Relational Data Mining

Springer
Berlin
Heidelberg
New York
Barcelona
Hong Kong
London
Milan
Paris
Tokyo

Sašo Džeroski • Nada Lavrač (Eds.)

Relational Data Mining

With 79 Figures and 69 Tables

Editors

Sašo Džeroski
Nada Lavrač

Jožef Stefan Institute
Jamova 39
1000 Ljubljana, Slovenia
E-mail: saso.dzeroski@ijs.si

Library of Congress Cataloging-in-Publication data applied for

Die Deutsche Bibliothek – CIP-Einheitsaufnahme

Relational data mining/Sašo Džeroski; Nada Lavrač (ed.). – Berlin; Heidelberg; New York; Barcelona; Hong Kong; London; Milan; Paris; Tokyo: Springer, 2001

ISBN 3-540-42289-7

ACM Subject Classification (1998): H.2.8, I.2.6, I.2.4, D.1.6, I.5

ISBN 3-540-42289-7 Springer-Verlag Berlin Heidelberg New York

Springer-Verlag Berlin Heidelberg New York
is a member of BertelsmannSpringer Science+Business Media GmbH

http://www.springer.de

Typesetting: Camera-ready by the editors
Cover Design: d&p, design & production, Heidelberg
Printed on acid-free paper SPIN 10843997 – 06/3142SR – 5 4 3 2 1 0

Foreword

The area of data mining, or knowledge discovery in databases, started to receive a lot of attention in the 1990s. Developments in sensing, communications and storage technologies made it possible to collect and store large collections of scientific and industrial data. The abilities to analyze such data sets had not developed as fast. Data mining research arouse at the intersection of several different research areas, notably statistics, databases, machine learning and algorithms. The area can loosely be defined as the analysis of large collections of data for finding models or patterns that are interesting or valuable.

The development of data mining methods requires the solution of several different types of problems. The data can have a very large number of dimensions, indicating that for example examining every pair of variables is impractical. The data can have hundreds of millions of observations, and therefore only a limited number of passes through the data can be done. The data can be observations of a process about which very little is known; hence there is no background knowledge available, and thus selection of appropriate models can be challenging. Or there can be heaps of background knowledge available, and methods that overlook it are destined to fail.

Most data mining methods have been developed for data in the traditional matrix form: rows represent observations, and columns represent variables. This representation has been the traditional one used in statistics, and it has many advantages. For example, matrix operations can be used to represent several data analytic procedures quite succintly, and these representations make it possible to devise efficient algorithms.

However, data about the real world is seldom of this form. Rather, the application domain contains several different types of entities, of which different types of data are known. Only recently has a large body of research aimed at data mining on such data emerged.

Relational data mining studies methods for knowledge discovery in databases when the database has information about several types of objects. This, of course, is usually the case when the database has more than one table. Hence there is little doubt as to the relevance of the area; indeed, one can wonder why most of data mining research has concentrated on the single table case.

Relational data mining has its roots in inductive logic programming, an area in the intersection of machine learning and programming languages. The early work in this area aimed at the synthesis of nontrivial programs from examples and background knowledge. The results were quite fascinating, but the true applicability of the techniques became clear only when the focus changed to the discovery of useful pieces of information from large collections of data, i.e., when the techniques started to be applied to data mining issues.

The present book "Relational Data Mining" provides a thorough overview of different techniques and strategies used in knowledge discovery from multi-relational data. The chapters describe a broad selection of practical inductive logic programming approaches to relational data mining and give a good overview of several interesting applications. I hope that the book will stimulate the interest for practical applications of relational data mining and further research in the development of relational data mining techniques.

Helsinki, June 2001 Heikki Mannila

Preface

Knowledge discovery in databases (KDD) is the process of identifying valid, novel, potentially useful, and ultimately understandable patterns in data. Data mining is the central step in this process, concerned with applying computational techniques to find patterns in data - other steps in the KDD process include data preparation and pattern evaluation. Most data mining approaches look for patterns in a single table of data. Since data in real databases typically reside in multiple tables, much thought and effort has to be invested in data preparation so as to squeeze as much relevant data as possible into a single table.

Relational data mining looks for patterns that involve multiple relations in a relational database. It does so directly, without transforming the data into a single table first and then looking for patterns in such an engineered table. The relations in the database can be defined extensionally, as lists of tuples, or intensionally, as database views or sets of rules. The latter allows relational data mining to take into account generally valid domain knowledge, referred to as background knowledge.

Relational data mining techniques have been mainly developed within the area of inductive logic programming (ILP). To learn patterns valid in multi-relational data, ILP approaches mainly use languages based on logic programming, a subset of the first-order predicate calculus or first-order logic. Relational algebra, the formalism of relational databases, is also a subset of first-order logic. To clarify the terminology, a predicate corresponds to a relation, arguments of a predicate correspond to attributes of a relation, and a relation defined by a view corresponds to a predicate defined intensionally. First-order logic is often referred to as predicate logic, but sometimes also as relational logic. In this spirit, we will sometimes use the term relational learning in the sense of relational data mining or ILP.

ILP is a research area at the intersection of machine learning and logic programming. It was initially concerned with the synthesis of logic programs from examples and background knowledge. This prototypical ILP task can be viewed as concept learning (inducing binary classifiers), but also as learning logical (intensional) definitions of relations. More recent developments, however, have expanded ILP to consider all of the main data mining tasks: classification, regression, clustering, and association analysis. The pattern languages used by single-table data mining approaches for these data mining tasks have been extended to the multiple-table case. Relational pattern languages now include relational association rules, relational classification rules, and relational decision trees, among others. The more expressive pattern languages that allow for multiple relations and the use of domain knowledge are the two distinguishing features of relational data mining.

Relational data mining algorithms have been developed to look for patterns expressed in relational pattern languages. Typically, data mining algorithms have been upgraded from the single-table case, i.e., propositional logic, to the multiple-table case, i.e., first-order logic. For example, distance-based algorithms for prediction and clustering have been upgraded from propositonal to first-order logic by defining a distance measure between examples/instances represented in first-order logic. Issues of efficiency and scaling-up to mine large datasets have been also addressed recently.

The number of successful applications of relational data mining has increased steadily over the recent years. In these, relational data mining approaches have clearly demonstrated improved performance over single-table approaches either in terms of performance or understandability of the discovered patterns, or both. Many successful applications come from life science domains, and in particular bioinformatics: these include the discovery of structural alerts for mutagenesis and genome-scale prediction of protein functional class. Other application areas include medicine, environmental sciences and engineering.

This book provides an introduction to relational data mining, a description of a representative sample of relational data mining approaches, and an overview of applications of and experiences with such approaches. The book is divided in four parts. The first part places relational data mining in the wider context of data mining and knowledge discovery. Part II provides a description of a number of relational data mining approaches. Part III shows how single table data mining approaches can be upgraded to or used as they are in a relational data mining context. The last part provides an overview of applications of and experiences with such approaches, as well as an overview of relevant Internet resources.

Part I starts with a brief overview of data mining (Chapter 1 by Džeroski) and proceeds with an overview of knowledge discovery in databases (KDD, Chapter 2 by Fayyad). Fayyad also lists some of the challenges faced by KDD and comments on how these could be addressed by relational data mining (and inductive logic programming (ILP)). Chapter 3 (by Džeroski and Lavrač) gives an introduction to inductive logic programming, which includes a brief introduction to logic programming and pointers to ILP literature. Wrobel (Chapter 4) gives a well-founded motivation for using relational data mining by providing a succint and illustrative account of the advantages of doing so — Sections 4.1 and 4.2 are not to be missed on a first reading of the book. He also presents an approach to discovering interesting subgroups in a relational context.

The chapters in Part II present a number of relational data mining approaches, including learning of relational decision trees, relational classification and association rules, and distance based approaches to relational learning and clustering. Chapter 5 (by De Raedt et al.) describes three data mining systems based on the ILP framework of learning from interpretations: these induce classification rules, decision trees and integrity constraints (clausal theories), respectively. Chapter 6 (by Kramer and Widmer) presents an approach to learning structural (relational) classification and regression trees. Chapter 7 (by Muggleton and Firth) describes the learning of relational (binary) classification rules, while Chapter 8 (by Dehaspe and Toivonen) describes the discovery of relational association rules. Chapter 9 (by Kirsten et al.) presents distance based approaches to relational learning and clustering, which include relational upgrades of the k-NN method, hierarchical agglomerative clustering and k-means clustering.

Part III presents a more detailed look at how a single table data mining approach can be upgraded to a relational data mining context or used in such a context after transforming the multi-table data to a single table. Chapter 10 (by Van Laer and De Raedt) presents a generic approach of upgrading single table data mining algorithms (propositional learners) to relational ones (first-order learners). Chapter 11 (by Kramer et al.) describes how a relational data mining problem can be transformed to a single table (propositional) data mining problem. Chapter 12 (by Quinlan) shows how the technique of boosting, increasingly often used in data mining, can be applied to improve the performance of a relational learner. Getoor et al. (Chapter 13) upgrade the language of probabilistic models (Bayesian networks) to probabilistic relational models and present techniques for finding such models from multi-relational data.

The last part is concerned with the practice of relational data mining. Chapter 14 (by Džeroski) gives an overview of applications of relational data mining (mostly ILP applications) in a number of areas, including drug design, protein structure and function, medicine, and engineering. Srinivasan (Chapter 15) gives a number of useful suggestions concerning the application of ILP, based on his experience with some of the most successful ILP applications. Finally, Chapter 16 (by Todorovski et al.) provides an overview of Internet resources on ILP for KDD, concerning ILP systems, applications, datasets and publications.

Ljubljana, June 2001 Sašo Džeroski and Nada Lavrač

Acknowledgments

The motivation for this book originates from the *International Summer School on Inductive Logic Programming and Knowledge Discovery in Databases* (ILP&KDD-97), held in Prague, Czech Republic, 15–17 September 1997, organized in conjunction with the *Seventh International Workshop on Inductive Logic Programming* (ILP-97). We wish to thank the lecturers and the participants of this exciting event. Our special thanks goes to Olga Stěpánkova and her colleagues from the Czech Technical University, who put a lot of effort into the local organization of the summer school and the workshop and did an excellent job.

Much of the research in inductive logic programming described in this volume has been supported by the Commission of the European Communities through the ESPRIT III Project ILP (*Inductive Logic Programming*, Basic Research Project 6020) and ESPRIT IV Project ILP2 (*Inductive Logic Programming II*, Long Term Research Project 2037). The Commission has also funded the PECO network ILPnet (*Inductive Logic Programming Pan-European Scientific Network*, CP93-94) and is funding the INCO network ILPnet2 (*Network of Excellence in Inductive Logic Programming*, WG-977102). The networks have greatly promoted the mobility of researchers and especially the creation of Internet resources in the area of ILP. We would like to thank all the participants in the above mentioned projects and networks for their invaluable cooperation.

We acknowledge the support of our local funding agency, the Slovenian Ministry of Education, Science and Sport (as of 2001, formerly the Slovenian Ministry of Science and Technology). Thanks also to our colleagues at the Department of Intelligent Systems at the Jožef Stefan Institute. Special thanks to Bernard Ženko for extensive help with LaTeX formatting.

A final word of thanks goes to the contributors to this volume. You have done an excellent job, some of you at short notice. You have also been patient with us. Thank you!

Table of Contents

Part I. Introduction

Part IV. Applications and Web Resources

List of Contributors

H. Blockeel
Department of Computer Science
Katholieke Universiteit Leuven
Celestijnenlaan 200A
B-3001 Leuven, Belgium
hendrik.blockeel@cs.kuleuven.ac.be

L. Dehaspe
Department of Computer Science
Katholieke Universiteit Leuven
Celestijnenlaan 200A
B-3001 Leuven, Belgium
luc.dehaspe@pharmadm.com

L. De Raedt
Institut für Informatik
Albert-Ludwigs-Universität Freiburg
Am Flughafen 17
D-79110 Freiburg, Germany
deraedt@informatik.uni-freiburg.de

S. Džeroski
Jožef Stefan Institute
Jamova 39
SI-1000 Ljubljana, Slovenia
saso.dzeroski@ijs.si

U. Fayyad
digiMine, Inc.
11250 Kirkland Way - Suite 201
Kirkland, WA 98033, USA
usama@digimine.com

J. Firth
Department of Computer Science
University of York
Heslington
York YO10 5DD, UK

P. Flach
Department of Computer Science
University of Bristol
Merchant Venturers Building
Woodland Rd
Bristol BS8 1UB, UK
peter.flach@bristol.ac.uk

N. Friedman
The School of Computer Science and Engineering
Hebrew University
Jerusalem 91904, Israel
nir@cs.huji.ac.il

L. Getoor
Computer Science Department
Stanford University
Stanford, CA 94305-9010, USA
getoor@cs.stanford.edu

T. Horváth
German National Research Center for Information Technology
GMD - AiS.KD
Schloß Birlinghoven
D-53754 Sankt Augustin, Germany
tamas.horvath@gmd.de

D. Kazakov
Department of Computer Science
University of York
Heslington
York YO10 5DD, UK
kazakov@cs.york.ac.uk

M. Kirsten
German National Research Center
for Information Technology
GMD - AiS.KD
Schloß Birlinghoven
D-53754 Sankt Augustin, Germany
mathias.kirsten@gmd.de

D. Koller
Computer Science Department
Stanford University
Stanford, CA 94305-9010, USA
koller@cs.stanford.edu

S. Kramer
Institut für Informatik
Albert-Ludwigs-Universität Freiburg
Am Flughafen 17
D-79110 Freiburg, Germany
skramer@
informatik.uni-freiburg.de

N. Lavrač
Jožef Stefan Institute
Jamova 39
SI-1000 Ljubljana, Slovenia
nada.lavrac@ijs.si

S. Muggleton
Department of Computer Science
University of York
Heslington
York YO10 5DD, UK
stephen@cs.york.ac.uk

A. Pfeffer
Division of Engineering
and Applied Sciences
Harvard University
Cambridge, MA 02138, USA
avi@eecs.harvard.edu

J. R. Quinlan
School of Computer Science
and Engineering
University of New South Wales
Sydney 2052, Australia
quinlan@cse.unsw.edu.au

A. Srinivasan
Computing Laboratory
Oxford University
Wolfson Building, Parks Road
Oxford OX1 3QD, UK
ashwin.srinivasan@
comlab.ox.ac.uk

O. Stěpánkova
Faculty of Electrical Engeneering
Department of Cybernetics
Czech Technical University
Technicka 2
166 27 Prague 6, Czech Republic
step@labe.felk.cvut.cz

L. Todorovski
Jožef Stefan Institute
Jamova 39
SI-1000 Ljubljana, Slovenia
ljupco.todorovski@ijs.si

H. Toivonen
Nokia Research Center
P.O. Box 407
FIN-00045 Nokia Group, Finland
hannu.tt.toivonen@nokia.com

W. Van Laer
Department of Computer Science
Katholieke Universiteit Leuven
Celestijnenlaan 200A
B-3001 Leuven, Belgium
wim.vanlaer@cs.kuleuven.ac.be

I. Weber
Institut für Informatik
Universität Stuttgart
Breitwiesenstr. 20–22
D-70565 Stuttgart, Germany
irene.weber@informatik.uni-stuttgart.de

G. Widmer
Austrian Research Institute for Artificial Intelligence
Schottengasse 3
A-1010 Vienna, Austria
gerhard@ai.univie.ac.at

S. Wrobel
School of Computer Science, IWS
University of Magdeburg
Universitätsplatz 2
D-39016 Magdeburg, Germany
wrobel@iws.cs.uni-magdeburg.de

D. Zupanič
Jožef Stefan Institute
Jamova 39
SI-1000 Ljubljana, Slovenia
darko.zupanic@ijs.si

Part I

Introduction

1. Data Mining in a Nutshell

Sašo Džeroski

Jožef Stefan Institute
Jamova 39, SI-1000 Ljubljana, Slovenia

Abstract

Data mining, the central activity in the process of knowledge discovery in databases, is concerned with finding patterns in data. This chapter introduces and illustrates the most common types of patterns considered by data mining approaches and gives rough outlines of the data mining algorithms that are most frequently used to look for such patterns. It also briefly introduces relational data mining, starting with patterns that involve multiple relations and laying down the basic principles common to relational data mining algorithms. An overview of the contents of this book is given, as well as pointers to literature and Internet resources on data mining.

1.1 Introduction

Knowledge discovery in databases (KDD) was initially defined as the "non-trivial extraction of implicit, previously unknown, and potentially useful information from data" [1.14]. A revised version of this definition states that "KDD is the non-trivial process of identifying valid, novel, potentially useful, and ultimately understandable patterns in data" [1.12]. According to this definition, data mining (DM) is a step in the KDD process concerned with applying computational techniques (i.e., data mining algorithms implemented as computer programs) to actually find patterns in the data. In a sense, data mining is the central step in the KDD process. The other steps in the KDD process are concerned with preparing data for data mining, as well as evaluating the discovered patterns (the results of data mining).

The above definitions contain very imprecise notions, such as knowledge and pattern. To make these (slightly) more precise, some explanations are necessary concerning data, patterns and knowledge, as well as validity, novelty, usefulness, and understandability. For example, the discovered patterns should be valid on new data with some degree of certainty (typically prescribed by the user). The patterns should potentially lead to some actions that are useful (according to user defined utility criteria). Patterns can be treated as knowledge: according to Frawley et al. [1.14], "a pattern that is interesting (according to a user-imposed interest measure) and certain enough (again according to the user's criteria) is called knowledge."

This chapter will focus on data mining and will not deal with the other aspects of the KDD process (such as data preparation). Since data mining is concerned with finding patterns in data, the notions of most direct relevance here are the notions of data and patterns. Another key notion is that of a data mining algorithm, which is applied to data to find patterns valid in the data. Different data mining algorithms address different data mining tasks, i.e., have different intended use for the discovered patterns.

Data is a set of facts, e.g., cases in a database (according to Fayyad et al. [1.12]). Most commonly, the input to a data mining algorithm is a single flat table comprising a number of attributes (columns) and records (rows). When data from more than one table in a database needs to be taken into account, it is left to the user to manipulate the relevant tables. Usually, this results in a single table, which is then used as input to a data mining algorithm.

The output of a data mining algorithm is typically a pattern or a set of patterns that are valid in the given data. A pattern is defined as a statement (expression) in a given language, that describes (relationships among) the facts in a subset of the given data and is (in some sense) simpler than the enumeration of all facts in the subset [1.14, 1.12]. Different classes of pattern languages are considered in data mining: they depend on the data mining task at hand. Typical representatives are equations; classification and regression trees; and association, classification, and regression rules. A given data mining algorithm will typically have a built-in class of patterns that it considers: the particular language of patterns considered will depend on the given data (the attributes and their values).

Many data mining algorithms come form the fields of machine learning and statistics. A common view in machine learning is that machine learning algorithms perform a search (typically heuristic) through a space of hypotheses (patterns) that explain (are valid in) the data at hand. Similarly, we can view data mining algorithms as searching, exhaustively or heuristically, a space of patterns in order to find interesting patterns that are valid in the given data.

In this chapter, we first look at the prototypical format of data and the main data mining tasks addressed in the field of data mining. We next describe the most common types of patterns that are considered by data mining algorithms, such as equations, trees and rules. We also outline some of the main data mining algorithms searching for patterns of the types mentioned above. Finally, we give an intuition of how these algorithms can be upgraded to look for patterns that involve multiple relations. Before summing up, we also give an overview of the contents of this book and some pointers to data mining literature and Internet resources.

1.2 Data mining tasks

This section first gives an example of what type of data is typically considered by data mining algorithms. It then defines the main data mining tasks addressed when such data is given. These include predictive modeling (classification and regression), clustering (grouping similar objects) and summarization (as exemplified by association rule discovery).

1.2.1 Data

The input to a data mining algorithm is most commonly a single flat table comprising a number of fields (columns) and records (rows). In general, each row represents an object and columns represent properties of objects. A hypothetical example of such a table is given in Table 1.1. We will use this example in the remainder of this chapter to illustrate the different data mining tasks and the different types of patterns considered by data mining algorithms.

Here rows correspond to persons that have recently (in the last month) visited a small shop and columns carry some information collected on these persons (such as their age, gender, and income). Of particular interest to the store is the amount each person has spent at the store this year (over multiple visits), stored in the field TotalSpent. One can easily imagine that data from a transaction table, where each purchase is recorded, has been aggregated over all purchases for each customer to derive the values for this field. Customers that have spent over 15000 in total are of special value to the shop. An additional field has been created (BigSpender) that has value yes if a customer has spent over 15000 and no otherwise.

Table 1.1. A single table with data on customers (table `Customer`).

CustomerID	Gender	Age	Income	TotalSpent	BigSpender
c1	Male	30	214000	18800	Yes
c2	Female	19	139000	15100	Yes
c3	Male	55	50000	12400	No
c4	Female	48	26000	8600	No
c5	Male	63	191000	28100	Yes
c6	Male	63	114000	20400	Yes
c7	Male	58	38000	11800	No
c8	Male	22	39000	5700	No
c9	Male	49	102000	16400	Yes
c10	Male	19	125000	15700	Yes
c11	Male	52	38000	10600	No
c12	Female	62	64000	15200	Yes
c13	Male	37	66000	10400	No
c14	Female	61	95000	18100	Yes
c15	Male	56	44000	12000	No
c16	Male	36	102000	13800	No
c17	Female	57	215000	29300	Yes
c18	Male	33	67000	9700	No
c19	Female	26	95000	11000	No
c20	Female	55	214000	28800	Yes

In machine learning terminology, rows are called examples and columns are called attributes (or sometimes features). Attributes that have numeric (real) values are called continuous attributes: Age, YearlyIncome and TotalSpent are continuous attributes. Attributes that have nominal values (such as Gender and BigSpender) are called discrete attributes.

1.2.2 Classification and regression

The tasks of classification and regression are concerned with predicting the value of one field from the values of other fields. The target field is called the class (dependent variable in statistical terminology). The other fields are called attributes (independent variables in statistical terminology).

If the class is continuous, the task at hand is called regression. If the class is discrete (it has a finite set of nominal values), the task at hand is called classification. In both cases, a set of data is taken as input, and a model (a pattern or a set of patterns) is generated. This model can then be used to predict values of the class for new data. The common term predictive modeling refers to both classification and regression.

Given a set of data (a table), only a part of it is typically used to generate (induce, learn) a predictive model. This part is referred to as the training set. The remaining part is reserved for evaluating the predictive performance of the learned model and is called the testing set. The testing set is used to estimate the performance of the model on new, unseen data, or in other words, to estimate the validity of the pattern(s) on new data.

1.2.3 Clustering

Clustering is concerned with grouping objects into classes of similar objects [1.18]. A cluster is a collection of objects that are similar to each other and are dissimilar to objects in other clusters. Given a set of examples, the task of clustering is to partition these examples into subsets (clusters). The goal is to achieve high similarity between objects within individual clusters (interclass similarity) and low similarity between objects that belong to different clusters (intraclass similarity).

Clustering is known as cluster analysis in statistics, as customer segmentation in marketing and customer relationship management, and as unsupervised learning in machine learning. Conventional clustering focusses on distance-based cluster analysis. The notion of a distance (or conversely, similarity) is crucial here: objects are considered to be points in a metric space (a space with a distance measure). In conceptual clustering, a symbolic representation of the resulting clusters is produced in addition to the partition into clusters: we can thus consider each cluster to be a concept (much like a class in classification).

1.2.4 Association analysis

Association analysis [1.16] is the discovery of association rules. Market basket analysis has been a strong motivation for the development of association analysis. Association rules specify correlations between frequent itemsets (sets of items, such as bread and butter, which are often found together in a transaction, e.g., a market basket).

The task of association analysis is typically performed in two steps. First, all frequent itemsets are found, where an itemset is frequent if it appears in at least a given percentage s (called support) of all transactions. Next, association rules are found of the form $X \to Y$, where X and Y are frequent itemsets and confidence of the rule (the percentage of transactions containing X that also contain Y) passes a threshold c.

1.2.5 Other data mining tasks

The above three data mining tasks receive by far the most attention within the data mining field and algorithms for performing such tasks are typically included in data mining tools. While classification and regression are of predictive nature, cluster analysis and association analysis are of descriptive nature. Subgroup discovery (Chapter 3) is at the boundary between predictive and descriptive tasks. Several additional data mining tasks [1.16] are of descriptive nature, including data characterization and discrimination, outlier analysis and evolution analysis. Below we give a short description of each of these, but we will not treat them further in this chapter.

Data characterization or summarization sums up the general characteristics or features of a target class of data: this class is typically collected by a database query. The summary data are typically generated using basic statistics or by aggregation in OLAP (On-line Analytical Processing) and can be presented in various graphical forms, such as pie charts or bar charts. Data discrimination compares the general features (as produced by characterization) of a target class with those of a given contrasting class (or each class from a set of contrasting classes).

Outlier detection is concerned with finding data objects that do not fit the general behavior or model of the data: these are called outliers. Outliers can be of interest, for example, for fraud detection. They can be found by looking for objects that are a substantial distance away from any of the clusters in the data or show large differences from the average characteristics of objects in a group.

Evolution analysis [1.16], describes and models regularities or trends whose behavior changes over time. It includes change and deviation detection [1.12], which focusses on discovering the most significant changes in the data from previously measured or normative values.

1.3 Patterns

Patterns are of central importance in data mining and knowledge discovery. Data mining algorithms search the given data for patterns. Discovered patterns that are valid, interesting and useful can be called knowledge.

Frawley et al. [1.14] define a pattern in a dataset as a statement that describes relationships in a subset of the dataset with some certainty, such that the statement is simpler (in some sense) than the enumeration of all facts in the dataset. A pattern thus splits the dataset, as it pertains to a part of it, and involves a spatial aspect which may be visualized.

This section introduces the most common types of patterns that are considered by data mining algorithms. Note that the same type of pattern may be used in different data mining algorithms addressing different tasks: trees can be used for classification, regression or clustering (conceptual), and so can distance-based patterns.

1.3.1 Equations

Statistics is one of the major scientific disciplines that data mining draws upon. A predictive model in statistics most commonly takes the form of an equation.

Linear models predict the value of a target (dependent) variable as a linear combination of the input (independent) variables. Three linear models that predict the value of the variable TotalSpent are represented by Equations 1.1, 1.2, and 1.3. These have been derived using linear regression on the data from Table 1.1.

$$\text{TotalSpent} = 189.5275 \times \text{Age} + 7146.89 \tag{1.1}$$

$$\text{TotalSpent} = 0.093 \times \text{Income} + 6119.74 \tag{1.2}$$

$$\text{TotalSpent} = 189.126 \times \text{Age} + 0.0932 \times \text{Income} - 2420.67 \tag{1.3}$$

Linear equations involving two variables (such as Equations 1.1 and 1.2) can be depicted as straight lines in a two-dimensional space (see Figure 1.1). Linear equations involving three variables (such as Equation 1.3) can be depicted as planes in a three-dimensional space. Linear equations, in general, represent hyper-planes in multidimensional spaces. Nonlinear equations are represented by curves, surfaces and hyper-surfaces.

Note that equations (or rather inequalities) can be also used for classification. If the value of the expression $0.093 \times \text{Income} + 6119.744$ is greater than 15000, for example, we can predict the value of the variable BigSpender to be "Yes". Points for which "Yes" will be predicted are those above the regression line in the left-hand part of Figure 1.1.

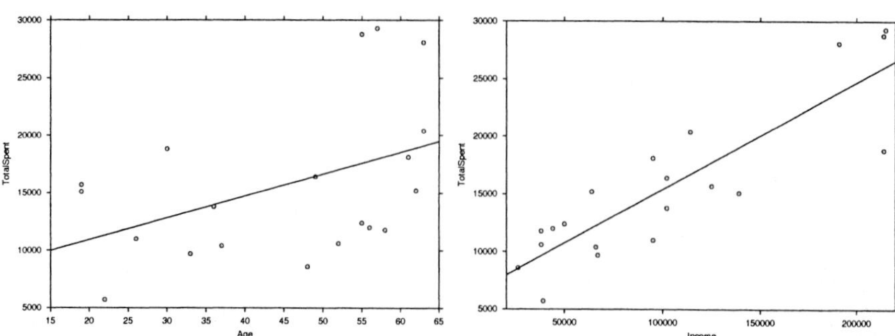

Fig. 1.1. Two regression lines that predict the value of variable TotalSpent from each of the variables Age and Income, respectively. The points correspond to the training examples.

1.3.2 Decision trees

Decision trees are hierarchical structures, where each internal node contains a test on an attribute, each branch corresponds to an outcome of the test, and each leaf node gives a prediction for the value of the class variable. Depending on whether we are dealing with a classification or a regression problem, the decision tree is called a classification or a regression tree, respectively. Two classification trees derived from the dataset in Table 1.1 are given in Figure 1.2. An example regression tree, also derived from the dataset in Table 1.1, is given in Figure 1.3.

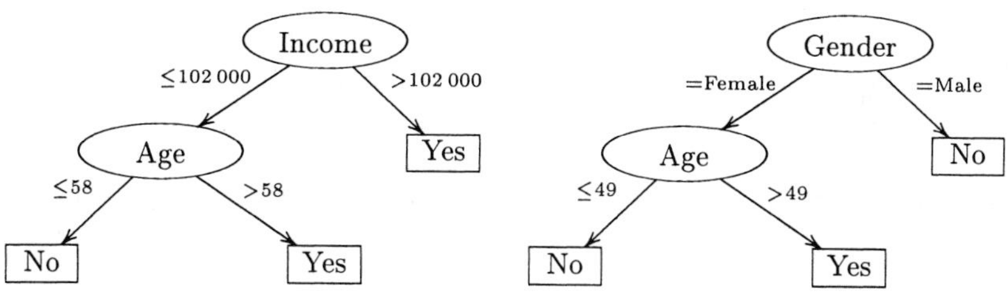

Fig. 1.2. Two classification trees that predict the value of variable BigSpender from the variables Age and Income, and Age and Gender, respectively.

Regression tree leaves contain constant values as predictions for the class value. They thus represent piece-wise constant functions. Model trees, where leaf nodes can contain linear models predicting the class value, represent piece-wise linear functions.

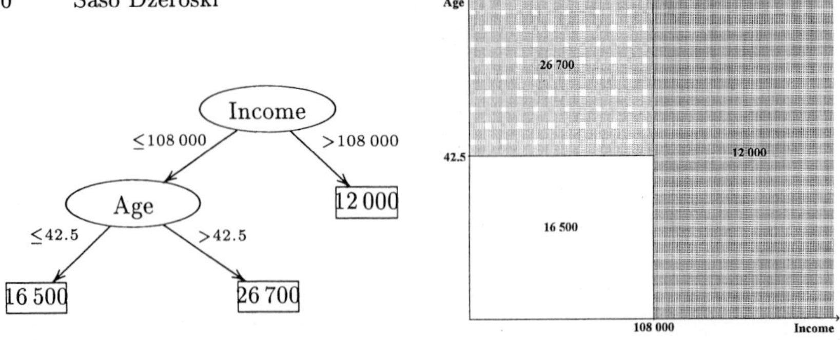

Fig. 1.3. A regression tree and the partition of the data space induced by the tree. The tree predicts the value of the variable TotalSpent from the variables Age and Income.

Note that decision trees represent total partitions of the data space, where each test corresponds to an axis-parallel split. This is illustrated in Figure 1.3. Most algorithms for decision tree induction consider such axis-parallel splits, but there are a few algorithms that consider splits along lines that need not be axis-parallel or even consider splits along non-linear curves.

1.3.3 Predictive rules

We will use the word rule here to denote patterns of the form "IF Conjunction of conditions THEN Conclusion." The individual conditions in the conjunction will be tests concerning the values of individual attributes, such as "Income $\leq$ 108000" or "Gender=Male". For predictive rules, the conclusion gives a prediction for the value of the target (class) variable.

If we are dealing with a classification problem, the conclusion assigns one of the possible discrete values to the class, e.g., "BigSpender=No". A rule applies to an example if the conjunction of conditions on the attributes is satisfied by the particular values of the attributes in the given example. Each rule corresponds to a hyper-rectangle in the data space, as illustrated in Figure 1.4.

Predictive rules can be ordered or unordered. Unordered rules are considered independently and several of them may apply to a new example that we need to classify. A conflict resolution mechanism is needed if two rules which recommend different classes apply to the same number of examples. A default rule typically exists, whose recommendation is taken if no other rule applies.

Ordered rules form a so-called decision list. Rules in the list are considered from the top to the bottom of the list. The first rule that applies to a given example is used to predict its class value. Again, a default rule with an empty precondition is typically found as the last rule in the decision list and is applied to an example when no other rule applies.

IF Income $\leq$ 102000
AND Age $\leq$ 58
THEN BigSpender=No
ELSE
DEFAULT BigSpender=Yes.

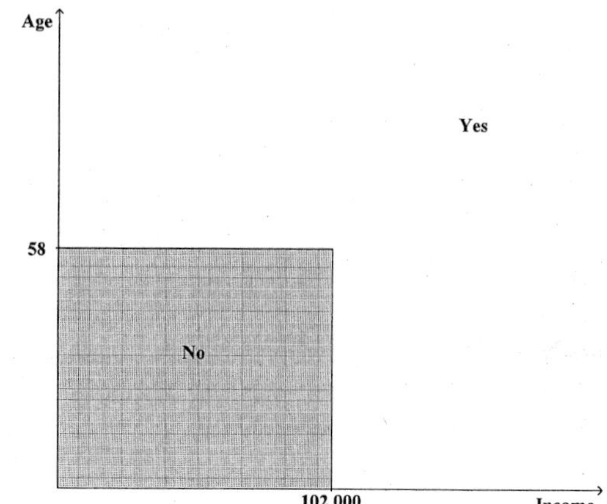

Fig. 1.4. A partition of the data space induced by an ordered list of rules, derived from the data in Table 1.1. The shaded box corresponds to the first rule in the list IF Income $\leq$ 102000 AND Age $\leq$ 58 THEN BigSpender=No, while the remainder of the data space is covered by the default rule BigSpender=Yes.

An ordered list and an unordered list of rules are given in Table 1.2. Both have been derived using a covering algorithm (see Section 4.1.3). The ordered list of rules in Figure 1.4, on the other hand, has been generated from the decision tree in the left-hand side of Figure 1.2. Note that each of the leaves of a classification tree corresponds to a classification rule. Although less common in practice, regression rules also exist, and can be derived, e.g., by transcribing regression trees into rules.

Table 1.2. An ordered (top) and an unordered (bottom) set of classification rules derived from the data in Table 1.1.

Ordered rules
IF Age $<$ 60 AND Income $<$ 81000 THEN BigSpender = No ELSE
IF Age $>$ 42 THEN BigSpender = Yes ELSE
IF Income $>$ 113500 THEN BigSpender = Yes ELSE
DEFAULT BigSpender=No

Unordered rules
IF Income $>$ 108000 THEN BigSpender = Yes
IF Age $\geq$ 49 AND Income $>$ 57000 THEN BigSpender = Yes
IF Age $\leq$ 56 AND Income $<$ 98500 THEN BigSpender = No
IF Income $<$ 51000 THEN BigSpender = No
IF 33 $<$ Age $\leq$ 42 THEN BigSpender = No
DEFAULT BigSpender=Yes

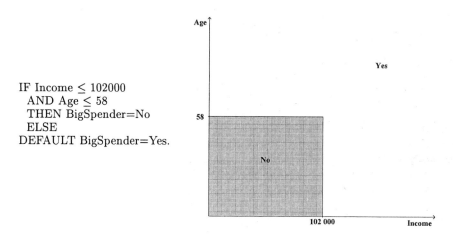

1.3.4 Association rules

Unlike predictive rules, association rules do not focus on predicting the value of a target class variable. This means that the conclusion parts of different association rules derived from the same data may concern different attributes/variables or even multiple variables. Association rules have the form $X \rightarrow Y$, where X and Y are conjunctions of conditions.

In the original formulation of association rules, X and Y are itemsets, such as $X = \{$ sausages, mustard $\}$ and $Y = \{$ beer $\}$, with $X \rightarrow Y$ meaning that a market basket (transaction) containing all items in X is likely to contain all items in Y. This assumes a relational sparse representation: each basket will contain only a few of the very large number of possible items. A tabular representation would have an attribute for every possible item, which would have value one if the item were in the basket and zero otherwise.

Table 1.3. A set of association rules derived from the data in Table 1.1. The continuous variables Age and Income have been discretized into 4 intervals each.

1. {Gender=Male, Income='(-inf-73250]'} → {BigSpender=No}
2. {Gender=Female, Age='(52-inf)'} → {BigSpender=Yes}
3. {Income='(167750-inf)'} → {BigSpender=Yes}
4. {Age='(52-inf)', BigSpender=No} → {Gender=Male, Income='(-inf-73250]'}
5. {Gender=Male, Age='(52-inf)', Income='(-inf-73250]'} → {BigSpender=No}
6. {Gender=Male, Age='(52-inf)', BigSpender=No} → {Income='(-inf-73250]'}
7. {Age='(52-inf)', Income='(-inf-73250]', BigSpender=No} → {Gender=Male}
8. {Age='(52-inf)', Income='(167750-inf)'} → {BigSpender=Yes}
9. {Age='(52-inf)', BigSpender=No} → {Income='(-inf-73250]'}
10. {Age='(52-inf)', BigSpender=No} → {Gender=Male}

Discovering association rules in a table with binary attributes is thus conceptually the same as discovering them in a relational sparse representation. The algorithms for association rule discovery can be easily extended to handle attributes with more than two values. Continuous attributes, however, should be discretized prior to applying association rule discovery.

A set of association rules discovered in the data from Table 1.1 after discretizing the attributes Age and Income is given in Table 1.3. All rules have a confidence of 100% and support of at least 15%: for each of the itemsets X and Y such that $X \rightarrow Y$ is in Table 1.3, Y appears in all transactions that X appears, and $X \cup Y$ appears in at least 3 (=15% × 20) transactions. Note that an itemset here corresponds to a conjunction of conditions and a transaction corresponds to an example: an itemset appearing in a transaction corresponds to a conjunction of conditions being true for an example.

Half of the association rules in Table 1.3 are classification rules that predict the value of the class BigSpender (rules 1,2,3,5, and 8). The remaining half, however, predict the values of variables Gender and Income, with rule 4 predicting the values of both. Rule 4 states that a person older than 52 who is not a big spender will be a male with income less than 73250.

1.3.5 Instances, distances, and partitions

Instance-based or nearest-neighbor methods for prediction simply store all the training examples and do not perform any generalization at training time. The data itself does not really qualify as a pattern, as patterns are required to be simpler than just enumerating all the facts in the data [1.14]. Of crucial importance to such methods is the notion of distance (or conversely, similarity) between two examples. Distance is also of crucial importance for clustering, where the patterns found by data mining algorithms are partitions of the training set.

If the examples only have continuous attributes, they can be viewed as points in a Euclidean space, and the Euclidean distance measure can be applied. Given two examples $x = (x_1, \ldots, x_n)$ and $y = (y_1, \ldots, y_n)$, their Euclidean distance is calculated as $d(x, y) = \sqrt{\sum_{i=1}^{n} (x_i - y_i)^2}$. Note that this does not allow for discrete attributes and also disregards differences in scale between attributes. A more general definition calculates the distance as

$$\text{distance}(x, y) = \sqrt{\sum_{i=1}^{n} w_i \times \text{difference}(x_i, y_i)^2}$$

where w_i is a non-negative weight value assigned to attribute A_i and the difference between attribute values is defined as follows

$$\text{difference}(x_i, y_i) = \begin{cases} |x_i - y_i| & \text{if attribute } A_i \text{ is continuous} \\ 0 & \text{if attribute } A_i \text{ is discrete and } x_i = y_i \\ 1 & \text{otherwise} \end{cases}$$

The weights allow to take into account the different importance of the attributes to the task at hand, after normalizing continuous attributes.

The notion of distance is used both for prediction (instance-based and nearest-neighbor methods) and clustering (especially distance-based clustering, but also conceptual clustering). The result of clustering is a partition of the training set into subsets. This can be a flat partition (i.e., a single partition with a fixed number of clusters) or a set of partitions hierarchically organized into a tree (where a cluster in one partition is split into several clusters in another partition). An example of the latter is given in Figure 1.5, where the top level partition contains one cluster only and the next level partition contains two clusters (one being { c1, c5, c17, c20 }). A graph such as the one in Figure 1.5 is called a dendrogram and results from hierarchical agglomerative clustering (see Section 1.4.5).

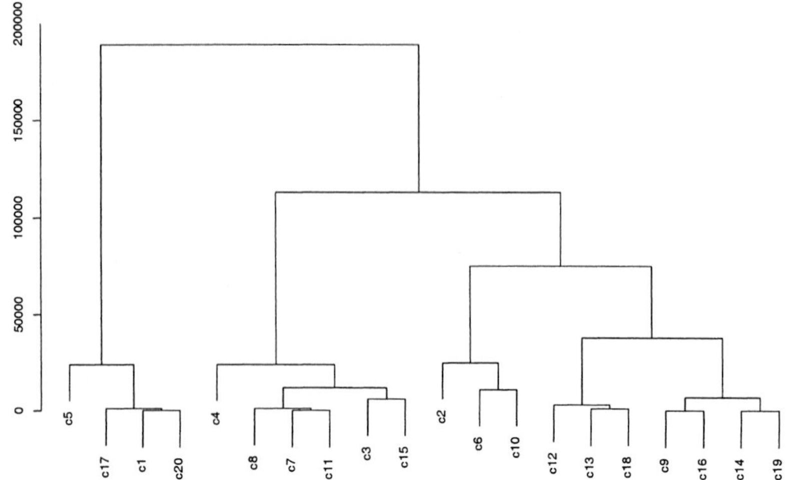

Fig. 1.5. A dendrogram representing a hierarchical clustering of the dataset from Table 1.1. The distance measure takes into account Gender, Age, and Income (non-normalized). The distance at which two clusters are joined can be read off the vertical axis.

1.3.6 Probabilistic models

Probabilistic models describe probabilistic dependencies among variables. Bayesian networks are currently a very popular class of probabilistic models, due to their powerful representation formalism and intuitive graphical representation. Bayesian networks are also known as Bayesian belief networks, belief networks, probabilistic networks, and probabilistic graphical models. Below we give a brief account of Bayesian networks (for a longer introduction we refer the reader to Chapter 13) and the naive Bayesian classifier, a special case of Bayesian networks, which has long been in widespread use.

A Bayesian network [1.22] is a directed acyclic graph, where nodes represent random variables and arcs represent probabilistic dependencies. If an arc is drawn from node A to B, then A is a parent of B. Each variable is probabilistically dependent on its descendants. Each variable is conditionally independent from its non-descendants, given the values of its parents. A conditional probability table is associated with each variable, specifying the conditional probability distribution $P(B|A)$, where $A = \{A_1, \ldots, A_n\}$ is the set of parents of B. The above two features of Bayesian networks allows for compact representation of joint probability distributions on a given set of variables, provided each variable has a relatively small number of parents.

Bayesian networks allow inferences to made for any subset of the variables involved, given evidence on any other subset. For classification, probabilistic models where inferences can be made only about the class variable given

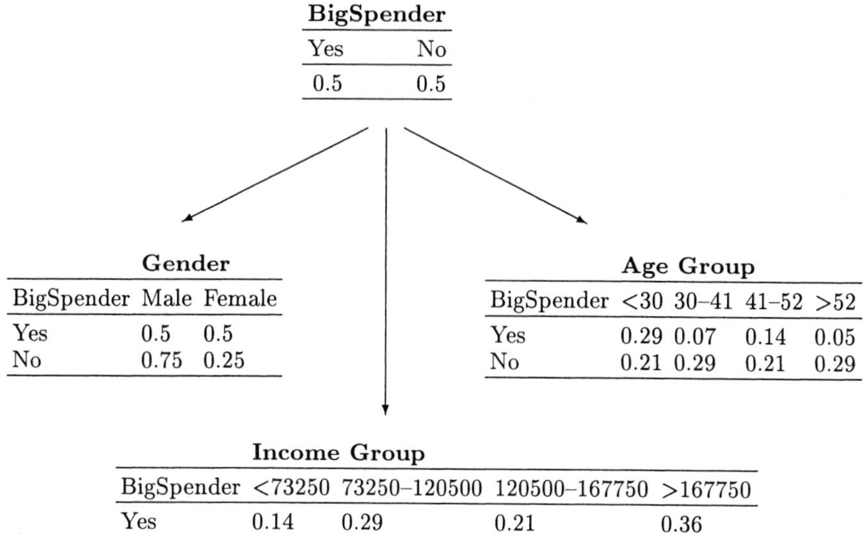

BigSpender

Yes	No
0.5	0.5

Gender

BigSpender	Male	Female
Yes	0.5	0.5
No	0.75	0.25

Age Group

BigSpender	<30	30–41	41–52	>52
Yes	0.29	0.07	0.14	0.05
No	0.21	0.29	0.21	0.29

Income Group

BigSpender	<73250	73250–120500	120500–167750	>167750
Yes	0.14	0.29	0.21	0.36
No	0.64	0.22	0.07	0.07

Fig. 1.6. A naive Bayesian classifier represented as a Bayesian network. The numbers in the BigSpender table represent the prior probabilities for the two class values. The numbers in the other tables represent the conditional probabilities of Gender, Age, and Income given the value of BigSpender. Age and Income have been discretized into 4 intervals each.

evidence on the attributes are used much more often. One such model is the naive Bayesian classifier. The naive Bayesian classifier corresponds to a Bayesian network where there are arcs from the class variable to each of the attributes, as depicted in Figure 1.6.

Figure 1.6 depicts the probability distributions of the class $P(C)$ and the conditional probability distributions $P(A|C)$ for the attributes given the class for the classification problem of predicting BigSpender from Gender, Age, and Income (see the dataset in Table 1.1) as estimated from the training data. The two classes (Yes and No) are equally likely by default (BigSpender table). A person that is a big spender is equally likely to be male or female, but a person that is not a big spender is much more likely to be male (0.75) than female (0.25) (Gender table).

For continuous attributes, we have two choices. One of is to discretize them and represent their conditional probability tables, as in Figure 1.6. An alternative is to make distributional assumptions, e.g., assume that the conditional probability distributions are normal and store the parameters (mean and standard deviation) of these distributions. Making this assumption, we find that for BigSpender=Yes, Age has mean μ =47.8 and deviation σ =18.1, while Income has μ =147300 and σ =56560; for BigSpender=No, Age has μ =42.3 and σ =13.1, while Income has μ =56500 and σ =25544.

1.4 Basic algorithms

The previous section described several types of patterns that can be found in data. This section outlines some basic algorithms that can be used to find such patterns in data. In most cases, this involves heuristic search through the space of possible patterns of the selected form.

1.4.1 Linear and multiple regression

Linear regression is the simplest form of regression [1.17]. Bivariate linear regression assumes that the class variable can be expressed as a linear function of one attribute, i.e., $C = \alpha + \beta \times A$. Given a set of data, the coefficients α and β can be calculated using the method of least squares, which minimizes the error $\sum_i (c_i - \alpha - \beta a_i)^2$ between the measured values for C (c_i), and the values calculated from the measured values for A (a_i) using the above equation. We have

$$\beta = \sum_i (a_i - \overline{a})(c_i - \overline{c}) / \sum_i (a_i - \overline{a})^2$$

$$\alpha = \overline{c} - \beta \overline{a},$$

where $\overline{a}$ is the average of $a_1, \ldots, a_n$ and $\overline{c}$ is the average of $c_1, \ldots, c_n$.

Multiple regression extends linear regression to allow the use of more than one attribute. The class variable can thus be expressed as a linear function of a multi-dimensional attribute vector, i.e., $C = \sum_{i=1}^{n} \beta_i \times A_i$. This form assumes that the dependent variable and the independent variables have mean values of zero (which is achieved by transforming the variables - the mean value of a variable is subtracted from each measured value for that variable). The method of least squares can also be applied to find the coefficients β_i. If we write the equation $C = \sum_{i=1}^{n} \beta_i \times A_i$ in matrix form $\underline{C} = \underline{\beta} \underline{A}$, where $\underline{C} = (c_1, \ldots, c_n)$ is the vector of measured values for the dependent variable and $\underline{A}$ is the matrix of measured values for the independent variables, we can calculate the vector of coefficients β as

$$\underline{\beta} = (\underline{A}^T \underline{A})^{-1} \underline{A}^T \underline{C}$$

where the operations of matrix transposition $\bullet^T$ and matrix inversion $\bullet^{-1}$ are used. The use of non-linear transformations, such as $A_i = A^i, i = 1, ..., n$, allows non-linear models to be found by using multiple regression: such models are linear in the parameters.

Note that both for linear and multiple regression, the coefficients α, β, and β_i can be calculated directly from a formula and no search through the space of possible equations takes place. Equation discovery approaches [1.11], which do not assume a particular functional form, search through a space of possible functional forms and look both for an appropriate structure and coefficients of the equation.

While linear regression is normally used to predict a continuous class, it can also be used to predict a discrete class. Generalized linear models can be used to this end, of which logistic regression is a typical representative. The fitting of generalized linear models is currently the most frequently applied statistical technique [1.26].

1.4.2 Top-down induction of decision trees

Finding the smallest decision tree that would fit a given data set is known to be computationally expensive (NP-hard). Heuristic search, typically greedy, is thus employed to build decision trees. The common way to induce decision trees is the so-called Top-Down Induction of Decision Trees (TDIDT) [1.25]). Tree construction proceeds recursively starting with the entire set of training examples (entire table). At each step, an attribute is selected as the root of the (sub)tree and the current training set is split into subsets according to the values of the selected attribute.

For discrete attributes, a branch of the tree is typically created for each possible value of the attribute. For continuous attributes, a threshold is selected and two branches are created based on that threshold. For the subsets of training examples in each branch, the tree construction algorithm is called recursively. Tree construction stops when the examples in a node are sufficiently pure (i.e., all are of the same class) or if some other stopping criterion is satisfied (there is no good attribute to add at that point). Such nodes are called leaves and are labeled with the corresponding values of the class.

Different measures can be used to select an attribute in the attribute selection step. These also depend on whether we are inducing classification or regression trees [1.7]. For classification, Quinlan [1.25] uses information gain, which is the expected reduction in entropy of the class value caused by knowing the value of the given attribute. Other attribute selection measures, however, such as the Gini index or the accuracy of the majority class, can and have been used in classification tree induction. In regression tree induction, the expected reduction in variance of the class value can be used.

An important mechanism used to prevent trees from over-fitting data is tree pruning. Pruning can be employed during tree construction (pre-pruning) or after the tree has been constructed (post-pruning). Typically, a minimum number of examples in branches can be prescribed for pre-pruning and a confidence level in accuracy estimates for leaves for post-pruning.

1.4.3 The covering algorithm for rule induction

In the simplest case of concept learning, one of the classes is referred to as positive (examples belonging to the concept) and the other as negative. For a classification problem with several class values, a set of rules is constructed for each class. When rules for class c_i are constructed, examples of this class are referred to as positive, and examples from all the other classes as negative.

The covering algorithm works as follows. We first construct a rule that correctly classifies some examples. We then remove the positive examples covered by the rule from the training set and repeat the process until no more examples remain. The pseudo code for this algorithm is given in Table 10.2.

Within this outer loop, different approaches can be taken to find individual rules. One approach is to heuristically search the space of possible rules top-down, i.e., from general to specific (in terms of examples covered this means from rules covering many to rules covering fewer examples) [1.9]. To construct a single rule that classifies examples into class c_i, we start with a rule with an empty antecedent (IF part) and the selected class c_i as a consequent (THEN part). The antecedent of this rule is satisfied by all examples in the training set, and not only those of the selected class. We then progressively refine the antecedent by adding conditions to it, until only examples of class c_i satisfy the antecedent. To allow for handling imperfect data, we may construct a set of rules which is imprecise, i.e., does not classify all examples in the training set correctly.

1.4.4 Finding frequent itemsets and association rules

The task of association rule discovery [1.2] is to find rules of the form $X \rightarrow Y$, where X and Y are itemsets, and the rule has a support and confidence above the user defined thresholds s and c, respectively. The support of a rule $X \rightarrow Y$ is the percentage of transactions in the database where the itemset $X \cup Y$ appears (the number of transactions where the itemset appears is called the frequency or support count of the itemset). The confidence of a rule $X \rightarrow Y$ is the percentage of transactions in the database containing X that also contain Y. In short, confidence and support can be defined as $support(X \rightarrow Y) = P(X \cup Y)$ and $confidence(X \rightarrow Y) = P(Y|X)$.

Association rule discovery typically proceeds in two phases. Frequent itemsets are found in the first phase. Frequent itemsets are itemsets with a support above s. In the second phase, strong association rules with confidence above c are formed from the discovered frequent itemset. For each frequent itemset Z, all non-empty subsets V of Z are generated and the rule $V \rightarrow Z - V$ is output if its confidence $support(V)/support(Z)$ is greater than the threshold c.

The search for frequent itemsets proceeds level-wise, with candidate itemsets being generated, pruned and eventually having their support counted at each level. The search starts with all singleton itemsets which have their support counted. At each level l, itemsets with l items are considered: candidate itemsets are generated by joining two frequent itemsets from the previous level which have all but one item in common. Candidate itemsets that have subsets that are not frequent are pruned: each subset of a frequent itemset

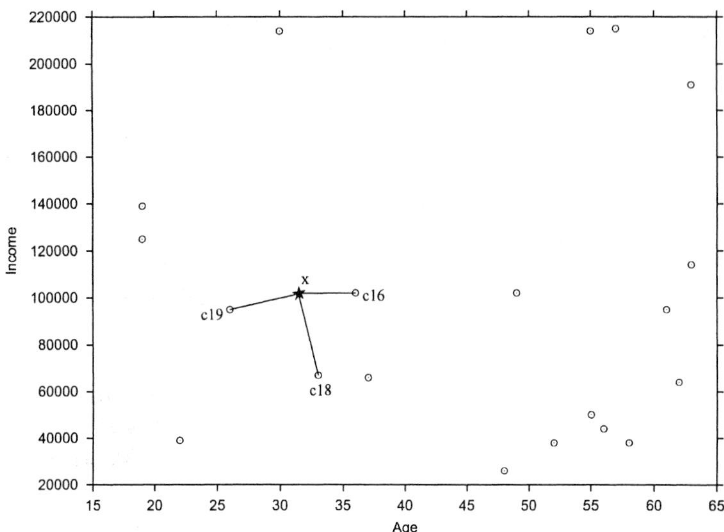

Fig. 1.7. 3-NN classification of a new example using the dataset from Table 1.1. The variables Age and Income are used to calculate distances.

is a frequent itemset. Candidate itemsets that are not eliminated have their support counted before proceeding to the next level. The search stops when no frequent itemsets are found at the current level.

1.4.5 Distance-based prediction and clustering

The nearest neighbor (NN) algorithm is one of the best known classification algorithms [1.10]. The NN algorithm treats attributes as dimensions of a metric space and examples as points in this space. In the training phase, the classified examples are stored without any processing. When classifying a new example, the distance between that example and all training examples is calculated and the class of the closest training example is assigned to the new example. This approach can be applied for both classification and regression.

The more general k-NN method takes the k nearest training examples when predicting the class for a new example. For classification, it determines the class of the new example by majority vote. For regression, it takes the average of the class values for the k training examples.

An example of 3-NN in action is given in Figure 1.7. A new person x who is 31 and has yearly income of 104000 (Age=31,Income=104000) is encountered. Persons c16, c18, and c19 from our dataset are closest to it and are used to predict the value of the class BigSpender (or TotalSpent). Since all three have the value No for class BigSpender, x is assigned class No. The predicted total spent for X is calculated as the average of the TotalSpent values for the three neighbors: TotalSpent$=(13800 + 9700 + 11000)/3 = 11500$.

In improved versions of k-NN, the votes/contributions of each of the k nearest neighbors can be weighted by the respective proximity to the new example. The contribution of each attribute to the distance may be weighted, in order to avoid problems caused by irrelevant features.

Hierarchical agglomerative clustering starts with every data point being a cluster. It then repeatedly aggregates the most similar (least dissimilar) groups together until there is just one big group. The result is a tree-like structure called dendrogram (see Figure 1.5) which shows which groups are joined and when. The final number of groups/clusters can be chosen subsequently by "cutting" the dendrogram at the desired height. Unlike k-means clustering, this kind of clustering can work with any kind of objects, as it only uses the matrix of distances between objects.

k-means clustering works on examples with continuous attributes only. The number of clusters k is an input to the algorithm. k of the training examples are randomly chosen as seeds or cluster prototypes. In each step of an iterative procedure, each of the training examples is assigned to one of the k prototypes, i.e., to the prototype closest to it. k new prototypes are then calculated as the average of the objects in each cluster. The procedure is repeated until there is no change of the assignment of examples to clusters in two consecutive steps.

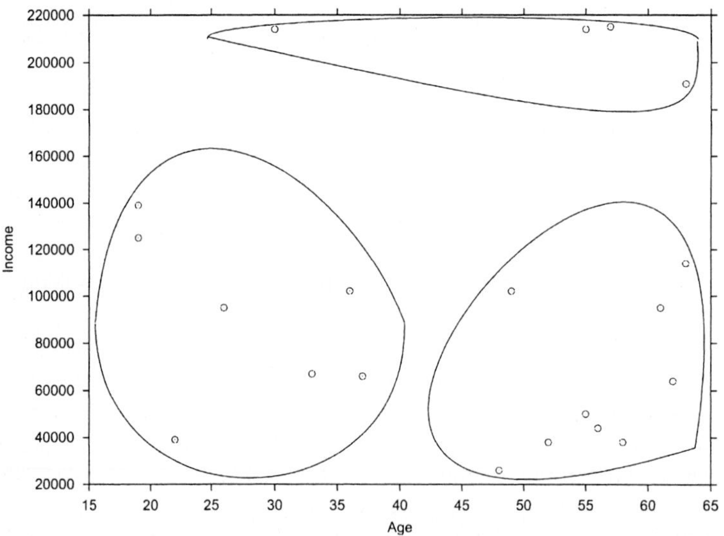

Fig. 1.8. A clustering produced by the k-means algorithm from the dataset in Table 1.1. The variables Age and Income are used to calculate distances.

An example clustering produced by the k-means algorithm from the dataset in Table 1.1 is shown in Figure 1.8. The distance measure employed only takes into account the attributes Age and Income, non-normalized. The three clusters contain the examples { c2, c8, c10, c13, c16, c18, c19 }, { c1, c5, c17, c20 }, and { c3, c4, c6, c7, c9, c11, c12, c14, c15 }, respectively. The prototypes for each of the three clusters are (27.43,90428.57), (51.25,208500.00), and (56.00,63444.44), respectively.

The k-means clustering method has two major drawbacks: it is sensitive to outliers and works on examples with continuous attributes only. The k-medoids method alleviates these drawbacks. It is similar to the k-means method, but differs in one important aspect. Instead of taking the mean value of the objects in a cluster as a prototype, it takes the medoid, i.e., the most centrally located object among the objects in the cluster.

1.4.6 Learning probabilistic models

This section describes how to learn Bayesian classifiers, and in particular naive Bayesian classifiers and Bayesian networks from data. Bayesian classification is based on the Bayes theorem $P(H|X) = P(X|H)P(H)/P(X)$, where $P(X)$ is the probability of X and $P(X|H)$ is the conditional probability of X given H. This theorem can be used to estimate the probability of an example belonging to each of the possible classes in a classification problem.

An example $X = (x_1, \ldots, x_n)$ is represented as conjunction of conditions $(A_i = x_i)$ for each of the n attributes A_i. If there are m classes, we let H range over $C_j \equiv C = c_j, j = 1, \ldots, m$, where C is the class and c_j is a possible class value. A Bayesian classifier will assign to a new example the class value c_k that maximizes $P(C_k|X)$, i.e., $P(C_k|X) \geq P(C_j|X)$ for $j = 1, \ldots, m$. According to the Bayes theorem, $P(C_j|X) = P(X|C_j)P(C_j)/P(X)$. Since $P(X)$ is constant across classes, we only need to maximize $P(X|C_j)P(C_j)$.

The key assumption of the naive Bayesian classifier is the assumption of class conditional independence (given the value of the class, the attributes are independent of one another). This allows the term $P(X|C_j)$ to be replaced by the product $\prod_i P(A_i = x_i|C_j)$. Unlike the joint probability $P(X|C_j)$, the individual conditional probabilities $P(A_i = x_i|C_j)$ are easy to estimate from training data (and so are the probabilities $P(C_j)$).

Learning a naive Bayes classifier consists of estimating the prior probabilities $P(Cj) = P(C = c_j)$ and $P(A_i = v_{ik}|C_j)$ for each of the possible values c_j for the class C and each of the attribute values v_{ik} of each attribute A_i. $P(C_j)$ is estimated by counting the number of examples $n_j = N(c_j)$ of class c_j and dividing this by n, the total number of training examples, i.e., $P(C_j) = n_j/n$. $P(A_i = v_{ik}|C_j)$ can be estimated as $N(A_i = v_{ik} \wedge C = c_j)$ divided by $N(C = c_j)$, where $N(X)$ is the number of examples where X is true. The above is a purely frequentist approach to estimating probabilities, which results in inaccurate estimates when few examples are used. Different probability estimates, such as the Laplace and the m-estimate [1.8] can result in more reliable estimations even when few examples are used.

Learning a Bayesian network from data involves two tasks. The first and easier task is estimating the probabilities in the conditional probability tables for a given dependency graph. This is very similar to estimating the probabilities for a naive Bayesian classifier and basically consists of counting.

Learning the structure, i.e., dependency graph of a Bayesian network is a harder task and involves a heuristic search through the space of possible structures. Operators on Bayesian networks that enable search algorithms to move through this space include adding an edge, deleting an edge, and reversing the direction of an edge. Candidate structures are evaluated by estimating the conditional probability tables and calculating how well the network fits the data. For a slightly more detailed description, we refer the reader to Chapter 13.

1.5 Relational data mining

Most existing data mining approaches look for patterns in a single table of data, as described in the previous sections. Relational data mining approaches, on the other hand, look for patterns that involve multiple relations from a relational database. The data taken as input by these approaches thus typically consists of several tables and not just one table. In the remainder of this section, we take a quick look at what data, patterns and algorithms look like in relational data mining.

1.5.1 Relational data

A relational database typically consists of several tables (relations) and not just one table. Staying with our example form Section 1.2.1, we might have the relation `MarriedTo` in addition to the table `Customer` from Table 1.1. An excerpt from the relation `MarriedTo` is given in Table 1.4.

Table 1.4. A relation providing information on the marital status of customers (table `MarriedTo`).

Spouse1	Spouse2
c1	c2
c2	c1
c3	c4
c4	c3
c5	c12
c6	c14
...	...
c12	c5
c14	c6
...	...

Note that relations can be defined extensionally (by tables as above) or intensionally through database views (as explicit logical rules). The latter typically represent relationships that can be inferred from other relationships. For example, having extensional representations of the relations mother and father, we can intentionally define the relations grandparent, grandmother, sibling, and ancestor, among others.

Intensional definitions of relations typically represent some general knowledge about the domain of discourse. For example, if we have extensional relations listing the atoms that make a compound molecule and the bonds between them, functional groups of atoms can be defined intentionally. Such general knowledge is called domain knowledge or background knowledge.

1.5.2 Relational patterns

Relational patterns involve multiple relations from a relational database. They are typically stated in a more expressive language than patterns defined on a single data table. The major types of relational patterns extend and are similar to the types of patterns outlined in Section 1.3. We can thus have relational classification rules, relational regression trees, and relational association rules, among others.

An example of a relational classification rule is given in Table 1.5. The rule involves the two relations `Customer` and `MarriedTo` from our customer database (Tables 1.1 and 1.4). It predicts a person to be a big spender if the person is married to somebody with high income (compare this to the rule that states a person is a big spender if he has high income, Table 1.2). Note that the two persons C1 and C2 are connected through the relation `MarriedTo`.

Table 1.5. A relational classification rule involving the relations `Customer` and `MarriedTo`) in the IF-THEN (top) and logic programming (bottom) notation.

```
IF Customer(C1,Age1,Income1,TotalSpent1,BigSpender1)
  AND MarriedTo(C1,C2)
  AND Customer(C2,Age2,Income2,TotalSpent2,BigSpender2)
  AND Income2 ≥ 108000
THEN BigSpender1 = Yes
```

```
big_spender(C1,Age1,Income1,TotalSpent1) ←
  married_to(C1,C2) ∧
  customer(C2,Age2,Income2,TotalSpent2,BigSpender2) ∧
  Income2 ≥ 108000.
```

Data mining algorithms search a given language of patterns to find valid patterns. The pattern language contains a very large number of possible patterns even in the single table case and is in practice limited by setting some parameters (e.g., the largest size of frequent itemsets for association rule discovery). For relational pattern languages, the number of possible patterns is even larger and it becomes necessary to limit the space of possible patterns by providing more explicit constraints. These typically specify what relations should be involved in the patterns, how the relations can be interconnected, and what other syntactic constraints the patterns have to obey. The explicit specification of the pattern language or constraints imposed upon it is known under the name of declarative bias.

1.5.3 Relational data mining algorithms

A relational data mining algorithm searches a language of relational patterns to find patterns that are valid in a given relational database. The search algorithms used here are very similar to those used in single table data mining: one can search exhaustively or heuristically (greedy search, best-first search, etc.). The essential ingredient of a relational data mining algorithm is the language of relational patterns.

Relational patterns are typically expressed in subsets of first-order logic, which is also called predicate logic or relational logic. A relation in a relational database corresponds to a predicate in predicate logic. Most commonly, logic programming (strongly related to deductive databases) is the formalism for expressing such patterns.

Just as many data mining algorithms come from the field of machine learning, most relational data mining algorithms come form the field of inductive logic programming (ILP) [1.21, 1.19]: situated at the intersection of machine learning and logic programming, ILP has been concerned with finding patterns expressed as logic programs. Initially, ILP focussed on automated program synthesis from examples, formulated as a binary classification task. In recent years, however, the scope of ILP has broadened to cover the whole spectrum of data mining tasks (classification, regression, clustering, association analysis).

The most common types of patterns have been extended to their relational versions (relational classification rules, relational regression trees, relational association rules) and so have the major data mining algorithms (decision tree induction, distance-based clustering and prediction, etc.). Van Laer and De Raedt (Chapter 10) present a generic approach of upgrading single table data mining algorithms (propositional learners) to relational ones (first-order learners).

Note that it is not trivial to extend a single table data mining algorithm to a relational one. Extending the key notions to, e.g., defining distance measures for multi-relational data requires considerable insight and creativity. Efficiency concerns are also very important, as it is often the case that even testing a given relational pattern for validity is computationally expensive, let alone searching a space of such patterns for valid ones. Such efficiency concerns also apply to the alternative approach of creating a single table from a multi-relational database in a systematic fashion, which in addition can have limited expressiveness.

1.6 Data mining literature and Internet resources

To learn more about data mining, a substantial body of literature and Internet resources can be consulted. Here we will focus on literature and resources devoted specifically to data mining and knowledge discovery, but bear in mind that many contributing disciplines can provide relevant material and knowledge (statistics, databases, machine learning).

Data mining literature now includes a dedicated journal titled *Data Mining and Knowledge Discovery* (DAMI), published by Kluwer Academic Publishers since 1997. The newsletter of the ACM SIGKDD (Special Interest Group on Knowledge Discovery in Databases) titled *SIGKDD Explorations* publishes short research and survey articles on various aspects of data mining and KDD.

After four workshops on knowledge discovery in databases, held from 1989 to 1994, a very successful conference series, the *International Conference on Knowledge Discovery and Data Mining* was started in 1995. Other conference series on this topic include the Pacific Asian Conference on Knowledge Discovery and Data Mining (PAKDD), the European Conference on Principles and Practice of Knowledge Discovery and Data Mining (PKDD), the Practical Application of Knowledge Discovery and Data Mining (PADD), the International Conference on Data Warehousing and Knowledge Discovery (DaWaK), and the IEEE International Conference on Data Mining (ICDM).

Edited books on data mining include *Knowledge Discovery in Databases*, edited by Piatetsky-Shapiro and Frawley [1.23] (an early collection of research papers), *Advances in Knowledge Discovery and Data Mining*, edited by Fayyad et al. [1.13] (a collection of more recent research results), and *Machine Learning and Data Mining: Methods and Applications*, edited by Michalski et al. [1.20].

Among the first authored books on data mining is *Data Mining* by Adriaans and Zantinge [1.1], published in 1996. This book was followed by quite a few books on the subject published in 1997, which include *Data Mining: A Hands-On Approach for Business Professionals* by Groth [1.15], *Predictive Data Mining: A Practical Guide* by Weiss and Indurkhya [1.27], *Data Warehousing, Data Mining and OLAP* by Berson and Smith [1.5], and *Data Mining Techniques for Marketing, Sales and Customer Support* by Berry and Linoff [1.3].

The most recent books on data mining include *Data Mining: Practical Machine Learning Tools and Techniques with Java Implementations* by Witten and Frank [1.28], *Mastering Data Mining: The Art and Science of Customer Relationship Management* by Berry and Linoff [1.4], *Intelligent Data Analysis*, edited by Berthold and Hand [1.6], *Data Preparation for Data Mining* by Pyle [1.24], and *Data Mining: Concepts and Techniques* by Han and Kamber [1.16]. New books on data mining appear with increasing frequency: for the latest news, check the Internet resources listed below.

Table 1.6. Links to Internet resources on data mining and knowledge discovery.

KDnuggets	`http://www.kdnuggets.com/`
KD Central	`http://www.kdcentral.com/`
The DAMI Journal	`http://www.wkap.nl/journalhome.htm/1384-5810`
The ACM SIGKDD	`http://www.acm.org/sigkdd/`

Table 1.6 list the links to some of the most important Internet resources on data mining. The KDnuggets site provides exhaustive information and lists of links in several categories, such as companies, publications, software, and solutions. KD Central is a WWW portal and resource center providing links to KDD related WWW sites, providing about a thousand links clustered in twenty-two categories.

1.7 Summary

This chapter provides an introduction to data mining, the central activity in the process of knowledge discovery in databases, which is concerned with finding patterns in data. It introduces and illustrates the most common types of patterns considered by data mining approaches and gave very brief descriptions of the major data mining algorithms. It also gives an intuition of what relational data mining is about, namely the discovery of patterns involving several relations. The remainder of this book provides a more detailed introduction to relational data mining, a description of a representative sample of relational data mining approaches, an overview of applications and experiences with such approaches, and an overview of Internet resources.

References

1.1 P. Adriaans and D. Zantinge. *Data Mining.* Addison-Wesley, Reading, 1996.
1.2 R. Agrawal, T. Imielinski, and A. Swami. Mining association rules between sets of items in large databases. In *Proceedings of the ACM SIGMOD Conference on Management of Data*, pages 207–216. ACM Press, New York, 1993.
1.3 M.J.A. Berry and G. Linoff. *Data Mining Techniques for Marketing, Sales and Customer Support.* John Wiley and Sons, New York, 1997.
1.4 M.J.A. Berry and G. Linoff. *Mastering Data Mining: The Art and Science of Customer Relationship Management.* John Wiley and Sons, New York, 1999.
1.5 A. Berson and S.J. Smith. *Data Warehousing, Data Mining and OLAP.* McGraw-Hill, New York, 1997.
1.6 M. Berthold and D.J. Hand, editors. *Intelligent Data Analysis: An Introduction.* Springer, Berlin, 1999.
1.7 L. Breiman, J. Friedman, R. Olshen, and C. Stone. *Classification and Regression Trees.* Wadsworth, Belmont, CA, 1984.
1.8 B. Cestnik. Estimating probabilities: A crucial task in machine learning. In *Proceedings of the Ninth European Conference on Artificial Intelligence*, pages 147–149. Pitman, London.

1.9 P. Clark and R. Boswell. Rule induction with CN2: Some recent improvements. In *Proceedings of the Fifth European Working Session on Learning*, pages 151–163. Springer, Berlin, 1991.
1.10 B.V. Dasarathy, editor. *Nearest Neighbor (NN) Norms: NN Pattern Classification Techniques.* IEEE Computer Society Press, Los Alamitos, CA, 1990.
1.11 S. Džeroski, L. Todorovski, I. Bratko, B. Kompare, and V. Križman. Equation discovery with ecological applications. In A.H. Fielding, editor, *Machine Learning Methods for Ecological Applications*, pages 185–207. Kluwer, Boston, 1999.
1.12 U. Fayyad, G. Piatetsky-Shapiro, and P. Smyth. From data mining to knowledge discovery: An overview. In U. Fayyad, G. Piatetsky-Shapiro, P. Smyth, and R. Uthurusamy, editors, *Advances in Knowledge Discovery and Data Mining*, pages 1–34. MIT Press, Cambridge, MA, 1996.
1.13 U. Fayyad, G. Piatetsky-Shapiro, P. Smyth, and R. Uthurusamy, editors. *Advances in Knowledge Discovery and Data Mining.* MIT Press, Cambridge, MA, 1996.
1.14 W. Frawley, G. Piatetsky-Shapiro, and C. Matheus. Knowledge discovery in databases: An overview. In G. Piatetsky-Shapiro and W. Frawley, editors, *Knowledge Discovery in Databases*, pages 1–27. MIT Press, Cambridge, MA, 1991.
1.15 R. Groth. *Data Mining: A Hands-On Approach for Business Professionals* Prentice Hall, Upper Saddle River, NJ, 1997.
1.16 J. Han and M. Kamber. *Data Mining: Concepts and Techniques.* Morgan Kaufmann, San Francisco, CA, 2001.
1.17 R.V. Hogg and A.T. Craig. *Introduction to Mathematical Statistics*, 5th edition. Prentice Hall, Englewood Cliffs, NJ, 1995.
1.18 L. Kaufman and P. J. Rousseeuw. *Finding Groups in Data: An Introduction to Cluster Analysis.* Wiley & Sons, New York, 1990.
1.19 N. Lavrač and S. Džeroski. *Inductive Logic Programming: Techniques and Applications.* Ellis Horwood, Chichester, 1994. Freely available at `http://www-ai.ijs.si/SasoDzeroski/ILPBook/`.
1.20 R.S. Michalski, I. Bratko, and M. Kubat, editors, *Machine Learning, Data Mining and Knowledge Discovery: Methods and Applications.* John Wiley and Sons, Chichester, 1997.
1.21 S. Muggleton. Inductive logic programming. *New Generation Computing*, 8(4): 295–318, 1991.
1.22 J. Pearl. *Probabilistic Reasoning in Intelligent Systems.* Morgan Kaufmann, San Mateo, 1988.
1.23 G. Piatetsky-Shapiro and W. Frawley, editors. *Knowledge Discovery in Databases..* MIT Press, Cambridge, MA, 1991.
1.24 D. Pyle. *Data Preparation for Data Mining.* Morgan Kaufmann, San Francisco, CA, 1999.
1.25 J. R. Quinlan. Induction of decision trees. *Machine Learning*, 1: 81–106, 1986.
1.26 P. Taylor. Statistical methods. In M. Berthold and D.J. Hand, editors, *Intelligent Data Analysis: An Introduction*, pages 67–127. Springer, Berlin, 1999.
1.27 S. Weiss and N. Indurkhya. *Predictive Data Mining: A Practical Guide.* Morgan Kaufmann, San Francisco, CA, 1997.
1.28 I.H. Witten and E. Frank. *Data Mining: Practical Machine Learning Tools and Techniques with Java Implementations.* Morgan Kaufmann, San Francisco, CA, 1999.

2. Knowledge Discovery in Databases: An Overview

Usama Fayyad

digiMine, Inc.
11250 Kirkland Way - Suite 201, Kirkland, WA 98033, USA

Abstract

Data Mining and Knowledge Discovery in Databases (KDD) promise to play an important role in the way people interact with databases, especially decision support databases where analysis and exploration operations are essential. Inductive logic programming can potentially play some key roles in KDD. We define the basic notions in data mining and KDD, define the goals, present motivation, and give a high-level definition of the KDD Process and how it relates to Data Mining. We then focus on data mining methods. Basic coverage of a sampling of methods will be provided to illustrate the methods and how they are used. We cover two case studies of successful applications in science data analysis, one of which is the classification of cataloging of a major astronomy sky survey covering 2 billion objects in the northern sky. The system can outperform human as well as classical computational analysis tools in astronomy on the task of recognizing faint stars and galaxies. We conclude with a listing of research challenges and we outline the areas where ILP could play some important roles in KDD.

2.1 Introduction

Data Mining and Knowledge Discovery in Databases (KDD) are rapidly evolving areas of research that draw on several disciplines, including statistics, databases, pattern recognition/AI, visualization, and high-performance and parallel computing. In this chapter, *which is intended to be strictly an introductory coverage*[1], we outline the basic notions in this area and show how data mining techniques can play an important role in helping humans analyse large databases.

We do not give an extensive coverage of applications, but rather describe two application case studies in science data analysis in this chapter. Applications are covered in [2.16], and in the special issue of *Communications of the ACM* [2.17]. A new technical journal, *Data Mining and Knowledge Discovery*, presents both new technical contributions and application case studies in this area [2.14].

---

[1] This chapter is a revised version of a companion reference to the invited talk given by the author at the ILP-97 workshop [2.11] and is intended to be general in its coverage, and not a presentation of new technical contributions.

2.2 From transactions to warehouses to KDD

With the widespread use of databases and the explosive growth in their sizes, individuals and organizations are faced with the problem of making use of this data. Traditionally, "use" of data has been limited to querying a reliable store via some well-circumscribed application or canned report-generating entity. While this mode of interaction is satisfactory for a wide class of well-defined processes, it was not designed to support data exploration and ad hoc querying of the data. Now that capturing data and storing it has become easy and inexpensive, certain questions begin to naturally arise: Will this data help my business gain an advantage? How can we use historical data to build models of underlying processes that generated such data? Can we predict the behavior of such processes? How can we "understand" the data? These questions become particularly important in the presence of massive data sets since they represent a large body of information that is presumed to be valuable, yet is far from accessible. The currently available interfaces between humans and machines do not support navigation, exploration, summarization, or modeling of large databases.

As transaction processing technologies were developed and became the mainstay of many business processes, great advances in addressing problems of reliable and accurate data capture were achieved. While transactional systems provided a solution to the problem of logging and book-keeping, there was little emphasis on supporting summarization, aggregation, and ad hoc querying over off-line stores of transactional data. A recent wave of activity in the database field has been concerned with turning transactional data into a more traditional relational database that can be queried for summaries and aggregates of transactions. Data warehousing also includes the integration of multiple sources of data along with handling the host of problems associated with such an endeavor. These problems include: dealing with multiple data formats, multiple database management systems (DBMS), distributed databases, unifying data representation, data cleaning, and providing a unified logical view of an underlying collection of non-homogeneous databases.

A data warehouse represents a large collection of data which in principle can provide views of the data that are not practical for individual transactional sources. For example, a supermarket chain may want to compare sales trends across regions at the level of products, broken down by weeks, and by class of store within a region. Such views are often precomputed and stored in special-purpose data stores that provide a multi-dimensional front-end to the underlying relational database and are sometimes called multi-dimensional databases.

Data warehousing is the first step in transforming a database system from a system whose primary purpose is reliable storage, i.e., On-Line Transaction Processing (OLTP) to one whose primary use is decision support, cf. Figure 2.1. A closely related area is called On-Line Analytical Processing (OLAP), named after principles first advocated by Codd [2.8].

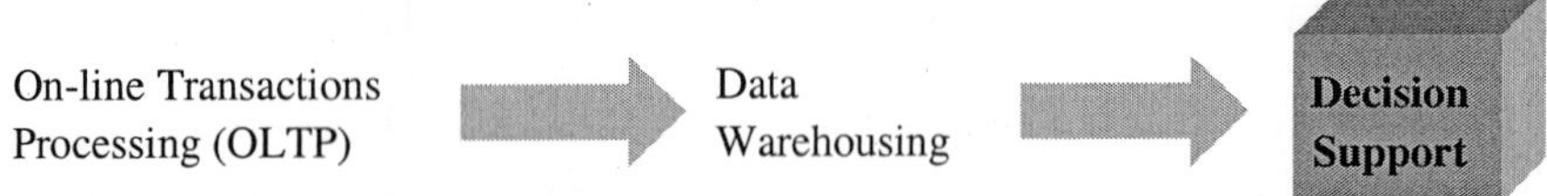

Fig. 2.1. The evolution of the view of a database system.

The current emphasis of OLAP systems is on supporting query-driven exploration of the data warehouse. Part of this entails precomputing aggregates along data "dimensions" in the multi-dimensional data store. Because the number of possible aggregates is exponential in the number of "dimensions", much of the work in OLAP systems is concerned with deciding which aggregates to pre-compute and how to derive other aggregates (or estimate them reliably) from the precomputed projections. Figure 2.2 illustrates one such example for data representing summaries of financial transactions in branches of a nationwide bank. The attributes (dimensions) have an associated hierarchy which defines how quantities are to be aggregated (rolled-up) as one moves to higher levels in the hierarchy.

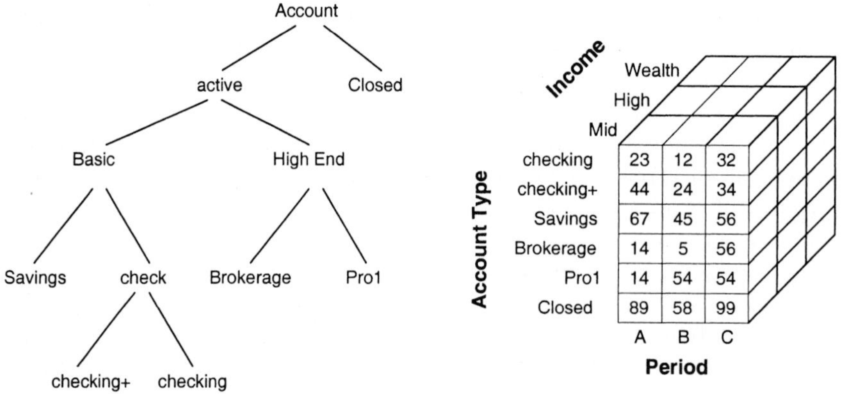

Fig. 2.2. An example multi-dimensional view of relational data.

Note that the multidimensional store may not necessarily be materialized as in principle it could be derived dynamically from the underlying relational database. For efficiency purposes, some OLAP systems employ "lazy" strategies in precomputing summaries and incrementally build up a cache of aggregates.

2.3 Why data mining?

Unlike OLAP, data mining techniques allow for the possibility of computer-driven exploration of the data. This opens up the possibility for a new way of interacting with databases: specifying queries at a much more abstract level than SQL permits. It also facilitates data exploration for problems that, due to high-dimensionality, would otherwise be very difficult to explore by humans, regardless of difficulty of use of, or efficiency issues with, SQL.

A problem that has not received much attention in database research is the *query formulation problem*: how can we provide access to data when the user does not know how to describe the goal in terms of a specific query? Examples of this situation are fairly common in decision support situations. For example, in a business setting, a credit card or telecommunications company would like to query its database of usage data for records representing fraudulent cases. In a science data analysis context, a scientist dealing with a large body of data would like to request a catalog of events of interest appearing in the data. Such patterns, while recognizable by human analysts on a case by case basis are typically very difficult to describe in a SQL query or even as a computer program in a stored procedure. A more natural means of interacting with the database is to state the query by example. In this case, the analyst would label a training set of examples of cases of one class versus another and let the data mining system build a model for distinguishing one class from another. The system can then apply the extracted classifier to search the full database for events of interest. This is typically more feasible because examples are usually easily available, and humans find it natural to interact at the level of cases.

Another major problem which data mining could help alleviate is the fact that humans find it particularly difficult to visualize and understand a large data set. Data can grow along two dimensions: the number of fields (also called dimensions or attributes) and the number of cases. Human analysis and visualization abilities do not scale to high-dimensions and massive volumes of data. A standard approach to dealing with high-dimensional data is to project it down to a very low-dimensional space and attempt to build models in this simplified subspace. As the number of dimensions grow, the number of choice combinations for dimensionality reduction explode. Furthermore, a projection to lower dimensions could easily transform an otherwise solvable discrimination problem into one that is impossible to solve. If the analyst is trying to explore models, then it becomes infeasible to go through the various ways of projecting the dimensions or selecting the right subsamples (reduction along columns and rows). An effective means to visualize data would be to employ data mining algorithms to perform the appropriate reductions. For example, a clustering algorithm could pick out a distinguished subset of the data embedded in a high-dimensional space and proceed to select a few dimensions to distinguish it from the rest of the data or from other clusters. Hence a much more effective visualization mode could be established: one that may enable an analyst to find patterns or models which may otherwise remain hidden in the high-dimensional space.

Another factor that is turning data mining into a necessity is that the rates of growth of data sets exceeds by far any rates that traditional "manual" analysis techniques could cope with. Hence, if one is to utilize the data in a timely manner, it would not be possible to achieve this goal in the traditional data analysis regime. Effectively this means that most of the data would remain unused. Such a scenario is not realistic in any competitive environment where those who do utilize data resources better will gain a distinct advantage. This sort of pressure is present in a wide variety of organizations, spanning the spectrum from business, to science, to government. It is leading to serious reconsideration of data collection and analysis strategies that are nowadays causing a build up of huge "write-only" data stores to accumulate.

2.4 KDD and data mining

The term *data mining* is often used as a synonym for the process of extracting useful information from databases. In this chapter, as in [2.15], we draw a distinction between the latter, which we call KDD, and "data mining". The term *data mining* has been mostly used by statisticians, data analysts, and the management information systems (MIS) communities. It has also gained popularity in the database field. The earliest uses of the term come from statistics and the usage in most cases was associated with negative connotations of blind exploration of data without a priori hypotheses to verify. However, notable exceptions can be found. For example, as early as 1978, [2.25] used the term in a positive sense in a demonstration of how generalized linear regression can be used to solve problems that are very difficult for humans and for traditional statistical techniques of that time to solve. The term KDD was coined at the first KDD workshop in 1989 [2.28] to emphasize that "knowledge" is the end product of a data-driven discovery.

In our view KDD refers to the overall *process* of discovering useful knowledge from data while *data mining* refers to a particular *step* in this process. Data mining is the application of specific algorithms for extracting patterns from data. The additional steps in the KDD process, such as data preparation, data selection, data cleaning, incorporating appropriate prior knowledge, and proper interpretation of the results of mining, are essential to ensure that useful knowledge is derived from the data. Blind application of data mining methods (rightly criticized as "data dredging" in the statistical literature) can be a dangerous activity easily leading to discovery of meaningless patterns. We give an overview of the KDD process in Figure 2.3. Note that in the KDD process, one typically iterates many times over previous steps and the process is fairly messy with plenty of experimentation. For example, one may select, sample, clean, and reduce data only to discover after mining that one or several of the previous steps need to be redone. We have omitted arrows illustrating these potential iterations to keep the figure simple.

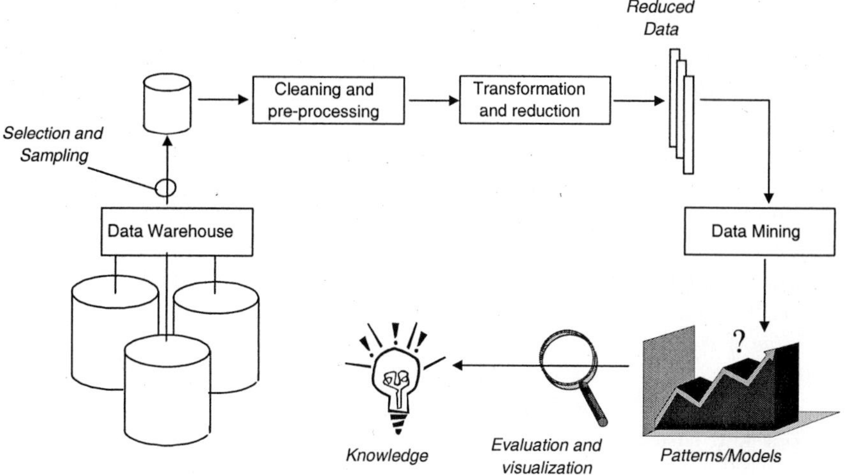

Fig. 2.3. An overview of the steps comprising the KDD process.

2.4.1 Basic definitions

We adopt the definitions of KDD and Data mining provided in [2.15].

> **Knowledge Discovery in Databases** is the *non-trivial process of identifying valid, novel, potentially useful, and ultimately understandable structure in data.*

Here *data* is a set of facts (e.g., cases in a database). *Structure* simply refers to patterns or models. A *pattern* is an expression in some language representing a parsimonious description of a subset of the data or a model applicable to that subset while a *model* is a characterization of the global data set. Hence, in our usage here, extracting *structure* designates finding patterns, fitting a model to data, finding structure from data, or in general any high-level description of a set of data. The term *process* implies that KDD comprises many steps, which involve data preparation, search for patterns, knowledge evaluation, and refinement, all repeated in multiple iterations. By *non-trivial* we mean that some search or inference is involved, i.e., it is not a straightforward computation of predefined quantities like computing the average value of a set of numbers. Since one can enumerate many (often infinitely) more patterns or models than there are data items, criteria for deciding what structure constitutes "knowledge" are needed. Classical measures such as *validity* (e.g., estimated prediction accuracy on new data) or *utility* (e.g., gain, perhaps in dollars saved due to better predictions or speed-up in response time of a system) are well-understood and can be adopted from decision analysis and statistics. However, other measures are much more difficult to define. How does one quantify *novelty* (at least to the system, and preferably to the user)? How does one quantify *understandability* of the patterns or model? In certain contexts understandability can be estimated by simplicity (e.g., the number of bits to describe a pattern). An important notion, called *interestingness* (e.g., see [2.29] and references within), is usually taken as an overall measure of pattern value, combining validity, novelty, usefulness, and simplicity. Interestingness functions can be explicitly defined or can be manifested implicitly via an ordering placed by the KDD system on the discovered patterns or models.

Hence, what we mean by the term "knowledge" in the KDD context can be characterized in a very simplified sense. A pattern can be considered as *knowledge* if it exceeds some interestingness threshold. This is by no means an attempt to define "knowledge" in the philosophical or even the popular view. It is purely user-oriented, domain-specific, and determined by whatever functions and thresholds the user chooses.

> **Data Mining** is *a step in the KDD process consisting of applying computational techniques that, under acceptable computational efficiency limitations, produce a particular enumeration of patterns (or models) over the data [2.15].*

Note that the space of patterns is often infinite, and the enumeration of patterns involves some form of search in this space. Practical computational constraints place severe limits on the subspace that can be explored by a data mining algorithm.

The data mining component of the KDD process is concerned with the algorithmic means by which patterns are extracted and enumerated from data. The overall KDD process (Figure 2.3) includes the *evaluation* and possible *interpretation* of the "mined" patterns to determine which patterns may be considered new "knowledge."

2.5 Data mining methods: An overview

Data mining techniques can be divided into five classes of methods. These methods are listed below. While much of these techniques have been historically defined to work over memory-resident data (typically read from flat files), some of these techniques are beginning to be scaled to operate on databases. Examples in classification (Section 2.5.1) include decision trees [2.27] and in summarization (Section 2.5.3) association rules [2.1]. For scalable clustering algorithms, see Section 2.5.2, [2.5, 2.6] and references within.

2.5.1 Predictive modeling

The goal is to predict some field(s) in a database based on other fields. If the field being predicted is a numeric (continuous) variable (such as a physical measurement of e.g., *height*) then the prediction problem is a *regression* problem. If the field is categorical then it is a *classification* problem. There is a wide variety of techniques for classification and regression. The problem in general is cast as determining the most likely value of the variable being predicted given the other fields (inputs), the training data (in which the target variable is given for each observation), and a set of assumptions representing one's prior knowledge of the problem.

Linear regression combined with non-linear transformation on inputs could be used to solve a wide range of problems. Transformation of the input space is typically the difficult problem requiring knowledge of the problem and quite a bit of "art". In classification problems this type of transformation is often referred to as "feature extraction".

In classification the basic goal is to predict the most likely state of a categorical variable (the class). This is fundamentally a density estimation problem. If one can estimate the probability that the class $C = c$, given the other fields $X = x$ for some feature vector x, then one could derive this probability from the joint density on C and X. However, this joint density is rarely known and very difficult to estimate. Hence one has to resort to various techniques for estimating. These techniques include:

1. Density estimation, e.g., kernel density estimators [2.9] or graphical representations of the joint density [2.21].
2. Metric-space based methods: define a distance measure on data points and guess the class value based on proximity to data points in the training set. An example of this approach is the k-nearest neighbor method [2.9].
3. Projection into decision regions: divide the attribute space into decision regions and associate a prediction with each. For example linear discriminant analysis finds linear separators, while decision tree or rule-based classifiers make a piecewise constant approximation of the decision surface. Neural nets find non-linear decision surfaces.

2.5.2 Clustering

Also known as segmentation, clustering does not specify fields to be predicted but targets separating the data items into subsets that contain items similar to each other. Since unlike classification we do not know the number of desired "clusters", clustering algorithms typically employ a two-stage search: An outer loop over possible cluster numbers and an inner loop to fit the best possible clustering for a given number of clusters. Given the number K of clusters, clustering methods can be divided into three classes:

1. Metric-distance based methods: a distance measure is defined and the objective becomes finding the best K-way partition such that cases in each block of the partition are closer to each other (or to a centroid) than to cases in other clusters.
2. Model-based methods: a model is hypothesized for each of the clusters and the idea is to find the best fit of that model to each cluster. If M_k is the model hypothesized for cluster k, then one way to score the fit of a model to a cluster is via the likelihood:
$$\text{Prob}(M_k|D) = \text{Prob}(D|M_k)\frac{\text{Prob}(M_k)}{\text{Prob}(D)}$$
The prior probability of the data D, $\text{Prob}(D)$ is a constant and hence can be ignored for comparison purposes, while $\text{Prob}(M_k)$ is the prior assigned to a model. In maximum likelihood techniques, all models are assumed equally likely and hence this term is ignored. A problem with ignoring this term is that more complex models are always preferred and this leads to overfitting the data.
3. Partition-based methods: basically enumerate various partitions and then score them by some criterion. The above two techniques can be viewed as special cases of this class. Many techniques in the AI literature fall under this category and utilize ad hoc scoring functions.

2.5.3 Data summarization

Sometimes the goal is to simply extract compact patterns that describe subsets of the data. There are two classes of methods which represent taking horizontal (cases) or vertical (fields) slices of the data. In the former, one would like to produce summaries of subsets: e.g., producing sufficient statistics, or logical conditions that hold for subsets. In the latter case, one would like to predict relations between fields. This class of methods is distinguished from the above in that rather than predicting a specified field (e.g., classification) or grouping cases together (e.g., clustering) the goal is to find relations between fields. One common method is called association rules [2.1]. Associations are rules that state that certain combinations of values occur with other combinations of values with a certain frequency and certainty. A common application of this is market basket analysis were one would like to summarize which products are bought with what other products.

2.5.4 Dependency modeling

Insight into data is often gained by deriving some causal structure within the data. Models of causality can be probabilistic (as in deriving some statement about the probability distribution governing the data) or they can be deterministic as in deriving functional dependencies between fields in the data [2.28]. Density estimation methods in general fall under this category, so do methods for explicit causal modeling (e.g., [2.18] and [2.21]).

2.5.5 Change and deviation detection

These methods account for sequence information, be it time-series or some other ordering (e.g., protein sequencing in genome mapping). The distinguishing feature of this class of methods is that ordering of observations is important and must be accounted for. Scalable methods for finding frequent sequences in databases, while in the worst-case exponential in complexity, do appear to execute efficiently given sparseness in real-world transactional databases [2.26].

2.6 Applications in science data analysis

We briefly review two case studies in order to illustrate the contribution and potential of KDD for science data analysis. For each case, the focus will primarily be on the impact of application, reasons why KDD systems succeeded, and limitations/future challenges. More expanded coverage of applications in science data analysis is given in [2.13].

2.6.1 Sky survey cataloging

The 2nd Palomar Observatory Sky Survey is a major undertaking that took over six years to complete. The survey produced 3 terabytes of image data containing an estimated 2 billion sky objects. The 3,000 photographic images are scanned into 16-bit/pixel resolution digital images at 23,040×23,040 pixels per image.

Classification. The basic problem is to generate a survey catalog which records the attributes of each object along with its class: star or galaxy. The attributes are defined by the astronomers. Once basic image segmentation is performed, 40 attributes per object are measured. The measurement process is fairly straightforward to automate. The problem is identifying the class of each object. Once the class is known, astronomers can conduct all sorts of scientific analyses like probing Galactic structure from star/galaxy counts, modeling evolution of galaxies, and studying the formation of large structure in the universe [2.31]. To achieve these goals the SKICAT system (Sky Image Cataloging and Analysis Tool) [2.32] was developed.

The problem of obtaining class labels is significant. The majority of objects in each image are faint objects whose class cannot be determined by visual inspection or classical computational approaches in astronomy. The goal was to classify objects that are at least one isophotal magnitude fainter than objects classified in previous comparable surveys. The problem was tackled using decision tree learning algorithms [2.12] to accurately predict the classes of objects. Accuracy of the procedure was verified by using a very limited set of high-resolution CCD images as ground truth.

By extracting rules via statistical optimization over multiple trees [2.12] it was possible to achieve 94% accuracy on predicting sky object classes. Since the approach could reliably classify objects that are one magnitude fainter than previous techniques, this resulted in tripling the size of data that is classified (usable for analysis). With the 300% increase in available data, astronomers were able to extract much more out of the data in terms of new scientific results [2.32], cf. below.

SKICAT was successful for the following reasons:

1. The astronomers solved the feature extraction problem: the proper transformation from pixel space to feature space. This transformation implicitly encodes a significant amount of prior knowledge to the system.
2. Within the 40 dimensional feature space, at least 8 dimensions are needed for accurate classification. Hence it was difficult for humans to discover which 8 of the 40 to use, let alone how to use them in classification. Data mining methods contributed in this case by solving the difficult classification problem.
3. Manual approaches to classification were simply not feasible. Astronomers needed an automated classifier to make the most out of the data.
4. Decision tree methods, although involving blind greedy search, proved to be an effective tool for finding the important dimensions for this problem.

Clustering: In search of rare clusters. An additional usage of the SKICAT catalog is in deriving new scientific results. Using the accurate classification of faint objects given by SKICAT helped a team of astronomers to discover 20 new high red-shift quasars [2.23]. These objects are extremely difficult to find, and are some of the farthest (hence oldest) objects in the universe. They provide valuable and rare clues about the early history of our universe. The search for quasars is an expensive operation requiring many observations. Since SKICAT provides accurate classifications of faint stars, the astronomers were able to use the classes to significantly narrow down the search. By combining classes and information from various color attributes, the new quasars were discovered using at least one order of magnitude less observations (and observation time) than was required by a comparable effort. The accurate classes translated into a small number of false alarms for the astronomers to cope with. The results after the first 5 quasars were discovered are detailed in [2.23].

However, what if one is interested in finding a new *unknown* class of objects, instead of a known target like quasars? Directions being pursued now involve the unsupervised learning (clustering) version of the problem. Unusual or unexpected clusters in the data might be indicative of new phenomena, perhaps even a new discovery. In a database of hundreds of millions of objects, automated analysis techniques are a necessity since browsing the feature vectors manually would only be possible for a small fraction of the survey. The idea is to pick out subsets of the data that look interesting, and ask the astronomers to focus their attention on those, perhaps perform further observations, and explain why these objects are different. A difficulty here is that new classes are likely to be rare in the data, so algorithms need to be tuned to looking for small interesting clusters rather than ignoring them. For example, the high-red-shift quasars mentioned above occur at a frequence of less than 10 per a million of objects.

In clustering problems, one typically needs to hold a large amount of data in memory, especially if one is searching for fairly rare clusters. The seemingly straightforward solution of randomly sampling the data set and building models from these random samples breaks down in case of rare clusters. Without an unreasonable amount of indexing, sampling would only work if random sampling (by row) is sufficient. In many cases, however, a stratified sample is required, in which case complexity of selection becomes a major issue. In the SKICAT application, for example, uniform random sampling would simply defeat the entire purpose. Clearly, members of a minority class could completely disappear from any small (or not so small) sample. One approach that can be adopted here is an iterative sampling scheme which exploits the fact that using a constructed model to classify the data scales linearly with the number of data points to be classified. The procedure goes as follows:

1. Generate a random sample S from the data set D.
2. Construct a model M_S based on S (based on probabilistic clustering or density estimation).
3. Apply the model to the entire set D, classifying items in D in the clusters with probabilities assigned by the model M_S.
4. Accumulate all the residual data points (members of D that do not fit in any of the clusters of M_S with high probability). Remove all data points that fit in M_S with high probability.
5. If a sample of residuals of acceptable size and properties is collected, go to step 7, else go to 6.
6. Let S be the set of residuals from step 4 mixed with a small sample (uniform) from D, return to step 2.
7. Perform clustering on the accumulated set of residuals, look for tight clusters as candidate new discoveries of minority classes in the data.

Other schemes for iteratively constructing a useful small sample via multiple efficient passes on the data are also possible [2.22]. The main idea is that sampling is not a straightforward matter. In more recent work, we have produced algorithms that scale clustering to very large databases [2.5, 2.6], but these were not available at the time of this application.

2.6.2 Finding volcanoes on Venus

The Magellan spacecraft orbited the planet Venus for over five years and used synthetic aperture radar (SAR) to map the surface of the planet penetrating the gas and cloud cover that permanently obscures the surface in the optical range. The resulting data set is a unique high-resolution global map of an entire planet. In fact, we have more of the planet Venus mapped at the 75m/pixel resolution than we do of our own planet Earth's surface (since most of Earth's surface is covered by water). This data set is uniquely valuable because of its completeness and because Venus is most similar to Earth in size. Learning about the geological evolution of Venus could offer valuable lessons about Earth and its history.

The sheer size of the data set prevents planetary geologists from effectively exploiting its content. For example, the first pass of Venus (completed in the first 2 years) using the left-looking radar resulted in over 30,000 images, each having 10^6 pixels (1000×1000). The data set was released on 100 CD-ROMs and is available to anyone who is interested. Lacking the proper tools to analyze this data, geologists did something very predictable: they simply examined browse images, looked for large features or gross structure, and cataloged/mapped the large-scale features of the planet. This means that the scientist operated at a much lower resolution, ignoring the potentially valuable high resolution data actually collected. Given that it took billions of dollars to design, launch, and operate the sensing instruments, it was a priority for NASA to insure that the data is exploited properly.

To help a group of geologists at Brown University analyze this data set [2.2], the JPL Adaptive Recognition Tool (JARtool) was developed [2.7]. The idea behind this system is to automate the search for an important feature on the planet, small volcanoes, by training the system via examples. The geologists would label volcanoes on a few (say 30-40) images, and the system would automatically construct a classifier that would then proceed to scan the rest of the image database and attempt to locate and measure the estimated 1 million small volcanoes.

Locating and cataloging the small volcanoes would enable the geologists to infer a lot about the geologic evolution of the planet and the processes it used to resurface itself. This is an excellent example of the wide gap between the raw collected data (pixels) and the level at which scientists operate (catalogs of objects). In this case, unlike in SKICAT, the mapping from pixels to features would have to be done by the system. Hence little prior knowledge is provided to the data mining system.

Using an approach based on matched filtering for focus of attention (triggering on any candidates with a high false alarm rate), followed by feature extraction based on projecting the data onto the dominant eigenvectors in the training data, and then classification learning to distinguish true detections from false alarms, JARtool can match scientist performance for certain classes of volcanoes (high probability volcanoes versus ones which scientists are not sure about) [2.7].

The use of data mining methods here was well-motivated because:

1. Scientists did not know much about image processing or about the SAR properties. Hence they could easily label images but programming clas-

sifiers was not really feasible. Hence a training-by-example framework is natural and justified.

2. Fortunately, as is often the case with cataloging tasks, there was little variation in illumination and orientation of objects of interest. Hence the mapping from pixels to features can be performed automatically.
3. The geologists did not have any other easy means for finding the small volcanoes, hence they were motivated to cooperate by providing training data and other help.
4. The result is to extract valuable data from an expensive data sets. Also, the adaptive approach (training by example) is flexible and would in principle allow us to reuse the basic approach on other problems.

With the proliferation of image databases and digital libraries, data mining systems that are capable of searching for content are becoming a necessity. In dealing with images, the train-by-example approach, i.e., querying for "things that look like this" is a natural interface since humans can visually recognize items of interest, but translating those visual intuitions into pixel-level algorithmic constraints is difficult to do. Future work on JARtool is proceeding to extend it to other applications like classification and cataloging of sun spots.

2.7 Research challenges for KDD

Successful KDD applications continue to appear, driven mainly by a glut in databases that have clearly grown to surpass raw human processing abilities. For examples of success stories about applications in industry see [2.4] and in science data analysis see [2.13]. More detailed case studies are found in [2.16]. Driving the healthy growth of this field are strong forces (both economic and social) that are a product of the data overload phenomenon. I view the need to deliver workable solutions to pressing problems as a very healthy pressure on the KDD field. Not only will it ensure our healthy growth as a new engineering discipline, but it will provide our efforts with a healthy dose of reality checks, ensuring that any theory or model that emerges will find its immediate real-world test environment.

The fundamental problems are still as difficult as they have been for the past few centuries as people considered difficulties of data analysis and how to mechanize it. The challenges facing advances in this field are formidable. Some of these challenges include:

1. Develop mining algorithms for classification, clustering, dependency analysis, and change/deviation detection that scale to large databases. There is a trade-off between performance and accuracy as one surrenders to the fact that data resides primarily on disk or on a server and cannot fit in main memory.

2. Develop schemes for encoding "metadata" (information about the content and meaning of data) over data tables so that mining algorithms can operate meaningfully on a database and so that the KDD system can effectively ask for more information from the user.
3. While operating with very large sample sizes is a blessing against overfitting problems, data mining systems need to guard against fitting models to data by chance. This problem becomes significant as a program explores a huge search space over many models for a given data set.
4. Develop effective means for data sampling, data reduction, and dimensionality reduction that operate on a mixture of categorical and numeric data fields. While large sample sizes allow us to handle higher dimensions, our understanding of high dimensional spaces and estimation within them is still fairly primitive. The curse of dimensionality is still with us.
5. Develop schemes capable of mining over non-homogeneous data sets (including mixtures of multimedia, video, and text modalities) and deal with sparse relations that are only defined over parts of the data.
6. Develop new mining and search algorithms capable of extracting more complex relationships between fields and able to account for structure over the fields (e.g., hierarchies, sparse relations), i.e., go beyond the flat file or single table assumption.
7. Develop data mining methods that account for prior knowledge of data and exploit such knowledge in reducing search, that can account for costs and benefits, and that are robust against uncertainty and missing data problems. Bayesian methods and decision analysis provide the basic foundational framework.
8. Enhance database management systems to support new primitives for the efficient extraction of necessary sufficient statistics as well as more efficient sampling schemes. This includes providing SQL support for new primitives that may be needed (e.g., [2.20]).
9. Scale methods to parallel databases with hundreds of tables, thousands of fields, and terabytes of data. Issues of query optimization in these settings are fundamental.
10. Account for and model comprehensibility of extracted models; allow proper trade-offs between complexity and understandability of models for purposes of visualization and reporting; enable interactive exploration where the analyst can easily provide hints to help the mining algorithm with its search.
11. Develop theory and techniques to model growth and change in data. Large databases, because they grow over a long time, do not typically grow as if sampled from a static joint probability density. The question of how does the data grow needs to be better understood (see articles by P. Huber, by Fayyad & Smyth, and by others in [2.24]) and tools for coping with it need to be developed.

12. Incorporate basic data mining methods with "any-time" analysis that can quantify and track the trade-off between accuracy and available resources, and optimize algorithms accordingly.

2.8 ILP and KDD: Prospects and challenges

ILP techniques address the problem of inferring models from data. Hence, they fit the definition of data mining methods. However, unlike standard statistical classification or clustering methods, ILP methods allow for the possibility of describing much richer structure in data. For example, one can learn a *descriptive* theory that describes the data or a logic program that serves as a *generative* model. Furthermore, one can utilize the power of inference during learning. For an excellent paper on how KDD and ILP fit together, the reader is referred to [2.10].

We summarize some aspects of ILP that makes its potential role in KDD exceedingly important:

Representation of data: Extends to include relations between items (training points) and allows for exploring data sets that are not limited to a single table of attribute value pairs or feature vectors. Representing general relations as attribute-value pairs results in blowing up data sizes on disk and in memory representations.

Representation of models: The expressive power of the model language is capable of representing much more complex model than the traditional propositional models used in decision trees and rules. In fact, the models can be as complex as full logic programs.

Feature extraction: ILP methods can explore many techniques for feature extraction prior to model learning. This includes using prior knowledge (domain theory) to dynamically derive features as well as deriving new features by making explicit implicit hidden facts via deriving logic programs to compute these.

Convenient representation of prior knowledge: A big challenge in KDD is the use of prior knowledge to guide discovery and mining. ILP provides a very natural mechanism for expressing (deterministic) background knowledge and its use during discovery. Rather than hand-encoding each specific piece of domain knowledge to be used by the mining algorithm, ILP makes it possible to derive such knowledge as part of the mining process.

Relational databases: There is a natural fit between the standard representation used in relational databases (and the query language SQL) and ILP representations and techniques. This allows for an easy way to define the interface between database and learning algorithm. In addition, it enables the algorithm to "browse" the tables of a database system as a natural part of its search. This alleviates the need of adding intermediate data representations between the database schema and the learning algorithm.

Support for deductive databases: Deductive databases [2.30] are able to aid in exploring a database and in executing more complex queries. There is a strong match between logic programming and the typical deductive database representation (several use Prolog in fact). This makes it convenient to allow the domain-specific knowledge encoded in the deductive database to be accessible by the learning program.

Understandability: The patterns derived via ILP are already symbolically expressed and hence should not be too difficult to understand by humans in many cases.

Post processors for derived patterns: If basic patterns are easier to derive via optimized traditional statistical pattern recognition techniques, then ILP can be used to enhance the derived models by using ILP techniques as a second stage. Hence the low-level interface to data can be via more traditional Data mining algorithms, while the second stage involves reasoning about the derived models and building more complex models. Hence traditional data mining techniques could serve as "data reduction" tools to reduce size of input to more expensive (search-intensive) ILP techniques

Data access costs dominate: When dealing with large databases, the dominant cost is by far the data access cost. Thus "in-memory" operations of search and inference may not necessarily dominate the search time. Sometimes a bit of knowledge might save many extraneous data scans.

Below we address two limitations/challenges for using ILP in KDD.

Minimize data scans: While the data representation scheme in ILP is well-matched to that of relational databases, the patterns of access required by ILP operations may not necessarily be efficient. A major challenge is to study the access patterns of ILP techniques and optimize the data scans so that the information necessary to drive a multitude of steps in the inductive search process are derived in as few scans of the data as possible. ILP algorithms, because they assume data is memory resident, perform many expensive joins over multiple relations without considering the true costs of such joins. This might require a rethinking of the basic design of some ILP methods.

Integrate with probabilistic techniques: The other major limitation and challenge for ILP is the need for a mechanism to represent uncertainty and probabilistic models. Many desired models in data mining applications are by their very nature probabilistic and not deterministic. Such models include characterization of probability density functions or explicit representation of uncertainty over derived models, missing values, and accounting for noise in the data. Without an integration with probabilistic techniques, it will be difficult to allow ILP methods to fulfill their potential promise in data mining.

A note from the editors: The above two challenges have recently received considerable attention and have been successfully addressed within several of the approaches presented in this volume. Reduction of the number of data scans has been achieved within the approaches for induction of relational decision trees (TILDE, Chapter 5, see [2.3] for details) and relational association rules (WARMR, Chapter 8). The two algorithms have been re-designed and turned inside-out, so that their outer counting loops are over data segments, while the inner ones are over patterns. Integration with probabilistic techniques has been elegantly achieved within the approach of learning probabilistic relational models (see Chapter 13).

2.9 Concluding remarks

KDD holds the promise of an enabling technology that could unlock the knowledge lying dormant in huge databases. Perhaps the most exciting aspect is the possibility of the birth of a new research area properly mixing statistics, databases, automated data analysis and reduction, and other related areas. The mix could produce a new breed of algorithms, tuned to working on large databases and scalable both in data set size and in parallelism.

In this chapter, we provided an overview of this area, defined the basic terms, and covered some sample case studies of applications in science data analysis. We also outlined prospects and challenges for the role of ILP in KDD. We concluded by listing future challenges for data mining and KDD. While KDD will draw on the substantial body of knowledge built up in its constituent fields, it is our hope that a new science will inevitably emerge as these challenges are addressed, and that suitable mixtures of ideas from each of the disciplines constituting the KDD field will greatly enhance our ability to exploit massive and ever-changing data sets.

References

2.1 R. Agrawal, H. Mannila, R. Srikant, H. Toivonen, and I. Verkamo. Fast discovery of association rules. In U. Fayyad, G. Piatetsky-Shapiro, P. Smyth, and R. Uthurusamy, editors, *Advances in knowledge Discovery and Data Mining*, pages 307–328. MIT Press, Cambridge, MA, 1996.

2.2 J. Aubele, L. Crumpler, U. Fayyad, P. Smyth, M. Burl, and P. Perona. Locating small volcanoes on Venus using a scientist-trainable analysis system. In *Proceedings of the Twenty-sixth Lunar and Planetary Science Conference*, page 1458. LPI/USRA, Houston, TX, 1995.

2.3 H. Blockeel, L. De Raedt, N. Jacobs, and B. Demoen. Scaling up inductive logic programming by learning from interpretations. *Data Mining and Knowledge Discovery*, 3(1): 59–93, 1999.

2.4 R. Brachman, T. Khabaza, W. Kloesgen, G. Piatetsky-Shapiro, and E. Simoudis. Mining business databases. *Communications of the ACM*, 39(11): 42–48, 1996.

2.5 P. Bradley, U. Fayyad, and C. Reina. Scaling clustering algorithms to large databases. *Proceedings of the Fourth International Conference on Knowledge Discovery and Data Mining*, pages 9–15. AAAI Press, Menlo Park, CA, 1998.
2.6 P. Bradley, U. Fayyad, and C. Reina. Scaling EM (Expectation Maximization) Clustering to Large Databases. *Technical Report MSR-TR-98-35*, Microsoft Research, Redmond, WA, 1998.
2.7 M. C. Burl, U. Fayyad, P. Perona, P. Smyth, and M. P. Burl. Automating the hunt for volcanoes on Venus. In *Proceedings of the Computer Vision and Pattern Recognition Conference*, pages 302–308. IEEE Computer Society Press, Los Alamitos, CA, 1994.
2.8 E. F. Codd, S. B. Codd, and C. T. Salley. *Providing OLAP (On-line Analytical Processing) to User-Analysts: An IT Mandate.* E. F. Codd and Associates, Toronto, Canada, 1993.
2.9 R. O. Duda and P. E. Hart. *Pattern Classification and Scene Analysis.* John Wiley & Sons, New York, 1973.
2.10 S. Džeroski. Inductive logic programming and knowledge discovery in databases. In Fayyad et al., editors, *Advances in Knowledge Discovery and Data Mining*, pages 117–152. MIT Press, Cambridge, MA, 1996.
2.11 U. Fayyad. Knowledge discovery in databases: An overview. In *Proceedings of the Seventh International Workshop on Inductive Logic Programming*, pages 3–16. Springer, Berlin, 1997.
2.12 U. Fayyad, S. G. Djorgovski, and N. Weir. Automating the analysis and cataloging of sky surveys. In U. Fayyad, G. Piatetsky-Shapiro, P. Smyth, and R. Uthurusamy, editors, *Advances in Knowledge Discovery and Data Mining*, pages 471–493, MIT Press, Cambridge, MA, 1996.
2.13 U. Fayyad, D. Haussler, and P. Stolorz. Mining scientific data. *Communications of the ACM*, 39(11): 51–57, 1996.
2.14 U. Fayyad, H. Mannila, and G. Piatetsky-Shapiro, editors-in-chief. *Data Mining and Knowledge Discovery.* Kluwer, Boston, 1997-2000.
2.15 U. Fayyad, G. Piatetsky-Shapiro, and P. Smyth. From data mining to knowledge discovery: An overview. In U. Fayyad, G. Piatetsky-Shapiro, P. Smyth, and R. Uthurusamy, editors, *Advances in Knowledge Discovery and Data Mining*, pages 1–34. MIT Press, Cambridge, MA, 1996.
2.16 U. Fayyad, G. Piatetsky-Shapiro, P. Smyth, and R. Uthurusamy, editors. *Advances in Knowledge Discovery and Data Mining.* MIT Press, Cambridge, MA, 1996.
2.17 U. Fayyad and R. Uthurusamy, editors. Special Issue on Data Mining. *Communications of the ACM*, 39(11), 1996.
2.18 C. Glymour, R. Scheines, P. Spirtes, and K. Kelly. *Discovering Causal Structure.* Academic Press, New York, 1987.
2.19 C. Glymour, D. Madigan, D. Pregibon, and P. Smyth. Statistical themes and lessons for data mining. *Data Mining and Knowledge Discovery*, 1(1): 11–28, 1997.
2.20 J. Gray, S. Chaudhuri, A. Bosworth, A. Layman, D. Reichart, M. Venkatrao, F. Pellow, and H. Pirahesh. Data Cube: A relational aggregation operator generalizing group-by, cross-tab, and sub-totals. *Data Mining and Knowledge Discovery*, 1(1): 29–53, 1997.
2.21 D. Heckerman. Bayesian networks for data mining. *Data Mining and Knowledge Discovery*, 1(1):79–119, 1997.
2.22 L. Kaufman and P. J. Rousseeuw. *Finding Groups in Data: An Introduction to Cluster Analysis.* Wiley & Sons, New York, 1990.
2.23 J. D. Kennefick, R. R. De Carvalho, S. G. Djorgovski, M. M. Wilber, E. S. Dickinson, N. Weir, U. Fayyad, and J. Roden. The discovery of five

quasars at z>4 using the second Palomar sky survey. *Astronomical Journal*, 110(1): 78–86, 1995.

2.24 J. Kettenring and D. Pregibon, editors. *Statistics and Massive Data Sets, Report to the Committee on Applied and Theoretical Statistics.* National Research Council, Washington, D.C., 1996.

2.25 E. Leamer. *Specification Searches: Ad-hoc Inference with Nonexperimental Data.* Wiley & Sons, New York, 1978.

2.26 H. Mannila, H. Toivonen, A. I. Verkamo. Discovery of frequent episodes in event sequences. *Data Mining and Knowledge Discovery*, 1(3): 259–289, 1997.

2.27 M. Mehta, R. Agrawal, and J. Rissanen. SLIQ: a fast scalable classifier for data mining. In *Proceedings of the Fifth International Conference on Extending Database Technology*, pages 18–32. Springer, Berlin, 1996.

2.28 G. Piatetsky-Shapiro and W. Frawley, editors. *Knowledge Discovery in Databases.* MIT Press, Cambridge, MA, 1991.

2.29 A. Silberschatz and A. Tuzhilin. On subjective measures of interestingness in knowledge discovery. In *Proceedings of the First International Conference on Knowledge Discovery and Data Mining*, pages 275–281. AAAI Press, Menlo Park, CA, 1995.

2.30 J. Ullman. *Principles of Database and Knowledge Base Systems*, volume 1. Computer Science Press, Rockville, MA, 1988.

2.31 N. Weir, S. G. Djorgovski, and U. Fayyad. Initial galaxy counts from digitized POSS-II. *Astronomical Journal*, 110(1): 1–20, 1995.

2.32 N. Weir, U. Fayyad, and S. G. Djorgovski. Automated star/galaxy classification for digitized POSS-II. *The Astronomical Journal*, 109(6): 2401–2412, 1995.

3. An Introduction to Inductive Logic Programming

Sašo Džeroski and Nada Lavrač

Jožef Stefan Institute
Jamova 39, SI-1000 Ljubljana, Slovenia

Abstract

Inductive logic programming (ILP) is concerned with the development of techniques and tools for relational data mining. Besides the ability to deal with data stored in multiple tables, ILP systems are usually able to take into account generally valid background (domain) knowledge in the form of a logic program. They also use the powerful language of logic programs for describing discovered patterns. This chapter introduces the basics of logic programming and relates logic programming terminology to database terminology. It then defines the task of relational rule induction, the basic data mining task addressed by ILP systems, and presents some basic techniques for solving this task. It concludes with an overview of other relational data mining tasks addressed by ILP systems.

3.1 Introduction

From a KDD perspective, we can say that inductive logic programming (ILP) is concerned with the development of techniques and tools for relational data mining. While typical data mining approaches find patterns in a given single table, relational data mining approaches find patterns in a given relational database. In a typical relational database, data reside in multiple tables. ILP tools can be applied directly to such multi-relational data to find patterns that involve multiple relations. This is a distinguishing feature of ILP approaches: most other data mining approaches can only deal with data that resides in a single table and require preprocessing to integrate data from multiple tables (e.g., through joins or aggregation) into a single table before they can be applied.

Integrating data from multiple tables through joins or aggregation can cause loss of meaning or information. Suppose we are given the relation $customer(CustID, Name, Age, SpendsALot)$ and the relation $purchase(CustID, ProductID, Date, Value, PaymentMode)$, where each customer can make multiple purchases, and we are interested in characterizing customers that spend a lot. Integrating the two relations via a natural join will give rise to a relation $purchase1$ where each row corresponds to a purchase and not to a customer. One possible aggregation would give rise to the relation $customer1(CustID, Age, NofPurchases, TotalValue, SpendsALot)$. In this case, however, some information has been clearly lost during the aggregation process (see Sections 4.1 and 4.2 for a detailed discussion).

The following pattern can be discovered by an ILP system if the relations *customer* and *purchase* are considered together.

$$\begin{aligned}&customer(CID, Name, Age, yes) \leftarrow\\&\qquad Age > 30 \wedge\\&\qquad purchase(CID, PID, D, Value, PM) \wedge\\&\qquad PM = credit\_card \;\wedge\; Value > 100.\end{aligned}$$

This pattern says: "a customer spends a lot if she is older than 30, has purchased a product of value more than 100 and paid for it by credit card." It would not be possible to induce such a pattern from either of the relations *purchase*1 and *customer*1 considered on their own.

Besides the ability to deal with data stored in multiple tables directly, ILP systems are usually able to take into account generally valid background (domain) knowledge in the form of a logic program. The ability to take into account background knowledge and the expressive power of the language of discovered patterns are also distinctive for ILP.

Note that data mining approaches that find patterns in a given single table are referred to as *attribute-value* or *propositional learning* approaches, as the patterns they find can be expressed in propositional logic. ILP approaches are also referred to as first-order learning approaches, or relational learning approaches, as the patterns they find are expressed in the relational formalism of first-order logic. A more detailed discussion of the single table assumption, the problems resulting from it and how a relational representation alleviates these problems can be found in the introductory sections (4.1 and 4.2) of Chapter 4.

The remainder of this chapter first introduces the basics of logic programming and relates logic programming terminology to database terminology. It then defines the task of relational rule induction, the basic data mining task addressed by ILP systems, and presents some basic techniques for solving this task. It concludes with an overview of other relational data mining tasks addressed by ILP systems and a survey of ILP-related literature.

3.2 Logic programming and databases

In this section, we introduce the basic logic programming and database terminology and explain the fundamentals of relational and deductive databases.

Patterns discovered by ILP systems are typically expressed as *logic programs*, an important subset of first-order (predicate) logic, also called relational logic. Logic programs consist of *clauses*. We can think of clauses as first-order rules, where the conclusion part is termed the *head* and the condition part the *body* of the clause. The head and body of a clause consist of atoms, where an *atom* is a predicate applied to some arguments (e.g., variables). A set of clauses is called a *clausal theory*.

As an example, consider the clause $father(X, Y) \vee mother(X, Y) \leftarrow parent(X, Y)$. This clause is read as follows: "if X is a parent of Y then X is the father of Y or X is the mother of Y" ($\vee$ stands for logical or). $parent(X, Y)$ is the body of the clause and $father(X, Y) \vee mother(X, Y)$ is the head. *parent*, *father* and *mother* are predicates, X and Y are variables, and $parent(X, Y)$, $father(X, Y)$ and $mother(X, Y)$ are atoms. Variables in clauses are implicitly universally quantified. Note that we adopt the Prolog [3.3] syntax in that capital letters denote variables.

As opposed to full clauses, *definite clauses* contain exactly one atom in the head. As compared to definite clauses, *program clauses* can also contain negated atoms in the body. *Logic programs* are sets of program clauses. While the clause in the paragraph above is a *full clause*, the clause $ancestor(X, Y) \leftarrow parent(Z, Y) \wedge ancestor(X, Z)$ is a definite clause ($\wedge$ stands for logical and). It is also a recursive clause, since it defines the relation *ancestor* in terms of itself and the relation *parent*. The clause $mother(X, Y) \leftarrow parent(X, Y) \wedge not\ male(X)$ is a program clause. A set of program clauses with the same predicate in the head is called a *predicate definition.* Most ILP approaches learn predicate definitions.

A *predicate* in logic programming corresponds to a *relation* in a relational database. A n-ary relation p is formally defined as a set of tuples [3.37], i.e., a subset of the Cartesian product of n domains $D_1 \times D_2 \times \ldots \times D_n$, where a *domain* (or a *type*) is a set of values. It is assumed that a relation is finite unless stated otherwise. A *relational database* (RDB) is a set of relations.

Thus, a predicate corresponds to a relation, and the arguments of a predicate correspond to the attributes of a relation. The major difference is that the attributes of a relation are *typed* (i.e., a domain is associated with each attribute).

Program clauses can be typed. For example, in the relation $lives\_in(X, Y)$, we may want to specify that X is of type *person* and Y is of type *city*.

The basic difference between program clauses and database clauses is in the use of types. *Database clauses* are *typed program clauses* of the form $T \leftarrow L_1 \wedge \ldots \wedge L_m$, where T is an atom and $L_1, \ldots, L_m$ are literals. A *literal* is either a positive literal (an atom), or a negative literal (the negation of an atom).

A *deductive database* is a set of database clauses. In deductive databases, relations can be defined *extensionally* as sets of tuples (as in RDBs) or *intensionally* as sets of database clauses (*views*). Database clauses use variables and function symbols in predicate arguments. As recursive types and recursive predicate definitions are allowed, the language is substantially more expressive than the language of relational databases [3.21, 3.37].

If we restrict database clauses to be non-recursive, we obtain the formalism of deductive hierarchical databases. A *deductive hierarchical database* (DHDB) is a deductive database restricted to non-recursive predicate definitions and non-recursive types. Non-recursive types determine finite sets of values which are constants or structured terms with constant arguments. While the expressive power of DHDB is the same as that of RDB, DHDB allows intensional relations which can be much more compact than a RDB representation.

Note that full clauses, such as $father(X, Y) \vee mother(X, Y) \leftarrow parent(X, Y)$ can be used to express integrity constraints on databases. While most ILP approaches are used to learn *views* (logical definitions or predicate definitions) from extensional relations [3.32], some ILP approaches can be used to learn integrity constraints [3.10] as well.

Table 3.1 relates basic database and logic programming terms. For a full treatment of logic programming, relational databases and deductive databases, we refer the reader to [3.21] and [3.37].

Table 3.1. Database and logic programming terms.

DB terminology	LP terminology
relation name p	predicate symbol p
attribute of relation p	argument of predicate p
tuple $\langle a_1, \ldots, a_n \rangle$	ground fact $p(a_1, \ldots, a_n)$
relation p -	predicate p -
a set of tuples	defined extensionally by a set of ground facts
relation q	predicate q
defined as a view	defined intensionally by a set of rules (clauses)

3.3 Logic programming in a nutshell

This section can be skipped upon first reading of the book. It is provided here for the sake of completeness, since some logic programming notions such as substitution, subsumption, logical (semantic) entailment, syntactic entailment, Herbrand models, soundness and completeness occur without explanation in some chapters of this volume. To understand the meaning of these terms, the reader should study this section. Notice, however, that most chapters can be understood also without devoting much effort to studying the fundamental issues of logic programming, using only the basic terminology of the previous section.

In this section, the basic concepts of logic programming are introduced. These include the language (syntax) of logic programs, as well as the basic notions from model and proof theory. The syntax defines what are legal the sentences/statements in the language of logic programs. Model theory is concerned with assigning meaning (truth values) to such statements. Proof theory focusses on (deductive) reasoning with such statements.

For a thorough treatment of logic programming we refer to the standard textbook of Lloyd [3.21]. The overview below is mostly based on the comprehensive and easily readable text by Hogger [3.15].

3.3.1 The language

A first-order *alphabet* consists of variables, predicate symbols and function symbols (which include constants). A *variable* is a term, and a *function symbol* immediately followed by a bracketed n-tuple of terms is a term. Thus $f(g(X),h)$ is a term when f, g and h are function symbols and X is a variable – strings starting with lower-case letters denote predicate and function symbols, while strings starting with upper-case letters denote variables. A *constant* is a function symbol of arity 0 (i.e., followed by a bracketed 0-tuple of terms, which is often left implicit). A *predicate symbol* immediately followed by a bracketed n-tuple of terms is called an *atomic formula* or *atom.* For example, $mother(maja, filip)$ and $father(X,Y)$ are atoms.

A *well-formed formula* (also called a sentence or statement) is either an atomic formula or takes one of the following forms: F, (F), $\overline{F}$, $F \vee G$, $F \wedge G$, $F \leftarrow G$, $F \leftrightarrow G$, $\forall X : F$ and $\exists X : F$, where F and G are well-formed formulae and X is a variable. $\overline{F}$ denotes the negation of F, $\vee$ denotes logical disjunction (or), and $\wedge$ logical conjunction (and). $F \leftarrow G$ stands for implication (F if G, $F \vee \overline{G}$) and $F \leftrightarrow G$ stands for equivalence (F if and only if G). $\forall$ and $\exists$ are the universal (for all X F holds) and the existential quantifier (there exists an X such that F holds). In the formulae $\forall X : F$ and $\exists X : F$, all occurrences of X are said to be *bound.* A *sentence* or a *closed formula* is a well-formed formula in which every occurrence of every variable symbol is bound. For example, $\forall Y \exists X father(X,Y)$ is a sentence, while $father(X, andy)$ is not.

The clausal form is a normal form for first-order sentences. A *clause* is a disjunction of *literals* – a *positive literal* is an atom, a *negative literal* the negation of an atom – preceded by a prefix of universal quantifiers, one for each variable appearing in the disjunction. In other words, a clause is a formula of the form $\forall X_1 \forall X_2 ... \forall X_s (L_1 \vee L_2 \vee ... L_m)$, where each L_i is a literal and $X_1, X_2, ..., X_s$ are all the variables occurring in $L_1 \vee L_2 \vee ... L_m$.

A clause can also be represented as a finite set (possibly empty) of literals. The set $\{A_1, A_2, ..., A_h, \overline{B_1}, \overline{B_2}, ..., \overline{B_b}\}$, where A_i and B_i are atoms, stands for the clause $(A_1 \vee ... \vee A_h \vee \overline{B_1} \vee ... \vee \overline{B_b})$, which is equivalently represented as $A_1 \vee ... \vee A_h \leftarrow B_1 \wedge ... \wedge B_b$. Most commonly, this same clause is written as $A_1, ..., A_h \leftarrow B_1, ..., B_b$, where $A_1, ..., A_h$ is called the *head* and $B_1, ..., B_b$ the *body* of the clause. Commas in the head of the clause denote logical disjunction, while commas in the body of the clause denote logical conjunction. A set of clauses is called a *clausal theory* and represents the conjunction of its clauses.

A clause is a *Horn clause* if it contains at most one positive literal; it is a *definite clause* if it contains exactly one positive literal. A set of definite clauses is called a *definite logic program.* A fact is a definite clause with an empty body, e.g., $parent(mother(X), X) \leftarrow$, also written simply as $parent(mother(X), X)$. A *goal* (also called a *query*) is a Horn clause with no positive literals.

A *program clause* is a clause of the form $A \leftarrow L_1, ..., L_m$ where A is an atom, and each of $L_1, ..., L_m$ is a positive or negative literal. A negative literal in the body of a program clause is written in the form *not* B, where B is an atom. A *normal program* (or *logic program*) is a set of program clauses. A *predicate definition* is a set of program clauses with the same predicate symbol (and arity) in their heads.

Let us now illustrate the above definitions with some examples. The clause

$$daughter(X, Y) \leftarrow female(X), mother(Y, X).$$

is a definite program clause, while the clause

$$daughter(X, Y) \leftarrow not\ male(X), father(Y, X).$$

is a normal program clause. Together, the two clauses constitute a predicate definition of the predicate *daughter*/2. This predicate definition is also a normal logic program. The first clause is an abbreviated representation of the formula

$$\forall X \forall Y : daughter(X, Y) \vee \overline{female(X)} \vee \overline{mother(Y, X)}$$

and can also be written in set notation as

$$\{daughter(X, Y), \overline{female(X)}, \overline{mother(Y, X)}\}$$

The set of *variables* in a term, atom or clause F, is denoted by $vars(F)$. A *substitution* $\theta = \{V_1/t_1, ..., V_n/t_n\}$ is an assignment of terms t_i to variables V_i. Applying a substitution θ to a term, atom, or clause F yields the instantiated term, atom, or clause $F\theta$ where all occurrences of the variables V_i are simultaneously replaced by the term t_i. A term, atom or clause F is called *ground* when there is no variable occurring in F, i.e., $vars(F) = \emptyset$. The fact $daughter(mary, ann)$ is thus ground.

A clause or clausal theory is called *function-free* if it contains only variables as terms, i.e., contains no function symbols (this also means no constants). The clause $daughter(X, Y) \leftarrow female(X), mother(Y, X)$ is function free and the clause $even(s(s(X)) \leftarrow even(X)$ is not. A *Datalog clause* (program) is a definite clause (program) that contains no function symbols of non-zero arity. This means that only variables and constants can be used as predicate arguments. The size of a term, atom, clause, or a clausal theory T is the number of symbols that appear in T, i.e., the number of all occurrences in T of predicate symbols, function symbols and variables.

3.3.2 Model theory

Model theory is concerned with attributing meaning (truth value) to sentences (well-formed formulae) in a first-order language. Informally, the sentence is mapped to some statement about a chosen domain through a process known as interpretation. An *interpretation* is determined by the set of ground facts (ground atomic formulae) to which it assigns the value true. Sentences involving variables and quantifiers are interpreted by using the truth values of the ground atomic formulae and a fixed set of rules for interpreting logical operations and quantifiers, such as "$\overline{F}$ is true if and only if F is false".

An interpretation which gives the value true to a sentence is said to satisfy the sentence; such an interpretation is called a *model* for the sentence. An interpretation which does not satisfy a sentence is called a *counter-model* for that sentence. By extension, we also have the notion of a model (counter-model) for a set of sentences (e.g., for a clausal theory): an interpretation is a model for the set if and only if it is a model for each of the set's members. A sentence (set of sentences) is *satisfiable* if it has at least one model; otherwise it is *unsatisfiable.*

A sentence F *logically implies* a sentence G if and only if every model for F is also a model for G. We denote this by $F \models G$. Alternatively, we say that G is a *logical* (or *semantic*) *consequence* of F. By extension, we have the notion of *logical implication* between sets of sentences.

A *Herbrand interpretation* over a first-order alphabet is a set of ground facts constructed with the predicate symbols in the alphabet and the ground terms from the corresponding Herbrand domain of function symbols; this is the set of ground atoms considered to be true by the interpretation. A Herbrand interpretation I is a model for a clause c if and only if for all substitutions θ such that $c\theta$ is ground $body(c)\theta \subset I$ implies $head(c)\theta \cap I \neq \emptyset$. In that case, we say c is true in I. A Herbrand interpretation I is a model for a clausal theory T if and only if it is a model for all clauses in T. We say that I is a *Herbrand model* of c, respectively T.

Roughly speaking, the truth of a clause c in a (finite) interpretation I can be determined by running the goal (query) $body(c), not\ head(c)$ on a database containing I, using a *theorem prover* such as Prolog [3.3]. If the query succeeds, the clause is false in I; if it fails, the clause is true. Analogously, one can determine the truth of a clause c in the *minimal (least) Herbrand model* of a theory T by running the goal $body(c), not\ head(c)$ on a database containing T.

To illustrate the above notions, consider the Herbrand interpretation $i = \{parent(saso, filip), parent(maja, filip), son(filip, saso), son(filip, maja)\}$. The clause $c = parent(X, Y) \leftarrow son(Y, X)$ is true in i, i.e., i is a model of c. On the other hand, i is not a model of the clause $parent(X, X) \leftarrow$ (which means that everybody is their own parent).

3.3.3 Proof theory

Proof theory focusses on (deductive) reasoning with logic programs. Whereas model theory considers the assignment of meaning to sentences, proof theory considers the generation of sentences (conclusions) from other sentences (premises). More specifically, proof theory considers the *derivability* of sentences in the context of some set of inference rules, i.e., rules for sentence derivation. Formally, an inference system consists of an initial set S of sentences (axioms) and a set R of inference rules.

Using the inference rules, we can derive new sentences from S and/or other derived sentences. The fact that sentence s can be derived from S is denoted $S \vdash s$. A proof is a sequence $s_1, s_2, \ldots.., s_n$, such that each s_i is either in S or derivable using R from S and $s_1, ..., s_{i-1}$. Such a proof is also called a *derivation* or *deduction*. Note that the above notions are of entirely syntactic nature. They are directly relevant to the computational aspects of automated deductive inference.

The set of inference rules R defines the derivability relation $\vdash$. A set of inference rules is *sound* if the corresponding derivability relation is a subset of the logical implication relation, i.e., for all S and s, if $S \vdash s$ then $S \models s$. It is *complete*if the other direction of the implication holds, i.e., for all S and s, if $S \models s$ then $S \vdash s$. The properties of *soundness* and *completeness* establish a relation between the notions of *syntactic* ($\vdash$) and *semantic* ($\models$) *entailment* in logic programming and first-order logic. When the set of inference rules is both sound and complete, the two notions coincide.

Resolution comprises a single inference rule applicable to clausal-form logic. From any two clauses having an appropriate form, resolution derives a new clause as their consequence. For example, the clauses $daughter(X,Y) \leftarrow female(X), parent(Y,X)$ and $female(sonja) \leftarrow$ resolve into the clause $daughter(sonja,Y) \leftarrow parent(Y,sonja)$. Resolution is sound: every resolvent is implied by its parents. It is also refutation complete: the empty clause is derivable by resolution from any set S of Horn clauses if S is unsatisfiable.

3.4 The basic ILP task: Relational rule induction

While logic programming (and in particular its proof theory) is concerned with deductive inference, inductive logic programming is concerned with inductive inference. It generalizes from individual instances/observations in the presence of background knowledge, finding regularities/hypotheses about yet unseen instances.

The most commonly addressed task in ILP is the task of learning logical definitions of relations [3.32], where tuples that belong or do not belong to the target relation are given as examples. From training examples ILP then induces a logic program (predicate definition) corresponding to a view that defines the target relation in terms of other relations that are given as background knowledge. This classical ILP task is addressed, for instance, by the seminal MIS system [3.35] (rightfully considered as one of the most influential ancestors of nowadays ILP) and one of the best known ILP systems FOIL [3.32] which is described also in Chapter 12. This ILP setting, aimed at learning concept descriptions in the form of first-order classification and prediction rules, nowadays called *predictive ILP*, is still the most common ILP setting. In the past, this setting was referred to as *normal ILP*, *explanatory induction*, *discriminatory induction*, and/or *strong ILP*. Despite many new developments in the area, much of this book is devoted to this most popular ILP framework (see Chapters 6, 7, 9, and 11).

Formally, given is a set of examples, i.e., tuples that belong to the target relation p (positive examples) and tuples that do not belong to p (negative examples). Given are also background relations (or background predicates) q_i that constitute the background knowledge and can be used in the learned definition of p. Finally, a hypothesis language, specifying syntactic restrictions on the definition of p is also given (either explicitly or implicitly). The task is to find a definition of the target relation p that is consistent and complete. Informally, it has to explain all the positive and none of the negative tuples.

More formally, given is a set of examples $E = P \cup N$, where P contains positive and N negative examples, and background knowledge B. The task is to find a hypothesis H such that $\forall e \in P : B \wedge H \models e$ (H is complete) and $\forall e \in N : B \wedge H \not\models e$ (H is consistent). This setting, introduced by Muggleton [3.25], is also called *learning from entailment*. In an alternative setting proposed by De Raedt and Džeroski [3.11], the requirement that $B \wedge H \models e$ is replaced by the requirement that H be true in the minimal Herbrand model of $B \wedge e$: this setting is called *learning from interpretations*.

In the most general formulation, each e, as well as B and H can be a clausal theory. In practice, each e is most often a ground example (tuple), B is a relational database (which may or may not contain views) and H is a definite logic program. Recall that $\models$ denotes logical implication (semantic entailment). The semantic entailment ($\models$) is in practice replaced with syntactic entailment ($\vdash$) or provability, where the resolution inference rule (as implemented in Prolog) is most often used to prove examples from a hypothesis and the background knowledge.

As an illustration, consider the task of defining relation $daughter(X, Y)$, which states that person X is a daughter of person Y, in terms of the background knowledge relations $female$ and $parent$. These relations are given in Table 3.2. There are two positive and two negative examples of the *target* relation *daughter*.

In the hypothesis language of definite program clauses it is possible to formulate the following definition of the target relation,

$$daughter(X, Y) \leftarrow female(X), parent(Y, X).$$

Table 3.2. A simple ILP problem: learning the *daughter* relation. Positive examples are denoted by ⊕ and negative examples by ⊖.

Training examples		Background knowledge	
$daughter(mary, ann)$.	⊕	$parent(ann, mary)$.	$female(ann)$.
$daughter(eve, tom)$.	⊕	$parent(ann, tom)$.	$female(mary)$.
$daughter(tom, ann)$.	⊖	$parent(tom, eve)$.	$female(eve)$.
$daughter(eve, ann)$.	⊖	$parent(tom, ian)$.	

which is consistent and complete with respect to the background knowledge and the training examples.

In general, depending on the background knowledge, the hypothesis language and the complexity of the target concept, the target predicate definition may consist of a set of clauses, such as

$$daughter(X,Y) \leftarrow female(X), mother(Y,X).$$
$$daughter(X,Y) \leftarrow female(X), father(Y,X).$$

if the relations *mother* and *father* were given in the background knowledge instead of the *parent* relation.

The hypothesis language is typically a subset of the language of program clauses. As the complexity of learning grows with the expressiveness of the hypothesis language, restrictions have to be imposed on hypothesized clauses. Typical restrictions are the exclusion of recursion and restrictions on variables that appear in the body of the clause but not in its head (so-called new variables).

From a data mining perspective, the task described above is a binary classification task, where one of two classes is assigned to the examples (tuples): ⊕ (positive) or ⊖ (negative). Classification tasks are most often addressed by decision tree approaches [3.33] within the data mining community. Rule induction, however, is also used occasionally. Rules can be generated from decision trees [3.33] or induced directly [3.22, 3.6].

ILP systems dealing with the classification task typically adopt the *covering approach* of rule induction systems. An outline of the covering algorithm can be found in Section 1.4.3. A more detailed description and pseudo-code can be found in Section 10.4, Tables 10.2 and 10.3. In a main loop, a covering algorithm constructs a set of clauses. Starting from an empty set of clauses, it constructs a clause explaining some of the positive examples, adds this clause to the hypothesis, and removes the positive examples explained. These steps are repeated until all positive examples have been explained (the hypothesis is complete).

In the inner loop of the covering algorithm, individual clauses are constructed by (heuristically) searching the space of possible clauses, structured by a specialization or generalization operator. Typically, search starts with a very general rule (clause with no conditions in the body), then proceeds to add literals (conditions) to this clause until it only covers (explains) positive examples (the clause is consistent). This search can be bound from below by using so-called *bottom clauses*, constructed by least general generalization or inverse resolution/entailment. We discuss this in detail in the remainder of this chapter.

When dealing with incomplete or noisy data, which is most often the case, the criteria of consistency and completeness are relaxed. Statistical criteria are typically used instead. These are based on the number of positive and negative examples explained by the definition and the individual constituent clauses.

3.5 Structuring the space of clauses

The previous section described how to learn sets of clauses by using the covering algorithm for clause/rule set induction. This section described some of the mechanisms underlying single clause/rule induction. Learning of single clauses is also the topic of Sections 3.6 and 3.7.

In order to search the space of relational rules (program clauses) systematically, it is useful to impose some structure upon it, e.g., an *ordering*. One such ordering is based on *θ-subsumption*, defined below.

Recall first (Section 3.3.1) that a *substitution* $\theta = \{V_1/t_1, ..., V_n/t_n\}$ is an assignment of terms t_i to variables V_i. Applying a substitution θ to a term, atom, or clause F yields the instantiated term, atom, or clause $F\theta$ where all occurrences of the variables V_i are simultaneously replaced by the term t_i.

Let c and c' be two program clauses. Clause c *θ-subsumes* c' if there exists a substitution θ, such that $c\theta \subseteq c'$ [3.31].

To illustrate the above notions, consider the clause c

$$c = daughter(X, Y) \leftarrow parent(Y, X).$$

Applying the substitution $\theta = \{X/mary, Y/ann\}$ to clause c yields

$$c\theta = daughter(mary, ann) \leftarrow parent(ann, mary).$$

Recall also that the clausal notation $daughter(X, Y) \leftarrow parent(Y, X)$ stands for

$$\{daughter(X, Y), \overline{parent(Y, X)}\}$$

where all variables are assumed to be universally quantified and the commas denote disjunction. According to the definition, clause c θ-subsumes c' if there is a substitution θ that can be applied to c such that every literal in the resulting clause occurs in c'. Clause c θ-subsumes the clause

$$c' = daughter(X, Y) \leftarrow female(X), parent(Y, X)$$

under the empty substitution $\theta = \emptyset$, since $\{daughter(X, Y), \overline{parent(Y, X)}\}$ is a proper subset of $\{daughter(X, Y), \overline{female(X)}, \overline{parent(Y, X)}\}$. Furthermore, under the substitution $\theta = \{X/mary, Y/ann\}$, clause c θ-subsumes the

clause $c' = daughter(mary, ann) \leftarrow$
$female(mary), parent(ann, mary), parent(ann, tom)$.

θ-subsumption introduces a syntactic notion of generality. Clause c *is at least as general as* clause c' ($c \leq c'$) if c *θ-subsumes* c'. Clause c *is more general than* c' ($c < c'$) if $c \leq c'$ holds and $c' \leq c$ does not. In this case, we say that c' *is a specialization of* c and c *is a generalization of* c'. If the clause c' is a specialization of c then c' is also called a *refinement* of c. The only clause refinements usually considered by ILP systems are the minimal (most general) specializations of the clause.

There are two important properties of θ-subsumption:

- If c θ-subsumes c' then c logically entails c', $c \models c'$. The reverse is not always true. As an example, [3.14] gives the following two clauses $c = list([V|W]) \leftarrow list(W)$ and $c' = list([X, Y|Z]) \leftarrow list(Z)$. Given the empty list, c constructs lists of any given length, while c' constructs lists of even length only, and thus $c \models c'$. However, no substitution exists that can be applied to c to yield c', since it should map W both to $[Y|Z]$ and to Z which is impossible. Therefore, c does not θ-subsume c'.
- The relation $\leq$ introduces a *lattice* on the set of reduced clauses [3.31]. This means that any two reduced clauses have a *least upper bound* (*lub*) and a *greatest lower bound* (*glb*). Both the *lub* and the *glb* are unique up to equivalence (renaming of variables) under θ-subsumption. Reduced clauses are the minimal representatives of the equivalence classes of clauses defined by θ-subsumption. For example, the clauses $daughter(X, Y) \leftarrow parent(Y, X), parent(W, V)$ and $daughter(X, Y) \leftarrow parent(Y, X)$ θ-subsume one another and are thus equivalent. The latter is reduced, while the former is not.

The second property of θ-subsumption leads to the following definition: The *least general generalization* (*lgg*) of two clauses c and c', denoted by $lgg(c, c')$, is the least upper bound of c and c' in the θ-subsumption lattice [3.31]. The rules for computing the *lgg* of two clauses are outlined later in this chapter.

Note that θ-subsumption and least general generalization are purely syntactic notions since they do not take into account any background knowledge. Their computation is therefore simple and easy to be implemented in an ILP system. The same holds for the notion of generality based on θ-subsumption. On the other hand, taking background knowledge into account would lead to the notion of *semantic generality* [3.29, 3.5], defined as follows: Clause c is *at least as general* as clause c' with respect to background theory B if $B \cup \{c\} \models c'$.

The syntactic, θ-subsumption based, generality is computationally more feasible. Namely, semantic generality is in general undecidable and does not introduce a lattice on a set of clauses. Because of these problems, syntactic generality is more frequently used in ILP systems.

θ-subsumption is important for inductive logic programming for the following reasons:

- As shown above, it provides a generality ordering for hypotheses, thus structuring the hypothesis space. It can be used to prune large parts of the search space.
- θ-subsumption provides the basis for the following important ILP techniques:
  - clause construction by top-down *searching of refinement graphs*,
  - bounding the search of refinement graphs from below by the bottom clause constructed as
    - the *least general generalizations* of two (or more) training examples, relative to the given background knowledge [3.28], or
    - the *most specific inverse resolvent* of an example with respect to the given background knowledge [3.25].

These techniques will be described in more detail in the following sections.

3.6 Searching the space of clauses

Most ILP approaches search the hypothesis space of program clauses in a top-down manner, from general to specific hypotheses, using a θ-subsumption-based *specialization operator*. A specialization operator is usually called a *refinement operator* [3.35]. Given a hypothesis language $\mathcal{L}$, a refinement operator ρ maps a clause c to a set of clauses $\rho(c)$ which are specializations (refinements) of c: $\rho(c) = \{c' \mid c' \in \mathcal{L},\ c < c'\}$.

A refinement operator typically computes only the set of minimal (most general) specializations of a clause under θ-subsumption. It employs two basic syntactic operations on a clause:

- apply a substitution to the clause, and
- add a literal to the body of the clause.

The hypothesis space of program clauses is a lattice, structured by the θ-subsumption generality ordering. In this lattice, a *refinement graph* can be defined as a directed, acyclic graph in which *nodes* are program clauses and *arcs* correspond to the basic refinement operations: substituting a variable with a term, and adding a literal to the body of a clause.

Figure 3.1 depicts a part of the refinement graph for the family relations problem defined in Table 3.2, where the task is to learn a definition of the *daughter* relation in terms of the relations *female* and *parent*.

At the top of the refinement graph (lattice) is the clause

$$c = daughter(X,Y) \leftarrow$$

where an empty body is written instead of the body *true*. The refinement operator ρ generates the refinements of c, which are of the form

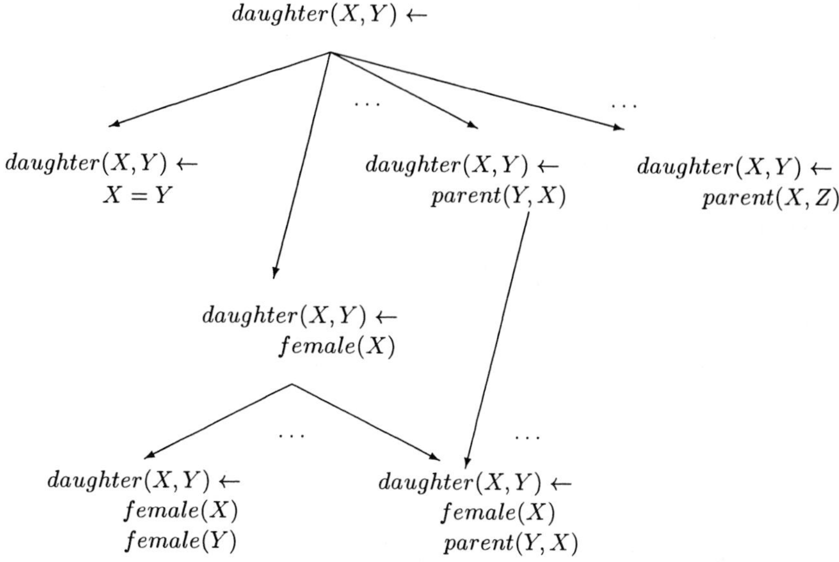

Fig. 3.1. Part of the refinement graph for the family relations problem.

$$\rho(c) = \{daughter(X,Y) \leftarrow L\}$$

where L is one of following literals:

- literals having as arguments the variables from the head of the clause: $X = Y$ (this corresponds to applying a substitution X/Y), $female(X)$, $female(Y)$, $parent(X,X)$, $parent(X,Y)$, $parent(Y,X)$, and $parent(Y,Y)$, and
- literals that introduce a new distinct variable Z ($Z \neq X$ and $Z \neq Y$) in the clause body: $parent(X,Z)$, $parent(Z,X)$, $parent(Y,Z)$, and $parent(Z,Y)$.

This assumes that the language is restricted to definite clauses, hence literals of the form *not* L are not considered, and non-recursive clauses, hence literals with the predicate symbol *daughter* are not considered.

The search for a clause starts at the top of the lattice, with the clause $d(X,Y) \leftarrow$ that covers all example (positive and negative). Its refinements are then considered, then their refinements in turn, an this is repeated until a clause is found which covers only positive examples. In the example above, the clause $daughter(X,Y) \leftarrow female(X), parent(Y,X)$ is such a clause. Note that this clause can be reached in several ways from the top of the lattice, e.g., by first adding $female(X)$, then $parent(Y,X)$ or vice versa.

The refinement graph is typically searched heuristically level-wise, using heuristics based on the number of positive and negative examples covered by a clause. As the branching factor is very large, greedy search methods are typically applied which only consider a limited number of alternatives at each level. Hill-climbing considers only one best alternative at each level, while beam search considers n best alternatives, where n is the beam width. Occasionally, complete search is used, e.g., A^* best-first search or breadth-first search.

3.7 Bounding the search for clauses

The branching factor of a refinement graph, i.e., the number of refinements a clause has, is very large. This is especially true for clauses deeper in the lattice that contain many variables. It is thus necessary to find ways to reduce the space of clauses actually searched.

One approach is to make the refinement graph smaller by making the refinement operator take into account the types of predicate arguments (discussed in Section 3.2), as well as input/output mode declarations. For example, we might restrict the $lives\_in(X, Y)$ predicate to only give us the city for a given person and not give us all persons that live in the city for a given city. This can be done with a mode declaration $lives\_in(+person, -city)$, specifying that the person has to be given when calling this predicate.

Type and mode declarations can be combined with the construction of a bottom clause that bounds the search of the refinement lattice from below. This is the most specific clause covering a given example (or examples). Only clauses on the path between the top and the bottom clause are considered, significantly improving efficiency.

The bottom clause can be constructed as the relative least general generalization of two (or more) examples [3.28] or the most specific inverse resolvent of an example [3.25], both with respect to given background knowledge. The bottom clause can be also derived using inverse entailment [3.26]. Below we discuss relative least general generalization and inverse resolution. Inverse entailment is discussed in detail in Chapter 7.

3.7.1 Relative least general generalization

Plotkin's notion of least general generalization (Plotkin 1969) forms the basis of cautious generalization: if two clauses c_1 and c_2 are true, it is very likely that $lgg(c_1, c_2)$ will also be true.

The least general generalization of two clauses c and c', denoted $lgg(c, c')$, is the least upper bound of c and c' in the θ-subsumption lattice. It is the most specific clause that θ-subsumes c and c'. If a clause d θ-subsumes c and c', it has to subsume $lgg(c, c')$ as well. To actually compute the lgg of two clauses, lgg of terms and literals need to be defined first [3.31].

The lgg of two terms $lgg(t_1, t_2)$ (where s, t, s_i, t_i denote terms and f and g are function symbols) is computed as

1. $lgg(t,t) = t$,
2. $lgg(f(s_1,..,s_n), f(t_1,..,t_n)) = f(lgg(s_1,t_1),..,lgg(s_n,t_n))$.
3. $lgg(f(s_1,..,s_m), g(t_1,..,t_n)) = V$, where $f \neq g$, and V is a variable which represents $lgg(f(s_1,..,s_m), g(t_1,..,t_n))$,
4. $lgg(s,t) = V$, where $s \neq t$ and at least one of s and t is a variable; in this case, V is a variable which represents $lgg(s,t)$.

For example, $lgg([a,b,c],[a,c,d]) = [a,X,Y]$, and $lgg(f(a,a), f(b,b)) = f(lgg(a,b), lgg(a,b)) = f(V,V)$, where V stands for $lgg(a,b)$. When computing lggs one must be careful to use the same variable for multiple occurrences of the lggs of subterms, i.e., $lgg(a,b)$ in this example. This holds for lggs of terms, atoms and clauses alike.

The lgg of two atoms $lgg(A_1,A_2)$ is computed as follows:

1. $lgg(p(s_1,..,s_n), p(t_1,..,t_n)) = p(lgg(s_1,t_1),..,lgg(s_n,t_n))$, if atoms have the same predicate symbol p,
2. $lgg(p(s_1,..,s_m), q(t_1,..,t_n))$ is undefined if $p \neq q$.

The lgg of two literals $lgg(L_1,L_2)$ is defined as follows:

1. if L_1 and L_2 are atoms, then $lgg(L_1,L_2)$ is computed as defined above,
2. if both L_1 and L_2 are negative literals, $L_1 = \overline{A_1}$ and $L_2 = \overline{A_2}$, then $lgg(L_1,L_2) = lgg(\overline{A_1},\overline{A_2}) = \overline{lgg(A_1,A_2)}$,
3. if L_1 is a positive and L_2 is a negative literal, or vice versa, $lgg(L_1,L_2)$ is undefined.

For example, $lgg(parent(ann,mary), parent(ann,tom)) = parent(ann,X)$., $lgg(\overline{parent(ann,mary)}, parent(ann,tom))$ is undefined, and $lgg(parent(ann,X), daughter(mary,ann))$ is undefined.

Taking into account that clauses are sets of literals, the lgg of two clauses is defined as follows. Let $c_1 = \{L_1,..,L_n\}$ and $c_2 = \{K_1,..,K_m\}$. Then $lgg(c_1,c_2) = \{M_{ij} = lgg(L_i,K_j) \mid L_i \in c_1, K_j \in c_2, lgg(L_i,K_j) \text{ is defined } \}$. If $c_1 = daughter(mary,ann) \leftarrow female(mary), parent(ann,mary)$ and $c_2 = daughter(eve,tom) \leftarrow female(eve), parent(tom,eve)$, then $lgg(c_1,c_2) = daughter(X,Y) \leftarrow female(X), parent(Y,X)$, where X stands for $lgg(mary,eve)$ and Y stands for $lgg(ann,tom)$.

The definition of *relative least general generalization(*lgg*) (rlgg)* is based on the semantic notion of generality. The *rlgg* of two clauses c_1 and c_2 is the least general clause which is more general than both c_1 and c_2 with respect (*relative*) to background knowledge B. The notion of *rlgg* was used in the ILP system GOLEM (Muggleton and Feng 1990). To avoid the problems with the semantic notion of generality, the background knowledge B in GOLEM is restricted to ground facts. If K denotes the conjunction of all facts from B, the *rlgg* of two ground atoms A_1 and A_2 (positive examples), relative to B can be computed as: $rlgg(A_1,A_2) = lgg((A_1 \leftarrow K),(A_2 \leftarrow K))$.

Given the positive examples $e_1 = daughter(mary,ann)$ and $e_2 = daughter(eve,tom)$ and the background knowledge B for the family example, the least general generalization of e_1 and e_2 relative to B is computed as: $rlgg(e_1, e_2) = lgg((e_1 \leftarrow K), (e_2 \leftarrow K))$ where K denotes the conjunction of the literals $parent(ann, mary)$, $parent(ann, tom)$, $parent(tom, eve)$, $parent(tom, ian)$, $female(ann)$, $female(mary)$, and $female(eve)$.

For notational convenience, the following abbreviations are used: *d-daughter, p-parent, f-female, a-ann, e-eve, m-mary, t-tom, i-ian.* The conjunction of facts from the background knowledge (comma stands for conjunction) is $K = p(a,m), p(a,t), p(t,e), p(t,i), f(a), f(m), f(e)$. The computation of $rlgg(e_1, e_2) = lgg((e_1 \leftarrow K), (e_2 \leftarrow K))$, produces the following clause

$$\underline{d(V_{m,e}, V_{a,t}) \leftarrow} p(a,m), p(a,t), p(t,e), p(t,i), f(a), f(m), f(e), \\ p(a, V_{m,t}), \underline{p(V_{a,t}, V_{m,e})}, p(V_{a,t}, V_{m,i}), p(V_{a,t}, V_{t,e}), \\ p(V_{a,t}, V_{t,i}), p(t, V_{e,i}), f(V_{a,m}), f(V_{a,e}), \underline{f(V_{m,e})}.$$

In the above clause, $V_{x,y}$ stands for $rlgg(x, y)$, for each x and y.

In general, a $rlgg$ of training examples can contain infinitely many literals or at least grow exponentially with the number of examples. Since such a clause can be intractably large, constraints are used on introducing new variables into the body of the $rlgg$. For example, literals in the body that are not connected to the head by a chain of variables are removed. In the above example, this yields the clause $\underline{d(V_{m,e}, V_{a,t}) \leftarrow p(V_{a,t}, V_{m,e})}, p(V_{a,t}, V_{m,i}), p(V_{a,t}, V_{t,e}), p(V_{a,t}, V_{t,i}), \underline{f(V_{m,e})}$. Also, nondeterminate literals (that can give more than one value of output arguments for a single value of the input arguments) are eliminated. The literals $p(V_{a,t}, V_{m,i}), p(V_{a,t}, V_{t,e}), p(V_{a,t}, V_{t,i})$ are nondeterminate since $V_{a,t}$ is the input argument and a parent can have more than one child. Eliminating these yields the bottom clause $d(V_{m,e}, V_{a,t}) \leftarrow p(V_{a,t}, V_{m,e}), f(V_{m,e})$, i.e., $daughter(X, Y) \leftarrow female(X), parent(Y, X)$. In this simple example, the bottom clause is our target clause. In practice, the bottom clause is typically very large, containing hundreds of literals.

3.7.2 Inverse resolution

The basic idea of *inverse resolution* introduced by Muggleton and Buntine [3.27], is to invert the *resolution* rule of deductive inference [3.34], i.e., to invert the SLD-resolution proof procedure for definite programs [3.21]. The basic resolution step in propositional logic derives the proposition $p \vee r$ given the premises $p \vee \overline{q}$ and $q \vee r$. In a first-order case, resolution is more complicated, involving substitutions. Let $res(c, d)$ denote the resolvent of clauses c and d.

To illustrate resolution in first-order logic, we use the family example. Suppose that background knowledge B consists of the clauses $b_1 = female(mary)$ and $b_2 = parent(ann, mary)$ and $H = \{c\} = \{daughter(X, Y) \leftarrow female(X), parent(Y, X)\}$. Let $T = H \cup B$. Suppose we want to derive the fact $daughter(mary, ann)$ from T. To this end, we proceed as follows:

- First, the resolvent $c_1 = res(c, b_1)$ is computed under the substitution $\theta_1 = \{X/mary\}$. This means that the substitution θ_1 is first applied to

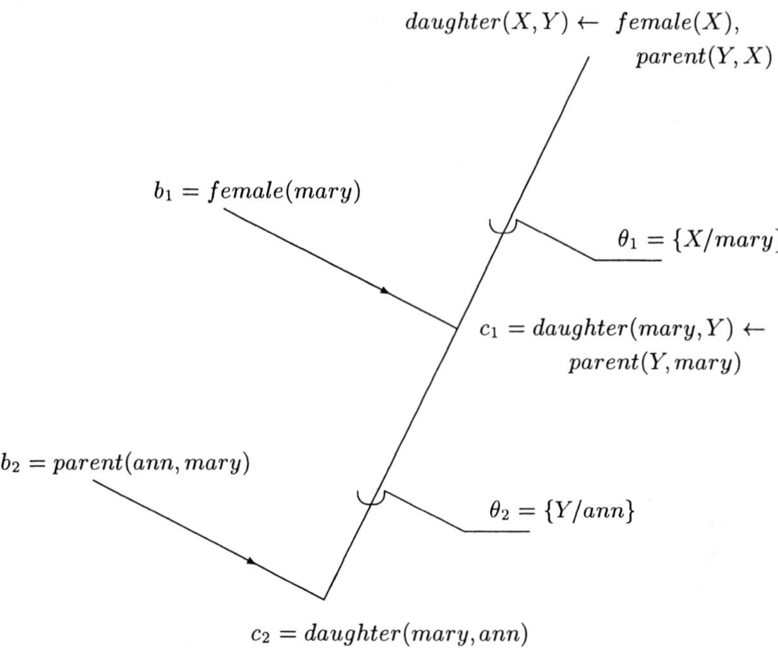

daughter(X,Y) ← female(X), parent(Y,X)
b1 = female(mary)
θ1 = {X/mary}
c1 = daughter(mary,Y) ← parent(Y,mary)
b2 = parent(ann,mary)
θ2 = {Y/ann}
c2 = daughter(mary,ann)

Fig. 3.2. A linear derivation tree.

clause c to obtain $daughter(mary, Y) \leftarrow female(mary), parent(Y, mary)$, which is then resolved with b_1 as in the propositional case. The resolvent of $daughter(X, Y) \leftarrow female(X), parent(Y, X)$ and $female(mary)$ is thus $c_1 = res(c, b_1) = daughter(mary, Y) \leftarrow parent(Y, mary)$.
- The next resolvent $c_2 = res(c_1, b_2)$ is computed under the substitution $\theta_2 = \{Y/ann\}$. The clauses $daughter(mary, Y) \leftarrow parent(Y, mary)$ and $parent(ann, mary)$ resolve in $c_2 = res(c_1, b_2) = daughter(mary, ann)$.

The linear derivation tree for this resolution process is given in Figure 3.2.

Inverse resolution, used in the ILP system CIGOL [3.27], inverts the resolution process using generalization operators based on inverting substitution [3.5]. Given a well-formed formula W, an *inverse substitution* θ^{-1} of a substitution θ is a function that maps terms in $W\theta$ to variables, such that $W\theta\theta^{-1} = W$.

Let $c = daughter(X, Y) \leftarrow female(X), parent(Y, X)$ and $\theta = \{X/mary, Y/ann\}$: then $c' = c\theta = daughter(mary, ann) \leftarrow female(mary), parent(ann, mary)$. By applying the inverse substitution $\theta^{-1} = \{mary/X, ann/Y\}$ to c_1, the original clause c is restored: $c = c'\theta^{-1} = daughter(X, Y) \leftarrow female(Y), parent(Y, X)$. In the general case, inverse substitution is substantially more complex. It involves the *places* of terms in order to ensure that the variables in the initial W are appropriately restored in $W\theta\theta^{-1}$. In fact, each occurrence of a term can be replaced by a different variable in an inverse substitution.

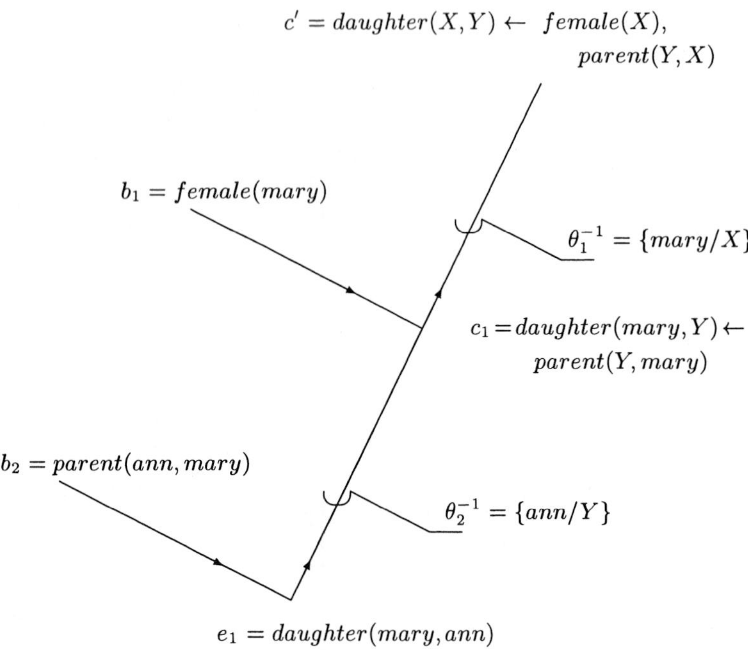
$c' = daughter(X,Y) \leftarrow female(X), parent(Y,X)$
$b_1 = female(mary)$
$\theta_1^{-1} = \{mary/X\}$
$c_1 = daughter(mary,Y) \leftarrow parent(Y,mary)$
$b_2 = parent(ann,mary)$
$\theta_2^{-1} = \{ann/Y\}$
$e_1 = daughter(mary,ann)$

Fig. 3.3. An inverse linear derivation tree.

We will not treat inverse resolution in detail, but will rather illustrate it by an example. Let $ires(c,d)$ denote the inverse resolvent of clauses c and d. As in the example above, let background knowledge B consist of the two clauses $b_1 = female(mary)$ and $b_2 = parent(ann,mary)$. The inverse resolution process might then proceed as follows:

- In the first step, inverse resolution attempts to find a clause c_1 which will, together with b_2, entail e_1. Using the inverse substitution $\theta_2^{-1} = \{ann/Y\}$, an inverse resolution step generates the clause $c_1 = ires(b_2,e_1) = daughter(mary,Y) \leftarrow parent(Y,mary)$.
- Inverse resolution then takes $b_1 = female(mary)$ and c_1. It computes $c' = ires(b_1,c_1)$, using the inverse substitution $\theta_1^{-1} = \{mary/X\}$, yielding $c' = daughter(X,Y) \leftarrow female(X), parent(Y,X)$.

The corresponding inverse linear derivation tree is illustrated in Figure 3.3.

Most specific inverse resolution uses only empty inverse substitutions. In the above example, this would yield the clause $daughter(mary, ann) \leftarrow female(mary), parent(ann, mary)$ as the final inverse resolvent. In practice, inverse substitutions replacing the occurrence of each constant with the same variable are considered to yield the most specific inverse resolvent, which can the be used as a bottom clause. In our case, this is again the target clause, while in practice the bottom clause will be much more specific/larger, containing a large number of literals. It should be noted that both inverse resolution and inverse entailment have problems when the target clause is recursive and is resolved with itself several times to derive a positive example.

3.8 Transforming ILP problems to propositional form

One of the early approaches to ILP, implemented in the ILP system LINUS [3.19] (see also Chapter 11), is based on the idea that the use of background knowledge can introduce new attributes for learning. The learning problem is transformed from relational to attribute-value form and solved by an attribute-value learner. An advantage of this approach is that data mining algorithms that work on a single table (and this is the majority of existing data mining algorithms) become applicable after the transformation.

This approach, however, is feasible only for a restricted class of ILP problems. Thus, the hypothesis language of LINUS is restricted to function-free program clauses which are *typed* (each variable is associated with a predetermined set of values), *constrained* (all variables in the body of a clause also appear in the head) and *nonrecursive* (the predicate symbol the head does not appear in any of the literals in the body), i.e., to function-free constrained DHDB clauses.

The LINUS algorithm which solves ILP problems by transforming them into propositional form consists of the following three steps:

- The learning problem is transformed from relational to attribute-value form.
- The transformed learning problem is solved by an attribute-value learner.
- The induced hypothesis is transformed back into relational form.

The above algorithm allows for a variety of approaches developed for propositional problems, including noise-handling techniques in attribute-value algorithms, such as CN2 [3.7], to be used for learning relations. It is illustrated on the simple ILP problem of learning family relations. The task is to define the target relation $daughter(X, Y)$, which states that person X is a daughter of person Y, in terms of the background knowledge relations $female$, $male$ and $parent$. All the variables are of the type $person$, defined as $person = \{ann, eve, ian, mary, tom\}$. There are two positive and two negative examples of the target relation. The training examples and the relations from the background knowledge are given in Table 3.2. However, since the LINUS approach can use non-ground background knowledge, let us assume that the background knowledge from Table 3.3 is given.

Table 3.3. Non-ground background knowledge for learning the *daughter* relation.

Training examples		Background knowledge		
daughter(mary, ann).	⊕	*parent(X, Y) ← mother(X, Y).*	*mother(ann, mary).*	*female(ann).*
daughter(eve, tom).	⊕		*mother(ann, tom).*	*female(mary).*
daughter(tom, ann).	⊖	*parent(X, Y) ← father(X, Y).*	*father(tom, eve).*	*female(eve).*
daughter(eve, ann).	⊖		*father(tom, ian).*	

The first step of the algorithm, i.e., the transformation of the ILP problem into attribute-value form, is performed as follows. The possible applications of the background predicates on the arguments of the target relation are determined, taking into account argument types. Each such application introduces a new attribute. In our example, all variables are of the same type *person*. The corresponding attribute-value learning problem is given in Table 3.4, where f stands for *female*, m for *male* and p for *parent*. The attribute-value tuples are generalizations (relative to the given background knowledge) of the individual facts about the target relation.

Table 3.4. Propositional form of the *daughter* relation problem.

Variables			Propositional features							
C	X	Y	$f(X)$	$f(Y)$	$m(X)$	$m(Y)$	$p(X,X)$	$p(X,Y)$	$p(Y,X)$	$p(Y,Y)$
⊕	*mary*	*ann*	*true*	*true*	*false*	*false*	*false*	*false*	*true*	*false*
⊕	*eve*	*tom*	*true*	*false*	*false*	*true*	*false*	*false*	*true*	*false*
⊖	*tom*	*ann*	*false*	*true*	*true*	*false*	*false*	*false*	*true*	*false*
⊖	*eve*	*ann*	*true*	*true*	*false*	*false*	*false*	*false*	*false*	*false*

In Table 3.4, *variables* stand for the arguments of the target relation, and *propositional features* denote the newly constructed attributes of the propositional learning task. When learning function-free clauses, only the new attributes (propositional features) are considered for learning.

In the second step, an attribute-value learning program induces the following if-then rule from the tuples in Table 3.4:

$$Class = \oplus \textbf{ if } [female(X) = true] \wedge [parent(Y, X) = true]$$

In the last step, the induced if-then rules are transformed into DHDB clauses. In our example, we get the following clause:

$$daughter(X, Y) \leftarrow female(X), parent(Y, X).$$

The LINUS approach has been extended to handle determinate clauses [3.12, 3.20], which allow the introduction of determinate new variables (which have a unique value for each training example). There also exist a number of other approaches to propositionalization, some of them very recent: an overview of these approaches can be found in Chapter 11.

Let us emphasize again, however, that it is in general not possible to transform an ILP problem into a propositional (attribute-value) form efficiently. De Raedt [3.8] treats the relation between attribute-value learning and ILP in detail, showing that propositionalization of some more complex ILP problems is possible in principle, but results in attribute-value problems that exponentially large, i.e., are of prohibitive size. This has also been the main reason for the development of a variety of new relational data mining (ILP) techniques, rather than using a transformation to propositional form and then standard data mining techniques.

3.9 Relational data mining tasks addressed by ILP

Initial efforts in ILP focussed on relational rule induction, more precisely on concept learning in first-order logic and synthesis of logic programs, cf. [3.25]. An overview of early work is given in the textbook on ILP by Lavrač and Džeroski [3.20]. Representative early ILP systems addressing this task are CIGOL [3.27], FOIL [3.32], GOLEM [3.28] and LINUS [3.19]. More recent representative ILP systems are PROGOL [3.26] and ALEPH [3.36].

State-of-the-art ILP approaches now span most of the spectrum of data mining tasks and use a variety of techniques to address these. The distinguishing features of using multiple relations directly and discovering patterns expressed in first-order logic are present throughout: the ILP approaches can thus be viewed as upgrades of traditional approaches. The Chapter 10 by Van Laer and De Raedt presents a case study of upgrading a propositional approach to classification rule induction to first order logic. Note, however, that upgrading to first-order logic is non-trivial: the expressive power of first-order logic implies computational costs and much work is needed in balancing the expressive power of the pattern languages used and the computational complexity of the data mining algorithm looking for such patterns. This search for a balance between the two has occupied much of the ILP research in the last ten years.

Present ILP approaches to multi-class classification involve the induction of relational classification rules (ICL De Raedt and Van Laer 1995), as well as first order logical decision trees in TILDE [3.2] and S-CART [3.17]. ICL upgrades the propositional rule inducer CN2 [3.6]. TILDE and S-CART upgrade decision tree induction as implemented in C4.5 [3.33] and CART [3.4]. A nearest-neighbor approach to relational classification is implemented in RIBL [3.13] and its successor RIBL2˙ICL is described in Chapters 5 and 10, TILDE in Chapter 5, and RIBL2 in Chapter 9.

Relational regression approaches upgrade propositional regression tree and rules approaches. TILDE and S-CART, as well as RIBL2 can handle continuous classes. FORS [3.16] learns decision lists (ordered sets of rules) for relational regression.

The main non-predictive or descriptive data mining tasks are clustering and discovery of association rules. These have been also addressed in a first-order logic setting. The RIBL distance measure has been used to perform hierarchical agglomerative clustering in RDBC, as well as k-means clustering (see Chapter 9). Chapter 8 describes a relational approach to the discovery of frequent queries and query extensions, a first-order version of association rules.

With such a wide arsenal of relational data mining techniques, there is also a variety of practical applications. ILP has been successfully applied to discover knowledge from relational data and background knowledge in the areas of molecular biology (including drug design, protein structure prediction and functional genomics), environmental sciences, traffic control and natural language processing. An overview of such applications is given in Chapter 14.

3.10 ILP literature

To learn more about ILP, a substantial body of literature and Internet resources can be consulted. Chapter 16 gives a detailed overview of the Internet resources on ILP, and especially those related to data mining and KDD. Here we will focus on literature devoted to ILP, and will only mention that information on ILP literature can be found in ILPnet2's on-line library (`http://www.cs.bris.ac.uk/~ILPnet2/Library/`).

While its origins can be traced back to Plotkin's work in the late 60's [3.31] and Shapiro's work in the early 80's [3.35], ILP started to claim its place under the sun as a separate branch of machine learning in 1991. The first *International Workshop on Inductive Logic Programming (ILP-91)* was organized in 1991 [3.23] and has continued to be organized yearly since then. The ILP workshops have since then become the main forum of publishing research on ILP theory, implementations and applications. Since 1996, the proceedings of the ILP workshops are published by Springer within the Lecture Notes in Artificial Intelligence subseries of the Lecture Notes in Computer Science series.

Papers on ILP regularly appear at a number of machine learning and artificial intelligence conferences. These include the European Conference on Machine Learning (ECML), the International Conference on Machine Learning (ICML), the European Conference on Artificial Intelligence (ECAI), and International Joint Conference on Artificial Intelligence (IJCAI). Recently, papers on ILP and applications of ILP to data mining problems have appeared at the European Conference on Principles and Practice of Knowledge Discovery and Data Mining (PKDD) and the International Conference on

Knowledge Discovery and Data Mining (KDD): a paper on the application of ILP to a data mining problem in functional genomics [3.18] won the best paper award at KDD-2000.

Papers on ILP regularly appear in a number of journals, including *Journal of Logic Programming*, *Machine Learning*, and *New Generation Computing*. Each of these has published several special issues on ILP. Extended versions of selected papers from ILP-97, ILP-98, and ILP-99, appear as special issues of the journals *New Generation Computing* (volume 17(1-2), 1999) and *Machine Learning* (volumes 43(1-2) and 44(3), 2001). Two journal special issues address specifically the topic of using ILP for KDD: *Applied Artificial Intelligence* (volume 12(5), 1998), and *Data Mining and Knowledge Discovery* (volume 3(1), 1999).

Selected papers from the ILP-91 workshop appeared as a book *Inductive Logic Programming*, edited by Muggleton [3.24], and selected papers from the ILP-95 workshop appeared as a book *Advances in Inductive Logic Programming*, edited by De Raedt [3.9]. The latter also includes an overview of the research results of the ESPRIT III Project ILP (*Inductive Logic Programming*, Basic Research Project 6020) supported by the Commission of the European Communities.

Authored books on ILP include *Inductive Logic Programming: Techniques and Application* by Lavrač and Džeroski [3.20], *Inductive Logic Programming: From Machine Learning to Software Engineering* by Bergadano and Gunetti [3.1], and *Foundations of Inductive Logic Programming* by Nienhuys-Cheng and de Wolf [3.30]. The first provides a practically oriented introduction to ILP, but is slightly dated now, given the fast development of ILP in the recent years. The other two deal with ILP from a software engineering and a theoretical perspective, respectively.

3.11 Summary

This chapter provided an introduction to inductive logic programming, a scientific discipline concerned with learning logic programs from examples and background knowledge. It introduced the basics of logic programming and related logic programming terminology to database terminology. The task of relational rule induction was the first data mining task addressed by ILP systems and was covered in detail in this chapter: both the space of relational rules and some basic techniques for searching this space were presented. An overview of other relational data mining tasks addressed by ILP systems and a survey of ILP-related literature was given.

The next chapter (Chapter 4 by Wrobel) gives a well-founded motivation for using relational data mining by providing a succint and illustrative account of the advantages of doing so — Sections 4.1 and 4.2 are not to be missed on a first reading of the book. The remainder of this book describes a variety of ILP approaches (Part II), shows how single table data mining approaches can be upgraded to or used as they are in a relational data mining context (Part II) and gives an overview applications of and experiences with such approaches, as well as an overview of relevant Internet resources (Part IV).

References

3.1 F. Bergadano and D. Gunetti. *Inductive Logic Programming: From Machine Learning to Software Engineering.* MIT Press, Cambridge, MA, 1995.
3.2 H. Blockeel and L. De Raedt. Top-down induction of first order logical decision trees. *Artificial Intelligence*, 101: 285–297, 1998.
3.3 I. Bratko. *Prolog Programming for Artificial Intelligence*, 3rd edition. Addison-Wesley, Harlow, England, 2001.
3.4 L. Breiman, J. H. Friedman, R. A. Olshen, and C. J. Stone. *Classification and Regression Trees.* Wadsworth, Belmont, 1984.
3.5 W. Buntine. Generalized subsumption and its applications to induction and redundancy. *Artificial Intelligence*, 36(2): 149–176, 1988.
3.6 P. Clark and R. Boswell. Rule induction with CN2: Some recent improvements. In *Proceedings of the Fifth European Working Session on Learning*, pages 151–163. Springer, Berlin, 1991.
3.7 P. Clark and T. Niblett. The CN2 induction algorithm. *Machine Learning*, 3(4): 261–283, 1989.
3.8 L. De Raedt. Attribute-value learning versus inductive logic programming: the missing links (extended abstract). In *Proceedings of the Eighth International Conference on Inductive Logic Programming*, pages 1–8. Springer, Berlin, 1998.
3.9 L. De Raedt, editor. *Advances in Inductive Logic Programming.* IOS Press, Amsterdam, 1996.
3.10 L. De Raedt and L. Dehaspe. Clausal discovery. *Machine Learning* 26: 99–146, 1997.
3.11 L. De Raedt and S. Džeroski. First order *jk*-clausal theories are PAC-learnable. *Artificial Intelligence*, 70: 375–392, 1994.
3.12 S. Džeroski, S. Muggleton, and S. Russell. PAC-learnability of determinate logic programs. In *Proceedings of the Fifth ACM Workshop on Computational Learning Theory*, pages 128–135. ACM Press, New York, 1992.
3.13 W. Emde and D. Wettschereck. Relational instance-based learning. In *Proceedings of the Thirteenth International Conference on Machine Learning*, pages 122–130. Morgan Kaufmann, San Mateo, CA, 1996.
3.14 P. Flach. Logical approaches to machine learning - an overview. *THINK*, 1(2): 25–36, 1992.
3.15 C. Hogger. *Essentials of Logic Pogramming.* Clarendon Press, Oxford, 1990.
3.16 A. Karalič and I. Bratko. First order regression. *Machine Learning* 26: 147-176, 1997.
3.17 S. Kramer. Structural regression trees. In *Proceedings of the Thirteenth National Conference on Artificial Intelligence*, pages 812–819. MIT Press, Cambridge, MA, 1996.
3.18 R.D. King, A. Karwath, A. Clare, and L. Dehaspe. Genome scale prediction of protein functional class from sequence using data mining. In *Proceedings of the Sixth International Conference on Knowledge Discovery and Data Mining*, pages 384–389. ACM Press, New York, 2000.

3.19 N. Lavrač, S. Džeroski, and M. Grobelnik. Learning nonrecursive definitions of relations with LINUS. In *Proceedings of the Fifth European Working Session on Learning*, pages 265–281. Springer, Berlin, 1991.
3.20 N. Lavrač and S. Džeroski. *Inductive Logic Programming: Techniques and Applications.* Ellis Horwood, Chichester, 1994. Freely available at `http://www-ai.ijs.si/SasoDzeroski/ILPBook/`.
3.21 J. Lloyd. *Foundations of Logic Programming*, 2nd edition. Springer, Berlin, 1987.
3.22 R. Michalski, I. Mozetič, J. Hong, and N. Lavrač. The multi-purpose incremental learning system AQ15 and its testing application on three medical domains. In *Proceedings of the Fifth National Conference on Artificial Intelligence*, pages 1041–1045. Morgan Kaufmann, San Mateo, CA, 1986.
3.23 Muggleton, S., editor. (1991) *Proceedings of the International Workshop on Inductive Logic Programming.* University of Porto, Portugal.
3.24 S.H. Muggleton, editor. *Inductive Logic Programming.* Academic Press, London, 1992.
3.25 S. Muggleton. Inductive logic programming. *New Generation Computing*, 8(4): 295–318, 1991.
3.26 S. Muggleton. Inverse entailment and Progol. *New Generation Computing*, 13: 245–286, 1995.
3.27 S. Muggleton and W. Buntine. Machine invention of first-order predicates by inverting resolution. In *Proceedings of the Fifth International Conference on Machine Learning*, pages 339–352. Morgan Kaufmann, San Mateo, CA, 1988.
3.28 S. Muggleton and C. Feng. Efficient induction of logic programs. In *Proceedings of the First Conference on Algorithmic Learning Theory*, pages 368–381. Ohmsha, Tokyo, 1990.
3.29 T. Niblett. A study of generalisation in logic programs. In *Proceedings of the Third European Working Session on Learning*, pages 131–138. Pitman, London, 1988.
3.30 S.-H. Nienhuys-Cheng and R. de Wolf. *Foundations of Inductive Logic Programming.* Springer, Berlin, 1997.
3.31 G. Plotkin. A note on inductive generalization. In B. Meltzer and D. Michie, editors, *Machine Intelligence 5*, pages 153–163. Edinburgh University Press, Edinburgh, 1969.
3.32 J. R. Quinlan. Learning logical definitions from relations. *Machine Learning*, 5(3): 239–266, 1990.
3.33 J. R. Quinlan. *C4.5: Programs for Machine Learning.* Morgan Kaufmann, San Mateo, CA, 1993.
3.34 J. Robinson. A machine-oriented logic based on the resolution principle. *Journal of the ACM*, 12(1): 23–41, 1965.
3.35 E. Shapiro. *Algorithmic Program Debugging.* MIT Press, Cambridge, MA, 1983.
3.36 A. Srinivasan. The Aleph Manual. Technical Report, Computing Laboratory, Oxford University, 2000. Available at `http://web.comlab.ox.ac.uk/oucl/research/areas/machlearn/Aleph/`
3.37 J. Ullman. *Principles of Database and Knowledge Base Systems*, volume 1. Computer Science Press, Rockville, MA, 1988.

4. Inductive Logic Programming for Knowledge Discovery in Databases

Stefan Wrobel

School of Computer Science, IWS, University of Magdeburg
Universitätsplatz 2, D-39016 Magdeburg, Germany

Abstract

Relational data mining algorithms and systems are capable of directly dealing with multiple tables or relations as they are found in today's relational databases. This reduces the need for manual preprocessing and allows problems to be treated that cannot be handled easily with standard single-table methods. This paper provides a tutorial-style introduction to the topic, beginning with a detailed explanation of why and where one might be interested in relational analysis. We then present the basics of Inductive Logic Programming (ILP), the scientific field where relational methods are primarily studied. After illustrating the workings of MIDOS, a relational methods for *subgroup discovery*, in more detail, we show how to use relational methods in one of the current data mining systems.

4.1 Introduction

Data Mining, or *Knowledge Discovery in Databases* (KDD), as it is referred to in the research world, has recently been gaining widespread attention. In one popular definition, KDD is seen as the "automatic extraction of novel, useful, and valid knowledge from large sets of data" [4.6]. As this definition indicates, KDD offers a general body of techniques that are capable of finding different kinds of "knowledge" in different kinds of data. A data mining task can only be defined precisely when we specify exactly what kind of knowledge we would like to find in which form, and in which way the data for analysis are available in a storage file or database system.

As for the knowledge that is to be discovered, rules, decision trees, cluster hierarchies, association rules or statistically unusual subgroups are just a few examples (see e.g., [4.6, 4.21]). Here, we will focus our attention on the form of the data that are available for analysis, and more precisely, on the question of using a single table or multiple tables for analysis. As we will see in the remainder of this section, using only a single table for analysis, despite being the standard in most current data mining systems, is restrictive and does not correspond to current database systems where relational technology with multiple tables has long been the standard.

We will then deal with the question of how to directly use multiple tables for analysis, introducing results from the field of *Inductive Logic Programming* (see e.g., [4.12, 4.19, 4.9]) which has been concerned with relational analysis technology for a number of years, and has now reached a state of maturity sufficient for practical applications. You will get to know a relational method for subgroup discovery named MIDOS [4.20], and will find out how to use this method in KEPLER [4.23], a data mining system that includes support for relational analysis technology, and includes a number of different ILP algorithms as plug-ins[1].

4.1.1 The single-table assumption

In the commercial world, KDD and Data Mining have up to now always used a very important simplifying assumption about the form of available data which we will refer to as the "single-table assumption": it is assumed that all data are stored in a single table (or "relation" in database terminology), and that there is one row (or "tuple") in this table for each object of interest. This is also referred to as a "propositional" or "rectangular table" representation. In a typical commercial application (e.g., database marketing), we would have a customer table with each line representing one of our customers, and the columns of the table representing different "attributes" of our customers (Table 4.1).

Table 4.1. Basic customer table.

ID	Name	First Name	Street	City	Zip	Sex	Social Status	Income	Age	Club Status	Res-ponse
...	...	...	...	...	...	...	...	...	...	...	...
3478	Smith	John	38 Lake Dr	Sampleton	34677	mal.	single	i60-70k	32	memb.	resp.
3479	Doe	Jane	45 Sea Ct	Invention	43666	fem.	married	i80-90k	45	nonm.	noresp.
...	...	...	...	...	...	...	...	...	...	...	...

The first field shows a customer's ID number, the subsequent fields his or her name and address. If we assume we require each new customer to fill in a small questionnaire (and assume our customers really do this), we also would have this kind of information, stored in the four fields. At the end, there is an attribute specifying whether this customer is a member of our buying club, and whether the customer responded to a recent direct mail campaign.

For analysis purposes, fields like name and exact address are uninteresting, since we are looking for general information about our customers, not specific information that applies only to one person. Thus for analysis, we would use a smaller table as shown in Table 4.2 (introducing a number of space-saving abbreviations for field names and field values).

From such data, traditional data mining software can produce different kinds of knowledge: decision tree algorithms induce models that predict

[1] The commercially available product version, marketed by Dialogis GmbH has recently been renamed to DIALOGIS D-MINER. We will stick to the name KEPLER for this chapter. All product and company names are used without explicitly marking their trademark status. This does not imply that these names are free.

Table 4.2. Customer table for analysis.

ID	Zip	Sex	SoSt	Income	Age	Club	Resp
....	...	...	...	...	...	...	...
3478	34677	m	si	60-70	32	me	nr
3479	43666	f	ma	80-90	45	nm	re
....	...	...	...	...	...	...	...

whether a certain customer will reply to a future mailing, clustering algorithms segment our customer base into homogeneous groups that can be treated together in marketing campaigns, etc.

4.1.2 Problems with the single-table assumption

In the customer table we can add as many attributes about our customers as we like, therefore the single-table assumption made by most systems seems to be no problem whatsoever. For the kind of general customer information recorded in Table 4.1, this is certainly correct. If we wanted to add information about a person's number of children, we would simply add another attribute.

However, for other kinds of valuable information about our customers, the single-table assumptions turns out to be a significant limitation. Assume we would like to add information about the orders placed by a customer, and would also like to include the delivery and payment modes and with which kind of store the order was placed (Size, Ownership, Location). For simplicity, we will not include information about goods ordered.

Table 4.3. Customer table including order and store information.

ID	Zip	Sex	SoSt	Income	Age	Club	Resp	Delivery Mode	Payment Mode	Store Size	Store Type	Store Location
....	...	...	...	...	...	...	...	...	...	...	...	...
3478	34677	m	si	60-70	32	me	nr	regular	cash	small	franchise	city
3479	43666	f	ma	80-90	45	nm	re	express	credit	large	indep	rural
....	...	...	...	...	...	...	...	...	...	...	...	...

We might consider augmenting our table by additional columns as in Table 4.3, using one column to indicate the delivery mode, a second to indicate payment mode, a third for store size, a fourth for store type and a fifth for store location. While this works well for once-only customers, what if our business, as any good business, has repeat customers who have ordered several times? If all our data mining method can handle is a single table, we are left with two equally unsatisfying options.

First, we could make one entry for each order in our customer table. Thus, if Mr. Smith (customer 3478) places three orders, he will be represented by three lines in the table, each duplicating his customer information fields (Table 4.4).

Table 4.4. Customer table with multiple orders.

ID	Zip	Sex	SoSt	Income	Age	Club	Resp	Delivery Mode	Payment Mode	Store Size	Store Type	Store Location
....	...	...	...	...	...	...	...	...	...	...	...	...
3478	34677	m	si	60-70	32	me	nr	regular	cash	small	franchise	city
3478	34677	m	si	60-70	32	me	nr	express	check	small	franchise	city
3478	34677	m	si	60-70	32	me	nr	regular	check	large	indep	rural
3479	43666	f	ma	80-90	45	nm	re	express	credit	large	indep	rural
3479	43666	f	ma	80-90	45	nm	re	regular	credit	small	franchise	city
....	...	...	...	...	...	...	...	...	...	...	...	...

Moreover, store information is also repeated for each customer and order. Thus, all the information is in our table, but it is stored redundantly. This not only wastes space, it also creates a number of problems. If there is an error in the data that represent Mr. Smith, it needs to be corrected in every row, not only in one single place. Even more importantly for analysis purposes, single-table analysis methods usually assume that each row represents one object of interest, i.e., one customer in our case. Since we now have one line per order, not per customer, analysis results will really be about orders, not customers, which is not what we might want.

The second option avoids redundancy at the expense of information detail. With a little creativity, the analyst might conclude that perhaps only the number of orders and the number of stores in which orders were placed are important. We could then create two new attributes as in Table 4.5.

Table 4.5. Customer table using summary attributes.

ID	Zip	Sex	SoSt	Income	Age	Club	Resp	No. of Orders	No. of Stores
....	...	...	...	...	...	...	...	...	...
3478	34677	m	si	60-70	32	me	nr	3	2
3479	43666	f	ma	80-90	45	nm	re	2	2
....	...	...	...	...	...	...	...	...	...

This avoids problems with redundancy and multiple rows, and thus allows analysis methods to operate properly on the resulting table. However, there is a lot less information in the new table. What if our analyst's opinion was wrong, and knowledge of the exact combinations of payment modes, delivery modes and store types *is* important after all? Our analysis methods cannot use this information, so again the results will not be as good as they could be if all information were used.

4.1.3 The solution: Relational representation

In the world of database systems, the problems inherent in the single-table representation have long been recognized and addressed by *relational* databases that are capable of representing information as a set of different interlinked tables. Today, even popular desktop databases have relational capabilities, and only the very simplest address managers are still restricted to single tables. In database terminology, a single table in the form described in our first solution above is said to be a "non-normal form" database and is considered bad database practice. Instead, database designers would represent the information in our sample problem as a set of tables as depicted in Figure 4.1.

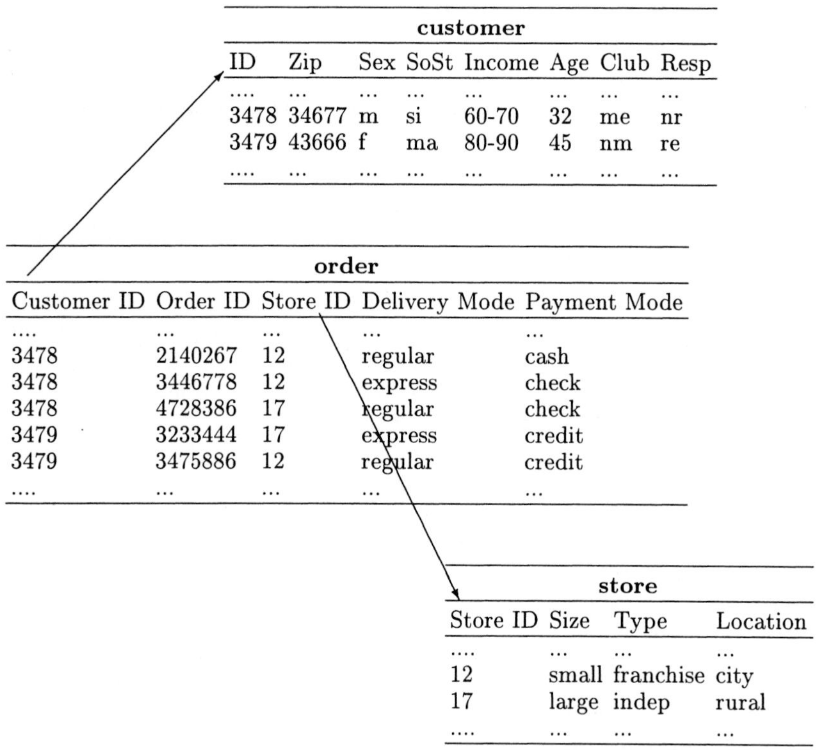

customer

ID	Zip	Sex	SoSt	Income	Age	Club	Resp
....	...	...	...	...	...	...	...
3478	34677	m	si	60-70	32	me	nr
3479	43666	f	ma	80-90	45	nm	re
....	...	...	...	...	...	...	...

order

Customer ID	Order ID	Store ID	Delivery Mode	Payment Mode
....	...	...	...	...
3478	2140267	12	regular	cash
3478	3446778	12	express	check
3478	4728386	17	regular	check
3479	3233444	17	express	credit
3479	3475886	12	regular	credit
....	...	...	...	...

store

Store ID	Size	Type	Location
....	...	...	...
12	small	franchise	city
17	large	indep	rural
....	...	...	...

Fig. 4.1. Relational representation of customers, orders and stores.

Here, we have one master table "customers" representing general information about each customer, and there is exactly one line per customer. Orders are described in a second table "orders", containing one line per order. The central element of relational technology is the use of identifiers that point to another table. In the "orders" relation, one field contains the customer ID, so if we want to know which customer placed an order, we simply go to the "customer" relation, find the row with this customer ID value, and get the required information. A field in a relation that points to a key field in another relation is often called a "foreign key", indicating that the value in this field is required to be a key value in the other relation. The arrows in Figure 4.1 represent the foreign key relations in our example. The same principle is used to represent the store information: the "order" table uses the store ID to refer to "store" which in turn describes each store.

4.1.4 Examples of relational representations

A car manufacturer's database. The relational representation thus quite naturally allows us to represent the required information in our simple example. In fact, it is so popular because it has turned out sufficient for almost any kind of data representation problem, and most commercial databases have the general structure of our little customer database. As a more complex example from a more technical domain, consider a car manufacturer who wants to represent information about each car manufactured, the manufacturing plant where the car was built, the parts built into the car, the dealer who sold the car, the customer who bought it etc. All of this could easily be represented in a relational database, perhaps the one shown in Figure 4.2.

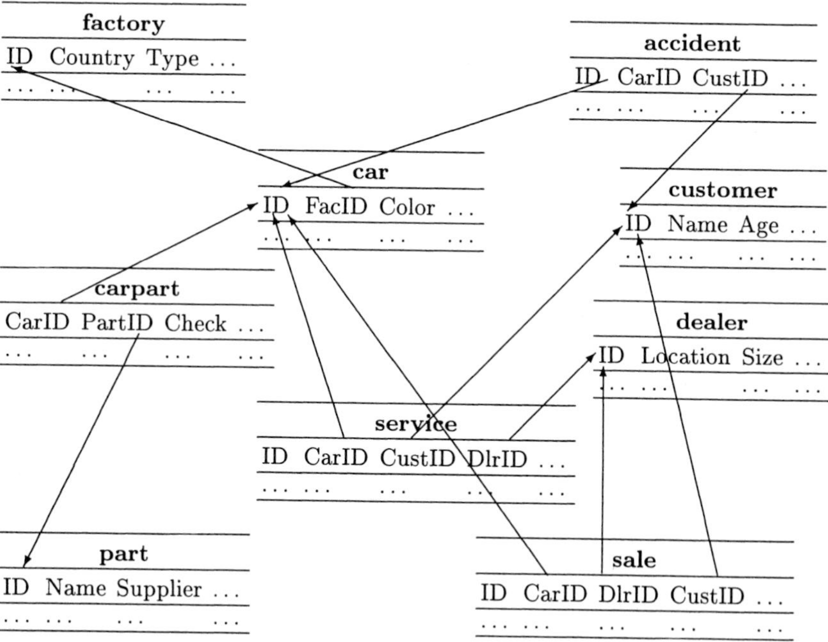

Fig. 4.2. A car manufacturer's relational database.

Non-rectangular questionnaire data. Another popular application for data mining is analysis of questionnaire data. Even though in principle a questionnaire can be represented as a single table, each column containing the answer for one question, in practice questionnaires are somewhat more complex. First, respondents may choose not to answer a question, creating an empty value. Such empty values (or "null" values) cannot be handled properly by most analysis algorithms — they would treat "null" as an ordinary value (just another possible answer), which is not appropriate in many circumstances. This is even more evident when we have different questions blocks ("If you marked *single* in question 10, jump directly to question 23, otherwise, please answer 11 to 22."). We could simply leave these attributes blank, but again, using them in the result of an analysis would not make sense for single persons. And last, if we allow multiple answers to each question ("which hobbies do you have?"), in a single table we need to create binary attributes, one for each possible hobby (meaning we have to fix the list of possible hobbies). In a relational representation, none of these present problems. If we use a master relation with one entry per respondent, we can have a separate relation for each question, containing zero (if the question was unanswered), one or many tuples per respondent, as shown in Figure 4.3.

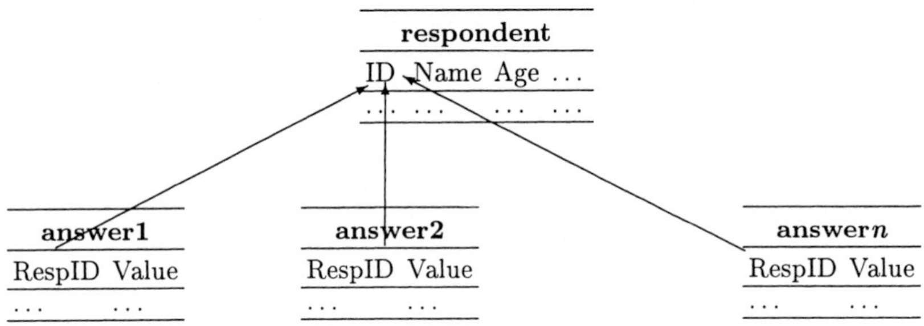

Fig. 4.3. Relational representation of non-rectangular questionnaire data.

Chemical applications. Finally, for a chemical application, assume we would like to represent chemical molecules and their component atoms. As shown in Figure 4.4, it is equally possible to represent this information in a relational form, using a relation "part-of" to indicate that an atom is a part of a molecule, and a relation "bond" to indicate that there is a bond between two atoms.

Note that even if we do not require the bond information, we cannot simply make a large number of columns in a single table called something like "atom 1" to "atom 23" (where 23 would be the number of atoms in the largest molecule). First, this would mean a lot of columns would be unused, meaning we need the problematic "null" values (whenever a molecule has less than 23 atoms). Worse, putting one of the atoms in the column "atom 1"

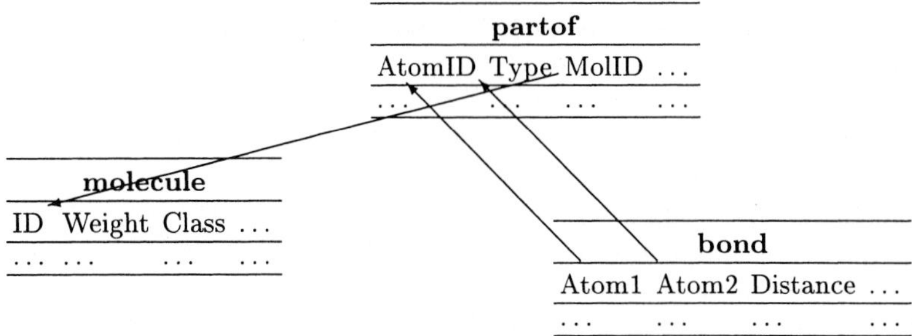

Fig. 4.4. Relational representation of molecules, atoms, and bonds.

would assign a numbering to the atoms that is not taken from reality, but assigned arbitrarily. Unfortunately, the way a single-table method works is that when trying to generate a model, only the values from the same column are compared to one another. Since the arbitrarily assigned "atom 1" elements from two different molecules are most likely not the functionally matching atoms at corresponding places in these two molecules, the analysis method will have no chance to discover the right model. Furthermore, our model might need to use not only information about particular atoms, but also information about their bonded neighbors and the neighbors of the neighbors, etc.

The structure of these chemical problems occurs in a similar fashion in technical applications, e.g., when we try to describe a power network or electrical circuit, where a faulty component has a non pre-determined number of neighbors, or when analyzing processes where predictions depend on predecessor states at varying temporal distances. None of these problems can be represented well in a single-table representation.

4.2 ILP: Relational analysis technology

We have thus seen that for almost all domains, it is very advantageous and often even necessary to use the power offered by relational technology and store available information in multiple relations. If we then want to analyze this information, current data mining tools based on the single-table assumption are ill-equipped to do the job. Either we invest a lot of thought and effort into pressing as much of the required information as necessary into a single table, perhaps producing redundancy and other problems, or we accept that these methods simply ignore a large part of the available information. Would it not be nice to have analysis methods and data mining systems capable of directly working with multiple relations as they are available in relational database systems ?

Fortunately, the research field called "Inductive Logic Programming" (ILP) has been asking exactly this question for a number of years now, and after a number of successful research prototypes, we are now seeing the first commercial data mining systems with ILP technology becoming available. These systems have already successfully solved applications of the types described above.

Especially in the field of chemistry/biology, ILP has had remarkable successes and has proven its superiority to other techniques. In the so-called "predictive toxicology evaluation" (PTE) challenge, the ILP system PROGOL [4.11] induced a classifier that was competitive with hand-crafted expert systems and classical propositional techniques like regression [4.16]. In another application, the ILP system RIBL [4.5] proved superior to propositional nearest neighbor and decision tree algorithms in the prediction of Diterpene classes from nuclear magnetic resonance (NMR) spectra [4.4]. Several other applications of this type have been reported, e.g., [4.10, 4.17]. Other applications outside of chemistry include e.g., analysis of survey data in medicine [4.20], analysis of traffic accident data [4.14], and prediction of ecological biodegradation rates [4.18, 4.3].

In this chapter, we cannot describe the theoretical foundations and main technological ingredients of ILP in an encompassing fashion, there are other papers, e.g., the introductory chapters of this volume, or [4.12, 4.19], that will give you a precise and reasonably formal introduction into what ILP is all about. The goal of the preceding sections of this chapter was to illustrate why you should be interested in relational technology and ILP even if you are not dealing with the problems in biochemistry or natural language for which ILP has become famous. In fact, we hope to have demonstrated that relational analysis technology is useful even when you are dealing with Data Mining and KDD problems in very down-to-earth domains like customers and their orders.

The main thing you need to know about ILP to use its methods is how data and analysis results are being represented — and of course, just what kind of knowledge is being discovered, i.e., which analysis task is being addressed. As for the knowledge representation, you are already familiar with it from the examples in the preceding section, except that ILP methods often use the notation popular from logic programming languages like Prolog (this is the "LP" in ILP). In logic programming, each row in a table or relation is represented by a fact (a positive *literal*) of the form

<tablename>(<v_1>, ..., <v_n>)

where <tablename> is called the *predicate*, and the <v_1>...<v_n>, the column values, are the *arguments*. Customer information about Mr. Smith (customer 3478) and Mrs. Doe (customer 3479) would thus be represented as two facts

customer(3478,34677,male,single,i60-70k,32,member,no_response).
customer(3479,43666,female,married,i80-90k,45,nonmember,response).

In fact, the language used in ILP is even more powerful than the language of relational databases, since arguments can be not only simple values, but nested structures called terms. Since terms go beyond what "classical" relational databases can do, and are not necessary for most applications, we will not describe them here, but see e.g., the Chapter 7 by Muggleton and Firth in this volume or [4.12, 4.19].

A little more complex for those not familiar with logic programming or Prolog is how ILP represents analysis results. Since we want to express general properties about unknown objects, we use *variables* to stand for unknown values. An expression with variables "stands for" all entries in a table for which matching values for all variables (and non-variables) can be found. As in Prolog, variable names start with a capital letter. A special variable "_" is used when we do not care which value was used. For example,

customer(_,_,female,_,_,_,_,_)

stands for all female customers in our customer relation. If we want to refer to groups of objects that require more complex descriptions involving multiple relations, we combine several such so-called literals with a logical "and" (conjunction), written as "," or "&". By using the same variable several times, we can easily represent the relational links between different relations. In contrast to all other variables, if "_" is used multiple times, each occurrence may refer to a different value. Also, two different variables may or may not refer to the same value. For example,

customer(C,_,female,_,_,_,_,_), order(C,_,_,_,credit_card)

stands for all female customers C who have ever placed an order paid for with a credit card. If relations have a large number of arguments (in real-life KDD applications, several dozens or even hundreds of arguments are not uncommon), this representation becomes very hard to read, since typically only a few field values are actually different from "_". Thus, an alternative representation is sometimes used, e.g., in the MIDOS analysis method (see below):

customer.Sex=female, customer.ID=order.Customer ID,
order.Paymt Mode=credit_card

or, if the base relation we are interested in ("customer") is always the same, simply

Sex=female, ID=order.Customer ID, order.Paymt Mode=credit_card

Often, we want to make predictions about each member of such a group. Assume we had another one-argument relation "good-customer" containing as its single argument the IDs of all customers that we classify as good customers. If indeed female credit-card customers are good customers, an ILP system like PROGOL (see the Chapter 7 by Muggleton and Firth in this volume or [4.11]) might induce the following if-then rule ("clause") from our database:

```
good_customer(C) :- customer(C,_,female,_,_,_,_,_),
                    order(C,_,_,_,credit_card)
```

As in Prolog, the two parts of the rule are separated by the symbol ":-", the left-hand part is called the "head", the right-hand part the "body" of the clause, to be read: "If a female customer has ever placed an order paid-for by credit card, then she is to be classified as a good customer." Often, rules are written in a more classical logical notation body-first with an implication arrow:

```
customer(C,_,female,_,_,_,_,_), order(C,_,_,_,credit_card) →
                                            good_customer(C)
```

Many ILP analysis algorithms offer an additional feature: they let you enter rules such as the one above, and automatically use them in their analysis process. This means you need not enter or generate every relation explicitly (as so-called "extensional" relations), but you can simply define new relations with clauses, and the analysis algorithm will use these so-called "intensional" relations as if they had been explicitly generated. Intensional definitions are not only more practical, they are actually even more powerful than extensionally given relations, e.g., for recursively defined concepts like "connected-to" in a network. If you work with an algorithm that cannot accept rules directly (or simply prefer to generate relations explicitly), the data mining platform that you use should offer facilities for taking rules and generating relations from them. Such rules, and generally all relations except the "target" relation, are often referred to as "background knowledge" to be used by the analysis method (see the Chapter 3 by Džeroski and Lavrač or [4.12, 4.19]).

4.3 ILP subgroup discovery: MIDOS

Even though we cannot dive into the detailed theory and mechanics behind ILP algorithms, this chapter would be incomplete without a short look at the mechanics of at least one ILP method. ILP systems for inducing predictive or descriptive rules are described in recent overview papers (e.g., [4.12, 4.19]), so for our purposes here we will illustrate a method for the task of subgroup discovery, one of the popular tasks in data mining [4.8]. The method is implemented in a software algorithm called MIDOS [4.20], available as part of the relational data mining platform KEPLER (see Section 4.4)). Even though the basic method is simple, we can use it to illustrate several techniques that are used in ILP to achieve good performance on larger data sets. (Further methods are illustrated in [4.22]).

But let us first define the task of subgroup discovery more precisely. In subgroup discovery, we assume we are given a so-called "population" of individuals (objects, customers, ...) and a property of those individuals that we are interested in. The task of subgroup discovery is then to discover the subgroups of the population that are statistically "most interesting", i.e., are as large as possible and have the most unusual statistical (distributional) characteristics with respect to the property of interest.

In our customer application, there are several subgroup discovery tasks in which we could be interested. We might want to find groups of orders with unusual distributions across stores, or look at groups of customers with unusual distributions across club membership or mailing response. If we pick club membership as our property of interest, one possible interesting subgroup returned by MIDOS could be (shown as it would appear in MIDOS' log file):

Target Type is: nominal([member,non_member])
Reference Distribution is: [66.1%, 33.9% - 1371 objects]

Sex=female, ID=order.Customer ID, order.Paymt Mode=credit_card
[69.9%, 30.1% - 478 objects] [1.53882%%]

We see that the entire population consisted of 1371 objects (i.e., customers) of which 66.1% are club members. In contrast, in the subgroup of female credit card buyers (478 customers), 69.9% are club members. This finding is assigned a quality value of 1.53882%% by MIDOS[2]. Normally, we ask the system for a certain number of interesting subgroups, and MIDOS will return a list ranked by the quality value (see Figure 4.10, Section 4.4.2 for a graphical rendering of these subgroups within KEPLER):

Target Type is: nominal([member,non_member])
Reference Distribution is: [66.1%, 33.9% - 1371 objects]

ID=order.Customer ID, order.Delivery Mode=express, order.Paymt Mode= credit_card [72.0%, 28.0% - 311 objects] [2.07221%%]
Age=a40_50, ID=order.Customer ID, order.Paymt Mode=check [57.9%, 42.1% - 152 objects] [1.67213%%]
Income=i50_60k, Response=no_response [60.4%, 39.6% - 270 objects] [1.60067%%]
Sex=female, ID=order.Customer ID, order.Paymt Mode=credit_card [69.9%, 30.1% - 478 objects] [1.53882%%]
ID=order.Customer ID, order.Store ID=store.Store ID, store.Size=large [61.5%, 38.5% - 353 objects] [1.47391%%]
ID=order.Customer ID, order.Delivery Mode=express [69.3%, 30.7% - 515 objects] [1.26099%%]
Income=i60_70k [70.6%, 29.4% - 289 objects] [1.08419%%]
Sex=male, Income=i60_70k [72.4%, 27.6% - 163 objects] [1.07433%%]

[2] See Section 4.3.1 for remarks on interpreting the significance of such a finding, and Section 4.3.2 for details on how this quality value could be computed.

Age=a60_70, Response=no_response [71.5%, 28.5% - 200 objects] [1.0023%%]
Sex=female, ID=order.Customer ID, order.Delivery Mode=express [70.6%, 29.4% - 252 objects] [0.933171%%]

More complex subgroup discovery tasks result when multi-valued or numerical properties are considered. So if we had stored the total amount of each order in our database, we might consider analyzing whether there are any subgroups that have an unusually high average order amount, or even whether one subgroup has an unusually high share in all purchases. We will not discuss these further here (but see [4.8]).

4.3.1 Use of subgroup discovery results

So how does one use subgroup discovery results, and how are they different from the results of predictive data mining algorithms? The typical use of subgroup discovery results is as a generator of ideas about phenomena in our application domain. For example, based on the analysis shown above, we might want to target the group of check payers aged 40 to 50 who show an unusually low rate of club membership.

Of course, a word of caution is important with respect to unreflected use of such subgroup discovery results (and in fact, with respect to unreflected use of any kind of data mining/knowledge discovery result). The subgroup discovery method *does* guarantee that a reported subgroup indeed possesses the distributional properties that are reported (this is simply a matter of counting correctly). Since no assumptions about the population and the distributions of the statistical phenomena that generate the data are used, however, the method *cannot* guarantee that the discovered subgroups represent general effects in reality that are statistically significant and can be reported and taken for true without further thought.

So if we find that check payers aged 40 to 50 are club members less frequently, we must carefully analyze whether this could be due to the way the sample was chosen — after all, we only have information about our own customers, so the effect could be due to the way we "recruit" our customers. A properly validated statistical statement like "check payers aged 40 to 50 are less likely to be club members" would require making distributional assumptions in order to determine the likelihood of error (i.e., the chance that check payers aged 40 to 50 are no different than the entire population when it comes to club membership).

Subgroup discovery thus functions as a convenient hypothesis generator for further analysis, not as a statistical oracle that can be blindly trusted. Its usefulness lies in the computer's power to quickly screen large numbers of hypotheses (hundreds of thousands of subgroups) and report only the most interesting ones, and its capability of considering hypotheses with a large number of factors where humans would consider only a handful of hypotheses with one or two factors (and thus might miss the truly surprising things that one would have never thought of).

The difference between the subgroups analysis task (on the simple binary property problem) and the prediction analysis task is twofold. On the downside, the discovered subgroups cannot be used for predictive purposes. If we know that the group of female credit card payers contains club members with a relative frequency of 69.9% vs. 66.1% in the entire population, this is not sufficient to predict that a given female credit card buyer will actually be or become a member. Predictive learning methods, in contrast, look only for subgroups with distributions that are almost "pure", i.e., groups where almost all members (except for noisy/erroneous cases) have the required property. On the upside, this means that subgroup discovery does useful work even in situations where such all-or-none groups cannot be found or in situations where the distribution is very unbalanced.

Similarly, the more complex average or market share analysis variants of subgroup discovery cannot be treated directly with predictive methods. If an exact numerical predictor could be learned by a predictive method, this predictor could be used to then estimate the average and market share properties. In most cases, however, such a predictor cannot be learned while subgroup discovery still yields interesting results. Subgroup discovery thus is a useful first approach in analyzing a problem and yields a "screening" of the statistical properties of the data. As further analysis steps, one could then try to see whether predictive models can also be induced.

4.3.2 MIDOS technological ingredients

So how does the MIDOS method work? As explained, the algorithm is to consider all possible subgroups and report the statistically "most interesting" ones. Thus, the key ingredients are the definition of interestingness or quality, and a clever way of searching so we find the most interesting results as quickly as possible.

Quality. To find a precise and technical definition of what "statistically interesting" means, two factors are considered: the size of a group and its so-called distributional unusualness. Distributional unusualness is of course central, since the more different the statistical properties of a group are when compared to the entire population, the more interesting this group is. We are more interested in groups with 73% club members than those with 68% club members if the population average is 66.1%. However, the size of the group is equally important, because smaller groups tend to be more statistically unusual simply by chance — in the extreme, a group consists of only one customer, and if this customer happened to be club member, we would have a group with 100% club members.

We thus need to balance the size of the group (usually referred to as factor g) with its distributional unusualness (usually referred to as factor

p). The properties of functions that combine these two factors have been extensively studied (the "p-g-space" [4.8]). As an example of a popular and typical function, here is one of the interestingness functions for the binary and multi-valued case that can be selected in MIDOS:

$$\frac{g}{1-g} \cdot \Sigma_{i=1..n}(p0_i - p_i)^2$$

Here, g is the relative size of the considered subgroup with respect to the population (between 0 and 100%), $p0_i$ is the relative frequency of value v_i in the entire population, p_i its relative frequency in the considered subgroup. As an example, in the population of 1371 customers used above of which 478 are female credit card buyers, g would be $\frac{478}{1371} = 0.348 = 34.8\%$. If we assume v_1 is "member" and v_2 is "nonmember", and 906 customers overall are members compared to 334 among female credit card buyers, $p0_1$ would be $\frac{906}{1371} = 0.661 = 66.1\%, p0_2$ would be $\frac{465}{1317} = 0.339 = 33.9\%, p_1$ would be $\frac{334}{478} = 0.699 = 69.9\%$, and p_2 would be $\frac{144}{478} = 0.301 = 30.1\%$. The interestingness value of female credit card buyers with respect to club membership would thus be:

$$\frac{0.348}{1-0.348} \cdot ((0.661 - 0.699)^2 + (0.339 - 0.301)^2) = 0.00153882 = 1.53882\%\%$$

(which is quite a low value, see Section 4.3.1 for comments on significance). MIDOS uses this quality value to rank possible subgroups.

Search. The search for subgroups in MIDOS proceeds in what is commonly referred to as "top-down" fashion. This means that MIDOS starts with the most general (largest) possible group, i.e., the entire population, and then adds restrictions to it in all possible ways, one by one until the groups have become too small. Whether a group is too small is determined by a user-given parameter called "minimal support" (see below). During the search, MIDOS always maintains a list of the n highest quality solutions found so far, where n is a user given parameter ("solution size"). When the entire space has been explored (or the user interrupts), this list of top solutions is returned to the user.

Restrictions are constructed from all the attributes of all available relations, combining relations wherever necessary. Thus, to reach the candidate "Sex=male, Income=i60_70k", MIDOS would add the restriction "Sex=male" to the group of all customers, and in a second step, add the restriction "Income=i60_70k". Figure 4.5 (page 89) graphically shows how MIDOS explores its search space for a single table.

In a similar fashion, MIDOS extends the single-table search space shown in Figure 4.5 to multiple relations by adding, instead of restrictions on individual attributes, new literals corresponding to one of the available relations. To see how this works, let us have a look at MIDOS' hypotheses in their original ILP-style form, where "Sex=male, Income=i60_70k" would be written as

customer(_,_,male,_,i60_70k,_,_,_)

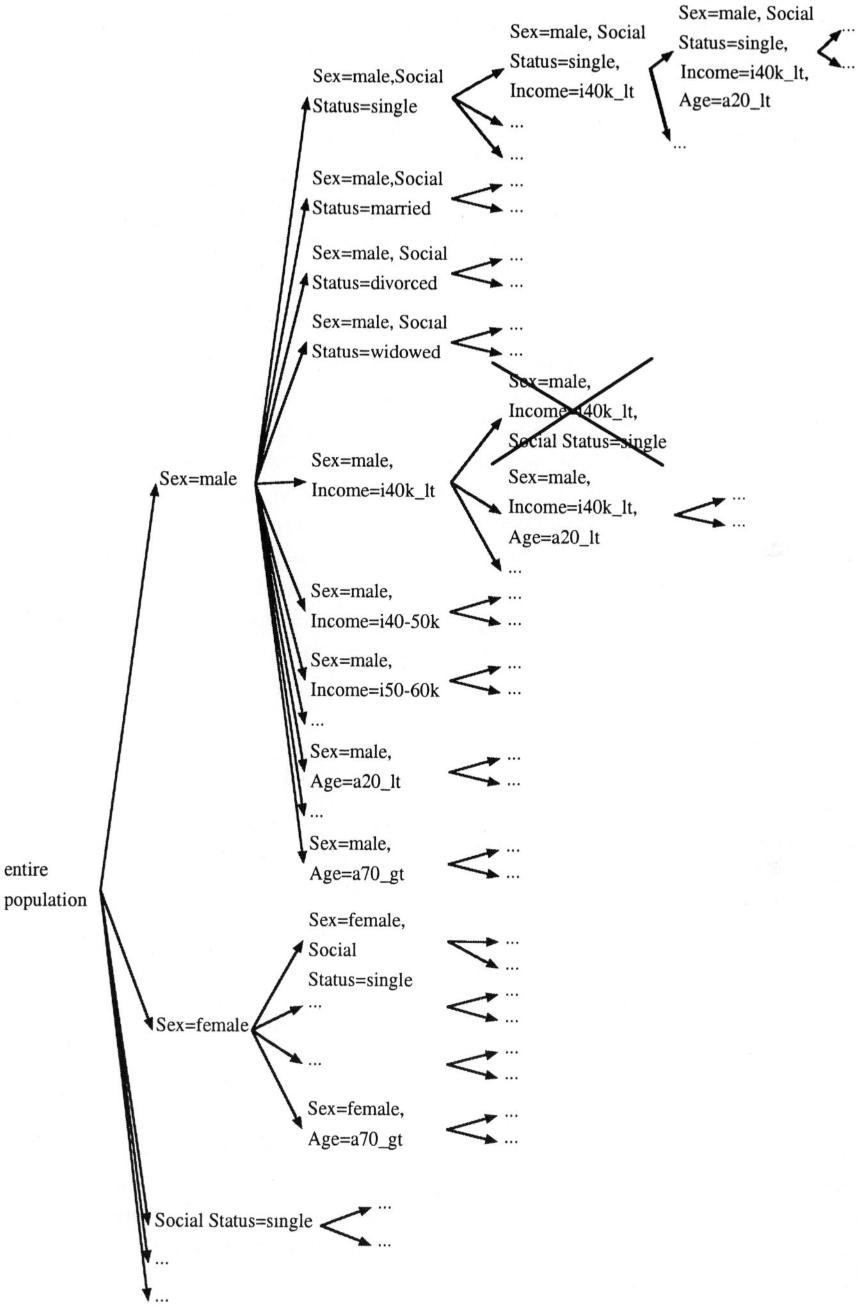

Fig. 4.5. MIDOS search space for a single table ("customer").

To bring in additional relations (or multiple copies of the same relations), Midos introduces a new literal that shares a variable with an existing literal. For example, to bring in the "order" relation, Midos would add a literal as follows:

customer(C,_,male,_,i60_70k,_,_,_), order(C,_,_,_,_)

using the shared variable "C" to indicate that the ordering customer should be the same as the customer referred to by the first literal (a "join" of the customer and order relations on the customer ID attribute). In Midos' notation, the above subgroup description would read

Sex=male, Income=i60_70k, ID=order.Customer ID

In our simple customer example, there are only two relations that can be brought in in addition to customer. In a fully developed real-life application, however, there might be dozens of relations that could be added, and several attributes for which it would make sense to link (join) these relations by shared variables. For any particular analysis, we might not actually be interested in all such combinations, so if Midos were to try all possible combinations, a lot of search effort would be needlessly wasted. To avoid this, Midos requires the user to provide a list of so-called *foreign links* that explicitly specify the "paths" along which new relations may be introduced with shared variables. To allow Midos to bring in "order" as in the example above, we would have to specify the foreign link

customer[1]->order[1]

telling Midos that it is allowed to bring in "order" if its first argument is linked to the first argument of "customer" with a shared variable. If we also wanted to use information about stores in our subgroups, we would also provide the foreign link

order[3]->store[1]

allowing Midos to form hypotheses like

customer(C,_,male,_,i60_70k,_,_,_), order(C,_,S,_,_), store(S,_,_,_)

or, in Midos notation,

Sex=male, Income=i60_70k, ID=order.Customer ID,
order.Store ID=store.store ID

Once brought in, the attributes of new relations themselves can again be restricted in the same fashion as shown in Figure 4.5 for "customer". Thus, at some point in its search, Midos would for example form the subgroup candidate

Sex=male, Income=i60_70k, ID=order.Customer ID,
order.Store ID=store.store ID, order.Paymt Mode=cash,
store.Size=small

The use of foreign links to allow the user to tell MIDOS where to search is an instance of a general approach known in ILP as "declarative bias", since it lets the user declaratively, i.e., without programming, control the search bias of the analysis algorithm. Foreign links owe their name to the close relationship to the concept of foreign keys in databases. Looking back at Figure 4.1, we see that the foreign links for analysis are almost the same as the foreign key relations between the three relations, except that the first arrow runs in the other direction (since we want to start with customers for the analysis). Nonetheless, the user is free to also use foreign links that do not correspond to a foreign key relation in the underlying database.

Organization and control of the search. As described above, MIDOS explores its search space top-down starting with the shortest descriptions and adding more and more restrictions. The user can influence the search with a number of parameters.

Minimal support. The minimal support specifies the minimal size of a subgroup as a fraction of the overall population. If we set the minimal support to 10%=0.1, MIDOS will not consider subgroups smaller than 10% of the population. This saves search time and also avoids findings that are uninteresting simply because the group is too small.

Search depth. Each restriction added by MIDOS to the initial group description (the entire population) is considered to be one step "deeper" in the search. For example,

Sex=male, Income=i60_70k

is at depth 2, and

Sex=male, Income=i60_70k, ID=order.Customer ID

is at depth 3 in the search (compare Figure 4.5 where deeper levels are to the right). By setting the desired search depth, the user can control how long and complex subgroup descriptions can be. As is obvious from Figure 4.5, increasing the depth of search greatly increases the number of subgroups that need to be considered, and thus requires significantly more execution time.

Search Mode. The default search mode of MIDOS is to explore the search space in a so-called "breadth-first" manner. This means MIDOS will first fully explore one level of the tree before starting the exploration of the next level. The user can change this to best first, meaning that MIDOS will always work on the subgroup that looks most promising at the time, independent of its depth. In most cases, this means that the most interesting hypotheses are found earlier in the search.

In addition to these-user controlled aspects of the search, MIDOS employs several optimizations to speed up the search.

Optimal refinement. This is a standard technique in search optimization: the search tree is structured in such a fashion that the same point can never be reached along more than one path. Consider the tree in Figure 4.5. The crossed-out node is already reachable along the very top path of the tree. By organizing the search in a way such that the crossed-out node is never even produced, we avoid having to check over and over again whether a new candidate perhaps already was considered in a previous branch.

Optimistic estimate pruning. When exploring the search space from top to bottom, we cannot predict how the quality of a subgroup's descendants will develop: by adding one more restriction, we might make the group more or less unusual depending on which individuals are excluded by the new restriction. We do know, however, that groups can only get smaller, and the maximally unusual groups are those with extreme distributions where certain values do not occur at all anymore. In fact, with mathematical transformations on the quality function explained above (and other quality functions that are used for the other search tasks), we can derive a so-called *optimistic estimate* function that will give us an upper limit on the quality that could potentially be reached by adding more restrictions to a subgroup description. Now, if this upper limit is lower than the qualities of solutions we have found so far, we need not consider this subgroup and its restrictions any further, since we know none of them can make it to our list of top candidates. This allows entire subtrees of the search space to be pruned. The optimistic estimate function is also used in best-first search (see above) to decide where to search next.

Sampling. For very large datasets, it is important to not only limit the number of subgroups that are considered, but also limit the time taken to test each subgroup description and compute its quality. Fortunately, statistics gives us a way to consider only a part of the population and subgroup and still be reasonably certain that we do not miss any interesting subgroup. Using basic statistical theorems, we can compute how many samples we need to draw to have a (say) 95 % chance that the quality of groups found is no more than (say) 5 % worse than the truly best subgroups (see [4.15] for details). In contrast to a priori sampling, where we can never be sure which results we might have lost, this approach allows a precise control over error probability during the run.

Subgroup suppression. Whenever MIDOS has found an interesting subgroup and adds restrictions to it, it is quite likely that the restricted subgroups will maintain a large part of the interestingness of the original group. Consider the case where the new restriction only excludes a small percentage of the members of the original group. There, the quality values of the two groups will not differ a whole lot. For the user, this means that the smaller subgroup is not very interesting, because the restriction does not make any new interesting things happen compared to the original group. For example, if we know that single males are twice as likely to be club members than the entire population, learning that single males aged 30 to 40 are also almost twice as likely is not very interesting. However, of course it would be interesting if single males aged 30 to 40 showed the same behavior as the entire population, since that is unexpected with respect to single males.

To account for this, MIDOS incorporates a subgroup suppression mechanism that works as follows. For a new hypothesis that is good enough to be on our top solutions list, MIDOS checks the solution list to see if there are any predecessors or descendants of the new candidate already on the solution list. For each of them, the algorithm computes a so-called *affinity* value [4.7] that increases with increasing overlap of the groups described by the two subgroup descriptions. If two hypotheses have high affinity, if one of them has a little higher quality (with respect to the entire population) than the other, the worse hypothesis will be suppressed *unless* the two hypotheses have a high quality with respect to each other. The lower the affinity value, the higher quality difference is required for suppression.

The user can influence suppression with two parameters.

Suppression factor. The higher the suppression factor, the smaller the quality difference that suffices for suppression (with equal affinity). A suppression factor of 0 turns off suppression.

Affinity influence. The higher this value, the more important it is how close two hypotheses are. For a value of 0, the higher quality hypothesis will suppress the lower quality hypothesis no matter how large the intersection of their groups is (so this is not a sensible setting).

Suppression is also useful for repeated analysis runs on the same or similar datasets, or in general whenever there is prior knowledge about certain subgroups. If such subgroups are given to MIDOS as previous solutions, any new subgroups that are too close to these already known solutions are suppressed, thus avoiding repeated reporting of known phenomena.

4.4 Using MIDOS and other ILP methods in KEPLER

For a practical look at some of the things discussed above, we will now describe how one would use MIDOS and other ILP methods in the data mining platform KEPLER, a commercial data mining system marketed by Dialogis GmbH[3]. For readers interested in trying the examples discussed in this chapter on their computer, a demonstration version of KEPLER is available on request free of charge from Dialogis at `http://www.dialogis.de`, and the customer dataset used for the examples, as well as other datasets that could be of interest, is available from `http://wwwiws.cs.uni-magdeburg.de/~wrobel/`. KEPLER is a client server system that needs Solaris/Sparc or Windows/Intel for the server; the Java clients runs wherever Java runs.

[3] KEPLER has recently been renamed to DIALOGIS D-MINER. All screenshots are taken from the 1999 release of the system. In D-MINER, the interface has been improved and reimplemented; but for our purposes, the basic principles still apply.

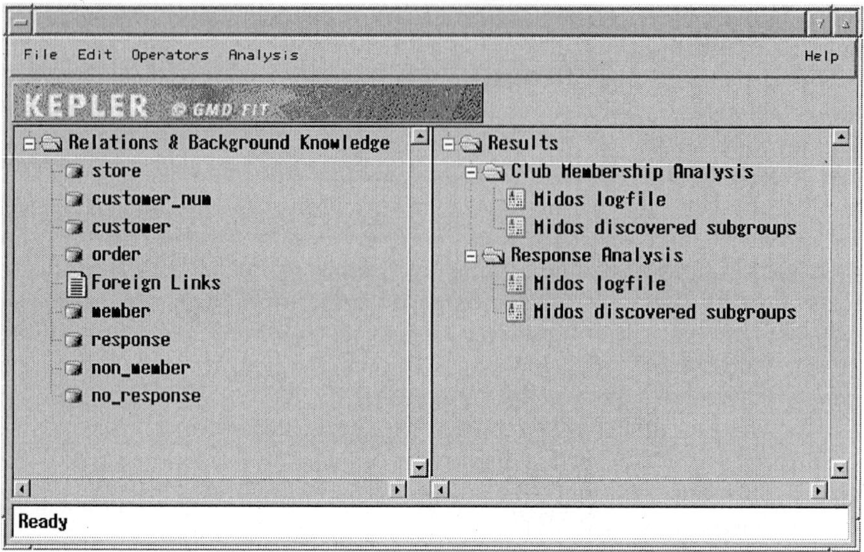

Fig. 4.6. KEPLER workspace window.

4.4.1 KEPLER basics

KEPLER is an extensible data mining platform [4.23] that was developed at GMD and is now available commercially. It was designed to support the entire process of knowledge discovery ranging from data access and import through data preparation and transformation to analysis and use of analysis results. The system features a plug-in interface that allows third party methods to be added with comparatively little effort, which has resulted in a large number of plug-in algorithms that are already available for the system.

More important for our purposes here, KEPLER was designed around relational technology from the start, so it can accommodate both single-table and relational plug-ins. As of now, plug-in wrappers for MIDOS, PROGOL [4.11], CLAUDIEN [4.2] and FOIL [4.13] are available, more will certainly follow.

Figure 4.6 shows the main window of KEPLER that represents the "workspace" in which all data and results are stored and organized. On the left, you see the available relations in our customer example; each relation can reside in KEPLER's own internal database or be linked to from an external database system. On the right, we find the analysis results obtained so far. Both relations and results can be structured into hierarchical folders as familiar from file system explorers.

The task concept. By double clicking on an item in a workspace, you bring up a tabbed-folder "task" window giving details of the item and of the way it was created. For example, Figure 4.7 shows the info window for the customer relation that was introduced in Figure 4.1, showing relation fields and their properties.

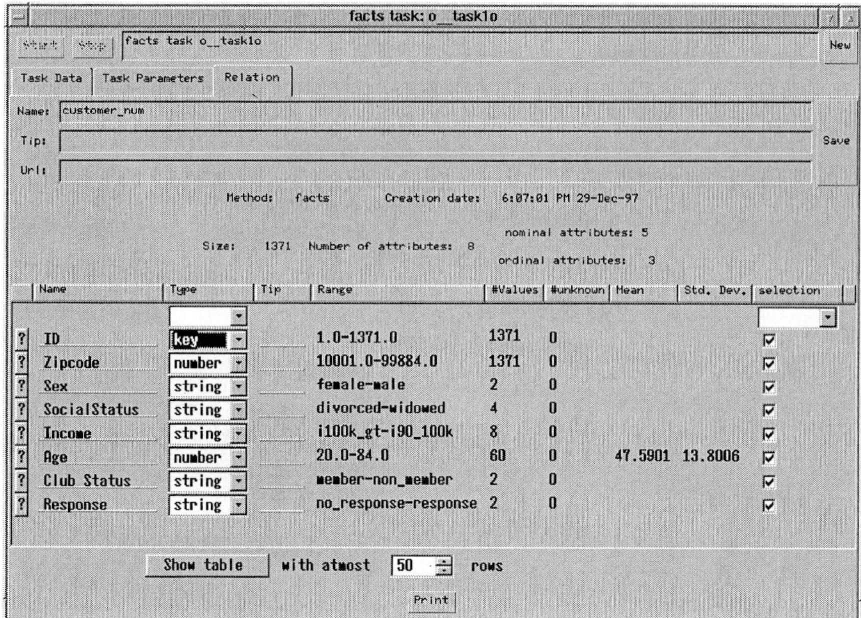

Fig. 4.7. Customer relation.

The info windows, besides information about the item proper, always contain panes that show how the item was created, i.e., by which "tool" with which inputs and which parameters. Thus, whenever you look at a result or relation and want to see how it was created, just switch to the appropriate panel. You can also click the "New" button to "clone" the task that created this result, and then re-run it with the same or modified parameters.

There are three kinds of tasks corresponding to the three kinds of plug-in tools possible in KEPLER: import, operator and analysis tasks, each available through menus and interfaces that are dynamically generated from the tools' descriptions. We will not describe import tasks in great detail; suffice it to say that for ILP purposes, the Prolog-like "facts" format is available as an import plug-in.

Operator tasks are responsible for data preprocessing, and there are plug-ins for different kinds of transformations and selections that one might want to apply to data. Perhaps most interesting in the ILP context are the operators for querying data and for applying user-defined Prolog rules. The query operator allows point-and-click construction of relational queries and suffices for a lot of the data transformations needed to adapt relational data to different ILP tools. The Prolog rule operator is for expert needs and allows Prolog rules to be used to define new relations; these Prolog rules can call user-defined predicates and thus allow complex processing. The rules are input and stored in KEPLER as "background knowledge" items which are

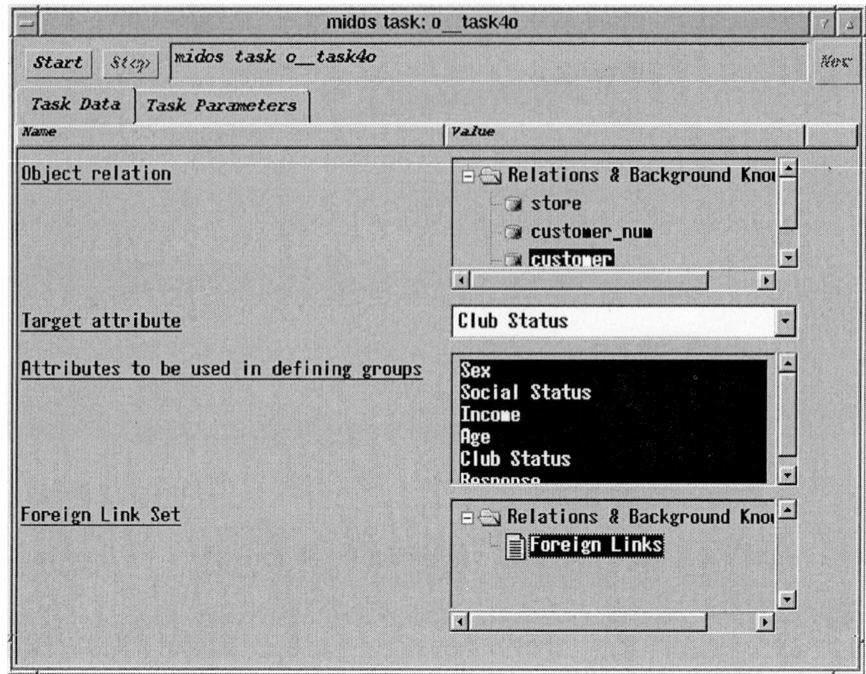

Fig. 4.8. MIDOS analysis task (inputs page).

capable of storing arbitrary non-relational information like Prolog rules or declarative bias specifications for analysis tools (see below).

Finally, analysis tasks are responsible for the data mining step proper, taking relations and background knowledge as input and producing analysis results.

4.4.2 Analysis tasks

Figure 4.8 shows the data page of a new MIDOS task window. This page is generated from input requirement declarations in MIDOS' plug-in wrapper, and allows you to select the start relation for subgroup discovery, the attributes of the start relation to be used in defining subgroups, and the foreign link set (entered as a KEPLER background knowledge item).

Figure 4.9 shows the parameter page of MIDOS. Once finished filling these pages, you start the task which is then executed in the background (so you can have as many tasks in parallel as your operating system and processor(s) will allow). Once finished, the results are retrieved and shown on the newly appearing result panes.

Figure 4.10 shows the result pane for the club member analysis task discussed above. On the left, you see a graphical image of the statistical proper-

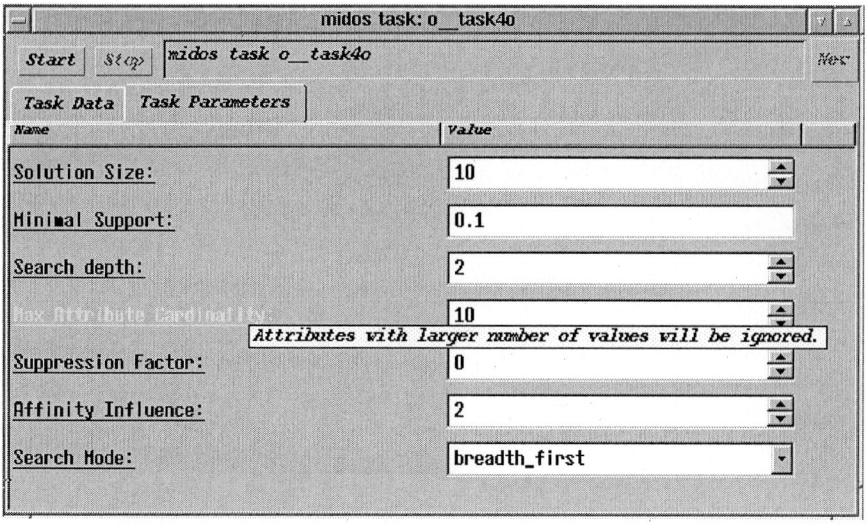

Fig. 4.9. MIDOS task (parameters page).

ties of the discovered subgroups. The horizontal position indicates the distributional differences from the population mean (shown as a vertical line). To the left of the vertical line, one finds groups with a smaller-than-usual share of group members; to the right those with a larger-than-usual share of group members. The colored areas show the size and distribution with respect to the entire population: the horizontally striped area represents all objects in the population with the currently selected value of the target attribute, the non-striped part the remaining objects of the population. The colored area represents all objects in the subgroup, and its overlap with the striped area indicates its proportion of objects with the chosen target value. On the right, you see for each group its subgroup description in the shorthand form described above. The columns preceding the group description show group size, number of elements possessing the target property, and the corresponding percentage value.

For other ILP algorithms, the visualizations are different, but the general principle is the same, organized around tasks with inputs, parameters and results.

Further details on this and on other aspects of KEPLER are of course available in the KEPLER manual; for the purposes of this chapter, we hope the above was sufficient to get an idea of how MIDOS and other ILP methods can be practically used. To really find out, you are encouraged to use the available data and try for yourself.

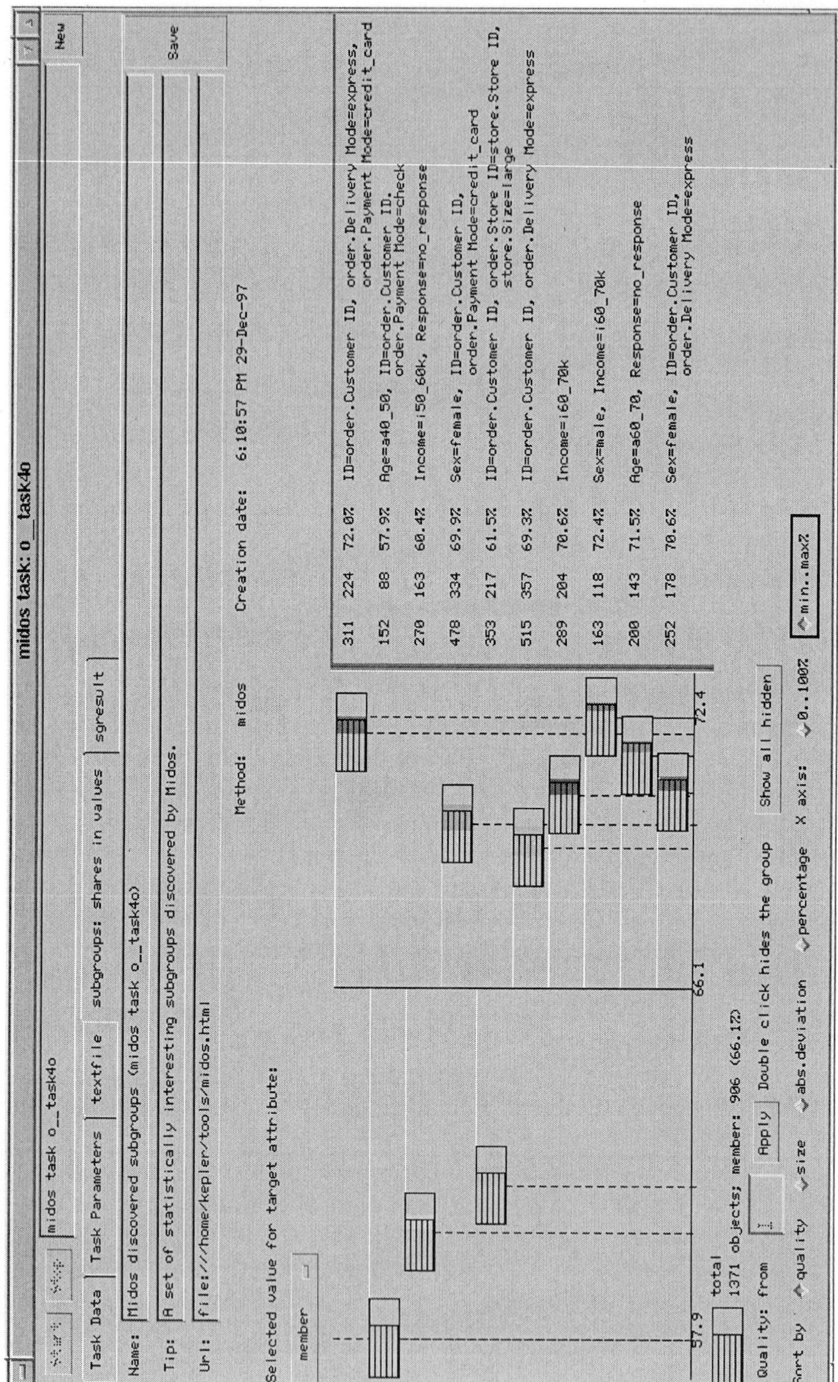

Fig. 4.10. MIDOS subgroup discovery result.

4.5 Conclusion

In this chapter, we have argued that relational analysis technology as developed in the field of ILP is useful not only in classical ILP domains like chemistry or natural language, but also in more everyday domains dealing with customers and orders, cars, parts and factories, or patients and their diagnoses and treatments. We have illustrated the analysis possibilities of ILP technology with a method for subgroup discovery and shown how it and other ILP methods can be used practically in a commercially available data mining system.

As discussed, the power of relational technology lies in its ability to work with multiple relations directly and its methods for automatically searching for the right combinations of these relations, saving you from having to think up and construct a single-table with its often problematic properties beforehand. However, we also pointed out that there is a price to be paid for this: Even though you do not have to think up the necessary joins beforehand, they do need to be computed during the run of a relational/ILP method. For smaller problems, this is usually no problem since all joins are done in main memory and ILP systems are smart in exploiting indexing and cutting off parts of their search spaces.

For larger problems, and in particular if databases are to be used directly (non-memory resident data), the cost of joins becomes the dominant factor. However, even for such large and very large problems, relational technology can usefully be applied based on a simple principle: *If* you know the answer or part of it, and in particular if you know which joins might be interesting and which ones might not, invest the extra work and make this knowledge available to the computer.

- For all joins that would fit in a single table without creating the problems discussed in the introduction, have your data mining platform or database precompute these joins. Make sure you use an ILP system that reaches the efficiency of a single table system when used on single table data (like MIDOS, CLAUDIEN [4.2], RIBL [4.5], TILDE [4.1]), or, if *all* necessary data can be sensibly joined into a single table, use a single-table method.
- For the remaining, more complex or alternative joins that could be interesting, use an available declarative bias language (e.g., MIDOS' foreign links) to describe as precisely as possible which joins are to be considered. But careful, knowledge discovery is about finding unexpected things, so make sure you do not exclude possibly interesting items just because you could not imagine they could be interesting.

If the above turns out impossible or insufficient to reach the runtimes you would like to see, use an ILP system capable of sampling (e.g., MIDOS, see Section 4.3.2) or use an ILP system on a sample of the data. Of course, if the sample is small (1% or less), solutions on the sample will not necessarily be the same as solutions on the entire dataset. One way out of this is to use the solutions on the sample as a guide for finding out what the important joins could be. You could then formulate a significantly more restricted search space and re-run ILP on the entire data set, or if possible, make a single table and run a propositional method. And of course, you can always at least verify your results on the entire dataset.

Either way, relational technology as developed in ILP is ready for practical applications and available now in commercial systems. Given the dynamic development of ILP methods, we can expect these techniques to have an even larger influence in the future.

Acknowledgments

This work was partially supported by ESPRIT LTR contract number 20237 ("Inductive Logic Programming 2"). A big thanks goes to my colleagues who read and commented on this chapter, and to Nada Lavrač and Sašo Džeroski for their excellent and detailed editorial comments.

References

4.1 H. Blockeel and L. De Raedt. Lookahead and discretization in ILP. In *Proceedings of the Seventh International Workshop on Inductive Logic Programming*, pages 77–84. Springer, Berlin, 1997.
4.2 L. De Raedt and L. Dehaspe. Clausal discovery. *Machine Learning*, 26: 99–146, 1997.
4.3 S. Džeroski, H. Blockeel, B. Kompare, S. Kramer, B. Pfahringer, and W. Van Laer. Experiments in Predicting Biodegradability. In *Proceedings of the Ninth International Workshop on Inductive Logic Programming*, pages 80–91. Springer, Berlin, 1999.
4.4 S. Džeroski, S. Schulze-Kremer, K. Heidtke, K. Siems, D. Wettschereck, and H. Blockeel. Diterpene structure elucidation from ^{13}C NMR spectra with Inductive Logic Programming. *Applied Artificial Intelligence*, 12: 363–383, 1998.
4.5 W. Emde and D. Wettschereck. Relational instance based learning. In *Proceedings of the Thirteenth International Conference on Machine Learning*, pages 122–130. Morgan Kaufmann, San Mateo, CA, 1996.
4.6 U. Fayyad, G. Piatetsky-Shapiro, and P. Smyth. From data mining to knowledge discovery: An overview. In U. Fayyad, G. Piatetsky-Shapiro, P. Smyth, and R. Uthurusamy, editors, *Advances in Knowledge Discovery and Data Mining*, pages 1–34. MIT Press, Cambridge, MA, 1996.
4.7 F. Gebhardt. Choosing among competing generalizations. *Knowledge Acquisition*, 3: 361–380, 1991.
4.8 W. Klösgen. Explora: A multipattern and multistrategy discovery assistant. In U. Fayyad, G. Piatetsky-Shapiro, P. Smyth, and R. Uthurusamy, editors, *Advances in Knowledge Discovery and Data Mining*, pages 249–271. MIT Press, Cambridge, MA, 1996.
4.9 N. Lavrač and S. Wrobel. Induktive Logikprogrammierung - Grundlagen und Techniken. *Künstliche Intelligenz*, 10(3): 46–54, 1996.

4.10 S. Muggleton, R. D. King, and M. J. E. Sternberg. Protein secondary structure prediction using logic-based machine learning. *Protein Engineering*, 5(7): 647–657, 1992.
4.11 S. Muggleton. Inverse entailment and Progol. In K. Furukawa, D. Michie, and S. Muggleton, editors, *Machine Intelligence 14*, pages 133–188. Oxford University Press, Oxford, 1995.
4.12 S. Muggleton and L. De Raedt. Inductive logic programming: Theory and methods. *Journal of Logic Programming*, 19/20: 629–679, 1994.
4.13 J. R. Quinlan. Learning logical definitions from relations. *Machine Learning*, 5(3): 239–266, 1990.
4.14 S. Roberts, W. Van Laer, N. Jacobs, S. Muggleton, and J. Broughton. A comparison of ILP and propositional systems on propositional traffic data. In *Proceedings of the Eighth International Conference on Inductive Logic Programming*, pages 291–299. Springer, Berlin, 1998.
4.15 T. Scheffer and S. Wrobel. A sequential sampling algorithm for a general class of utility criteria. In *Proceedings of the Sixth International Conference on Knowledge Discovery and Data Mining*, pages 330–334. ACM, New York, 2000.
4.16 A. Srinivasan, R. D. King, S. Muggleton, and M. J. E. Sternberg. The predictive toxicology evaluation challenge. In *Proceedings of the Fifteenth International Joint Conference on Artificial Intelligence*, pages 1–6. Morgan Kaufmann, San Mateo, CA, 1997.
4.17 A. Srinivasan, S. Muggleton, R. King, and M. Sternberg. Mutagenesis: ILP experiments in a non-determinate biological domain. In *Proceedings of the Fourth International Workshop on Inductive Logic Programming*, pages 217–232. GMD, Sankt Augustin, Germany, 1994.
4.18 W. Van Laer, L. De Raedt, and S. Džeroski. On multi-class problems and discretization in inductive logic programming. In *Proceedings of the Tenth International Symposium on Foundations of Intelligent Systems*, pages 277–286. Springer, Berlin, 1997.
4.19 S. Wrobel. Inductive logic programming. In G. Brewka, editor, *Advances in Knowledge Representation and Reasoning*, pages 153–189. CSLI-Publishers, Stanford, CA, 1996.
4.20 S. Wrobel. An algorithm for multi-relational discovery of subgroups. In *Proceedings of the First European Symposium on Principles of Data Mining and Knowledge Discovery*, pages 78–87. Springer, Berlin, 1997.
4.21 S. Wrobel. Data Mining und Wissensentdeckung in Datenbanken. *Künstliche Intelligenz*, 12(1): 6–10, 1998.
4.22 S. Wrobel. Scalability issues in Inductive Logic Programming. In *Proceedings of the Ninth International Conference on Algorithmic Learning Theory*, pages 11–30. Springer, Berlin, 1998.
4.23 S. Wrobel, D. Wettschereck, E. Sommer, and W. Emde. Extensibility in data mining systems. In *Proceedings of the Second International Conference on Knowledge Discovery and Data Mining*, pages 214–219. AAAI Press, Menlo Park, CA, 1996.

Part II

Techniques

5. Three Companions for Data Mining in First Order Logic

Luc De Raedt[1], Hendrik Blockeel[2], Luc Dehaspe[2], and Wim Van Laer[2]

[1] Institut für Informatik, Albert-Ludwigs-Universität Freiburg
Am Flughafen 17, D-79110 Freiburg, Germany

[2] Department of Computer Science, Katholieke Universiteit Leuven
Celestijnenlaan 200A, B-3001 Leuven, Belgium

Abstract

Three companion systems, CLAUDIEN, ICL and TILDE, are presented. They use a common representation for examples and hypotheses: each example is represented by a relational database. This contrasts with the classical inductive logic programming systems such as PROGOL and FOIL. It is argued that this representation is closer to attribute value learning and hence more natural. Furthermore, the three systems can be considered first order upgrades of typical data mining systems, which induce association rules, classification rules or decision trees respectively.

5.1 Introduction

Typical data mining algorithms employ a limited attribute value representation, where each example consists of a single tuple in a relational database. This representation is inadequate for problem-domains that require reasoning about the structure of objects in the domain and relations among such objects, such as in bio-chemistry [5.8], natural language processing [5.15], and traffic control [5.18]. This paper presents three companion systems, where each example corresponds to a small relational database (or Prolog knowledge base). Hence, examples consist of multiple relations and each example can have multiple tuples for these relations. This setting is known in the literature as *learning from interpretations*. It contrasts with the classical inductive logic programming setting employed by the systems PROGOL [5.20] and FOIL [5.22], cf. [5.11]. We will show that our representation is a natural upgrade of attribute value representations that is effective for inductive logic programming.

Current data mining approaches are often distinguished on the basis of their predictive or descriptive nature. In predictive data mining one is given a set of examples or observations that are classified into a finite number of classes. Typically, there are two (or more) classes, one that is called positive, and the other that is negative. The aim then is to induce a hypothesis that correctly classifies all the given (and unseen) examples. Consider Figure 5.1(a), where one is given two types of examples (+ and –). H is a correct hypothesis as it correctly discriminates the positives from the negatives. The purpose of predictive data mining is thus to generate hypotheses that can be used for classification. Common predictive data mining techniques include decision tree induction (e.g., C4.5 [5.21]) and rule induction (e.g., CN2 [5.10, 5.9] or AQ [5.19]).

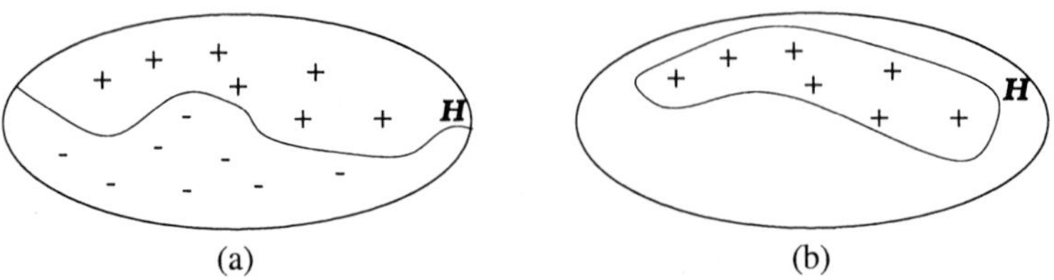

Fig. 5.1. Predictive versus descriptive induction.

In descriptive data mining one is given a set of unclassified examples and the aim is to find regularities within these examples. Furthermore, it is the aim to characterize as much as possible the given examples. Therefore as many properties or regularities as possible are derived. Together they form a kind of most specific hypothesis that covers (explains) all the examples. E.g., in Figure 5.1(b), the hypothesis H characterizes all the given examples. The most popular descriptive data mining technique is that of discovering association rules [5.2, 5.1, 5.24].

The three companion systems, CLAUDIEN [5.13, 5.12], ICL [5.14] and TILDE [5.5, 5.6], can be considered first order upgrades of existing attribute-value data mining approaches. CLAUDIEN upgrades the descriptive association rule approach, ICL upgrades the predictive production rule approach as incorporated in, e.g., CN2 [5.10, 5.9] and AQ [5.19], and TILDE is a recent upgrade of Quinlan's popular predictive C4.5 algorithm for decision tree induction [5.21].

This chapter is organized as follows: Section 2 introduces the representational framework and its relation to attribute value learning. In Sections 3, 4, and 5 the basic algorithms of the three companion systems CLAUDIEN, ICL and TILDE are introduced. Section 6 illustrates how we can use these three systems on an application and discusses some practical aspects. In Sections 7 and 8, we illustrate briefly the three companions on the mutagenesis problem, and give an exercise on an artificial problem from a series of problems called *Bongard problems*. Section 9 concludes.

5.2 Representation

5.2.1 Attribute value representations

Imagine that you have just been hired by a professional seminar organizer PSO in order to discover new knowledge about the activities of PSO that is to be used for commercial purposes. PSO also informs you that they have a database about past activities. Part of this database is listed in Figure 5.2.

PARTICIPANT Table

NAME	JOB	SENIORITY	COMPANY	PARTY	R_NUMBER
adams	researcher	junior	scuf	no	23
blake	president	junior	jvt	yes	5
king	manager	junior	ucro	no	78
miller	manager	senior	jvt	yes	14
scott	researcher	senior	scuf	yes	94
turner	researcher	junior	ucro	no	81

Fig. 5.2. Table of participants of a seminar.

It contains information about participants in a recent Seminar on Data Mining. For each of the participants, a number of attributes are stored: Name, Job, Seniority, Company, Party (whether the person attend the party or not), and Rnumber (Registration number). From talking to the marketing people you are informed that the variable Party is important. So, you decide to find out what type of people attend the parties at such seminars. (This is useful in order to set the price of the party as well as to decide upon the activities at parties).

Being familiar with attribute value approaches to data mining, you decide to carry out experiments with the predictive decision tree and production rule algorithms as well as with the descriptive association rule approach. Within the attribute value representation, each participant corresponds to an example. In predictive approaches, each example will be described by a vector *participant(Job,Seniority,Company)* in addition to the Class-variable Party as you decide that Name and Rnumber are bad indicators for the variable Party. Running your favorite rule-induction algorithm could then result in the following production rules:

Party=*yes* if Seniority=*senior*
Party=*yes* if Job=*president*

Using a Prolog representation, one could use:

party(*yes*) :- participant(_J,*senior*,_C)
party(*yes*) :- participant(*president*,_S,_C)

Using the latter representation, in order to predict whether a participant will attend the party or not, you simply assert the tuple describing the participant together with the above hypothesis in the Prolog-database and run the query

?- party(*yes*). If the query succeeds[1], the participant will attend the party; otherwise she won't.

Alternatively, you may induce a decision tree (see Figure 5.3) for this predictive data mining problem. For the purposes of this chapter, we assume that all tests in the decision trees are boolean (or binary). To use the tree for

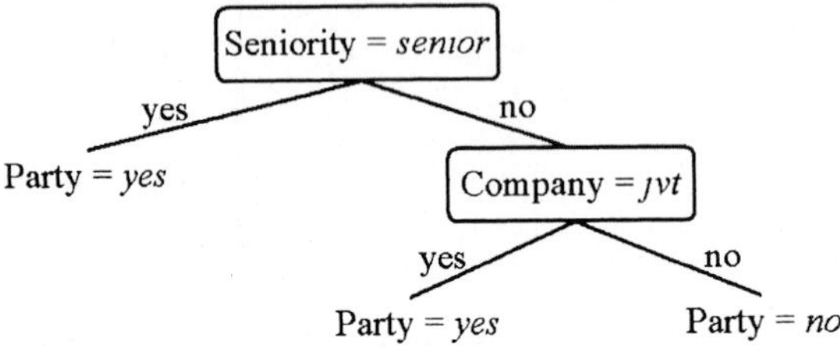

Fig. 5.3. A possible decision tree for the Participant table in Figure 5.2.

classifying a participant, one first tests whether Seniority=*senior*; if it is, one classifies the participant as going to the party; otherwise one tests whether Company=*jvt*. If it is, one classifies as Party=*yes*, otherwise as Party=*no*. Again, we could represent the decision tree by a Prolog program :

party(*yes*) :- participant(_J,*senior*,_C), !.
party(*yes*) :- participant(_J,_S,*jvt*), !.
party(*no*).

To classify an example one then again has to assert the tuple corresponding to the participant in the Prolog knowledge base containing the above hypothesis, and then run the query ?-party(_X). Prolog will then compute the value of _X : *yes* or *no*.

Alternatively, you may decide to use the association rule approach to discover all interesting characteristics of people that go to the party. Running a descriptive data mining algorithm on this subset of the original dataset then results in rules such as the following:

false if Company=*ucro*
Job=*president* or Seniority=*senior*
...

or in Prolog:

[1] A query succeeds if and only if values for its variables exist such that all the conditions in the query are true.

```
false :- participant(_J,_S,ucro)
_J=president;_S=senior :- participant(_J,_S,_C)
...
```

The first rule reveals no one from company *ucro* will attend the party. The second rule expresses the constraint that, within the subgroup going to the party, all participants are either presidents or senior staff. Therefore PSO may decide to increase the price of the party (since it will be attended anyway by the higher personnel) or decrease the price (in order to attract also the lower ranked personnel).

The representations introduced above have two drawbacks. Firstly, there is only a fixed feature-value representation of examples. This makes it impossible to represent structural information about the examples (cf. Exercise on Bongard Problem in Section 5.8). Secondly, no additional background information about the examples was provided. We discuss each of these drawbacks in turn, and show how they can be alleviated.

5.2.2 First order representations

Examples. A slightly more realistic version of our previous illustration is shown in Figure 5.4 (the attribute Seniority is deleted in table Participant

PARTICIPANT Table

NAME	JOB	COMPANY	PARTY	R_NUMBER
adams	researcher	scuf	no	23
blake	president	jvt	yes	5
king	manager	ucro	no	78
miller	manager	jvt	yes	14
scott	researcher	scuf	yes	94
turner	researcher	ucro	no	81

SUBSCRIPTION Table

NAME	COURSE
adams	erm
adams	so2
adams	srw
blake	cso
blake	erm
king	cso
king	erm
king	so2
king	srw
miller	so2
scott	erm
scott	srw
turner	so2
turner	srw

Fig. 5.4. The registration database with two tables.

and an extra table Subscription is added). The main distinction between the previous database and the current one is that properties of each person are now scattered around the whole database. In particular, information about each person is contained in multiple tables of the database. Furthermore, some tables contain multiple tuples for some persons. The question now is how this database can be turned into a set of examples. To obtain a set of examples, one merely has to partition the database into examples, each of which corresponds to a single person. This is shown in Figure 5.5(a) and (b). E.g., the partition corresponding to *adams* can be obtained by first selecting all tuples having Name=*adams* (see (a)), and then projecting on all attributes (excluding Name). Each partition thus obtained constitutes a *model* or an interpretation (see (b))[2].

PARTICIPANT Table

NAME	JOB	COMPANY	PARTY	R_NUMBER
adams	researcher	scuf	no	23
blake	president	jvt	yes	5
king	manager	ucro	no	78
miller	manager	jvt	yes	14
scott	researcher	scuf	yes	94
turner	researcher	ucro	no	81

SUBSCRIPTION Table

NAME	COURSE
adams	erm
adams	so2
adams	srw
blake	cso
blake	erm
king	cso
king	erm
king	so2
king	srw
miller	so2
scott	erm
scott	srw
turner	so2
turner	srw

(a)

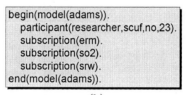

```
begin(model(adams)).
  participant(researcher,scuf,no,23).
  subscription(erm).
  subscription(so2).
  subscription(srw).
end(model(adams)).
```

(b)

Fig. 5.5. Partition corresponding to Adams.

Given the new representation, we could again run our favorite algorithms. However, before doing so, we will first consider the issue of background knowledge.

Background knowledge. It is often useful to employ additional knowledge in the induction process. Let us start, e.g., with global information. Consider

[2] The generation of these models from the database is intuitively straightforward. It can also be automated; we do not go into the details of the algorithm here but refer to [5.4].

Figure 5.6, which is an extension to the database in Figure 5.4. The information in the tables Company and Course could be helpful when generalizing from examples. However, the information contained in these tables is not

PARTICIPANT Table

NAME	JOB	COMPANY	PARTY	R_NUMBER
adams	researcher	scuf	no	23
blake	president	jvt	yes	5
king	manager	ucro	no	78
miller	manager	jvt	yes	14
scott	researcher	scuf	yes	94
turner	researcher	ucro	no	81

SUBSCRIPTION Table

NAME	COURSE
adams	erm
adams	so2
adams	srw
blake	cso
blake	erm
king	cso
king	erm
king	so2
king	srw
miller	so2
scott	erm
scott	srw
turner	so2
turner	srw

COMPANY Table

COMPANY	TYPE
jvt	commercial
scuf	university
ucro	university

COURSE Table

COURSE	LENGTH	TYPE
cso	2	introductory
erm	3	introductory
so2	4	introductory
srw	3	advanced

Fig. 5.6. The complete registration database with several tables.

only meaningful to a single example, it contains knowledge relevant to all examples. This kind of knowledge will be included as a separate background knowledge file (see the `company` and `course` facts in Figure 5.7(a)), and will be considered part of each example (or partition).

Secondly, it may be useful to encode derived knowledge using inference rules (such as those offered by Prolog or using, e.g., views in a relational database). Such knowledge can take various forms: e.g., abstraction of specific values into a taxonomy or interval (e.g., course-weight in Figure 5.7(b)), deriving new properties from a combination of existing ones, summarizing or aggregating values of several tuples into a single value (e.g., sum_due in Figure 5.7(b)), views on a specific attribute (e.g., job, party,... in Figure 5.7(a)) etc. Rules for *course_weight* (which is *hi* if the course is advanced or longer than 3, otherwise it is *low*) and *sum_due* (the amount of money a participant must pay, depending on the courses followed) can be found in Figure 5.7(b). Again this form of background knowledge will be included in the background knowledge file. To reason about examples, one then asserts the background theory and the partition in a Prolog knowledge base, and asks queries in Prolog. E.g., to reason about *adams*, one would assert the model(*adams*) in Prolog together with background. Then *adams* would have the following property: *sum_due*(*1000*).

Taken all together, the database in Figure 5.6 can be partitioned as in Figure 5.8. The background is common to all examples.

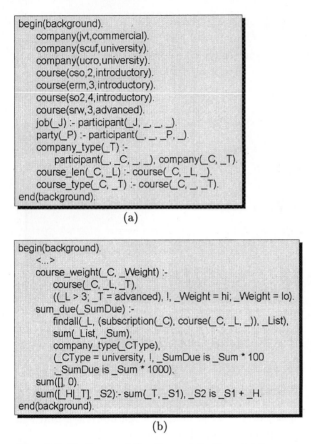

```
begin(background).
    company(jvt,commercial).
    company(scuf,university).
    company(ucro,university).
    course(cso,2,introductory).
    course(erm,3,introductory).
    course(so2,4,introductory).
    course(srw,3,advanced).
    job(_J) :- participant(_J, _, _, _).
    party(_P) :- participant(_, _, _P, _).
    company_type(_T) :-
        participant(_, _C, _, _), company(_C, _T).
    course_len(_C, _L) :- course(_C, _L, _).
    course_type(_C, _T) :- course(_C, _, _T).
end(background).
```

(a)

```
begin(background).
    <...>
    course_weight(_C, _Weight) :-
        course(_C, _L, _T),
        ((_L > 3; _T = advanced), !, _Weight = hi; _Weight = lo).
    sum_due(_SumDue) :-
        findall(_L, (subscription(_C), course(_C, _L, _)), _List),
        sum(_List, _Sum),
        company_type(_CType),
        (_CType = university, !, _SumDue is _Sum * 100
        ;_SumDue is _Sum * 1000).
    sum([], 0).
    sum([_H|_T], _S2):- sum(_T, _S1), _S2 is _S1 + _H.
end(background).
```

(b)

Fig. 5.7. Background for the registration database.

First order hypotheses. In the previous section, we merely adapted the representation of examples (and background knowledge) to first order logic. Given the representation of examples and background knowledge it is however easy to upgrade the representations for production rules, association rules and decision trees. One merely has to allow for condition parts of Prolog rules that allow for conjunctions of literals (or relations).

E.g., to predict the class *party*, one could induce: (\+ stands for Prolog's *not* operator)

party(*yes*) :- company_type(*commercial*).
party(*yes*) :- course_len(_S, 4), \+subscription(_S).

Similarly, for decision trees we have the tree in Figure 5.9.

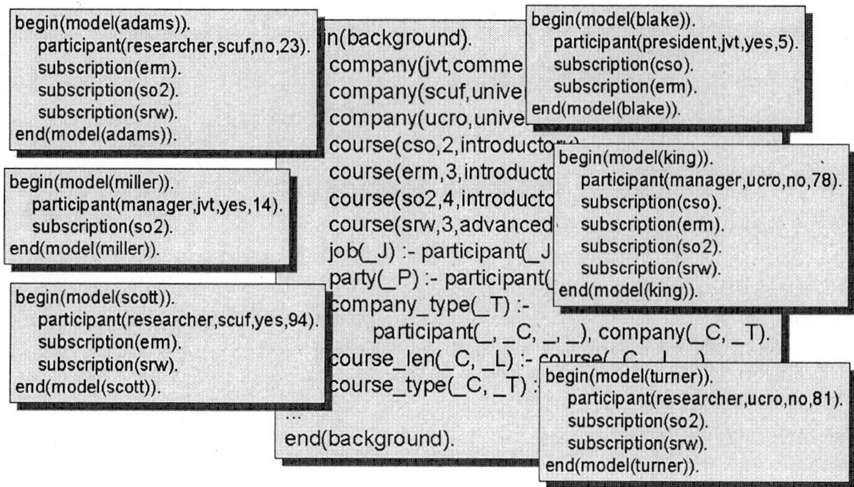

begin(model(adams)).
participant(researcher,scuf,no,23).
subscription(erm).
subscription(so2).
subscription(srw).
end(model(adams)).
begin(model(miller)).
participant(manager,jvt,yes,14).
subscription(so2).
end(model(miller)).
begin(model(scott)).
participant(researcher,scuf,yes,94).
subscription(erm).
subscription(srw).
end(model(scott)).
n(background).
company(jvt,comme
company(scuf,unive
company(ucro,unive
course(cso,2,introducto
course(erm,3,introductc
course(so2,4,introductc
course(srw,3,advanced
job(_J) :- participant(_J
party(_P) :- participant(
company_type(_T) :-
participant(_, _C, _, _), company(_C, _T).
course_len(_C, _L) :-
course_type(_C, _T) :
...
end(background).
begin(model(blake)).
participant(president,jvt,yes,5).
subscription(cso).
subscription(erm).
end(model(blake)).
begin(model(king)).
participant(manager,ucro,no,78).
subscription(cso).
subscription(erm).
subscription(so2).
subscription(srw).
end(model(king)).
begin(model(turner)).
participant(researcher,ucro,no,81).
subscription(so2).
subscription(srw).
end(model(turner)).

Fig. 5.8. The registration database in Prolog.

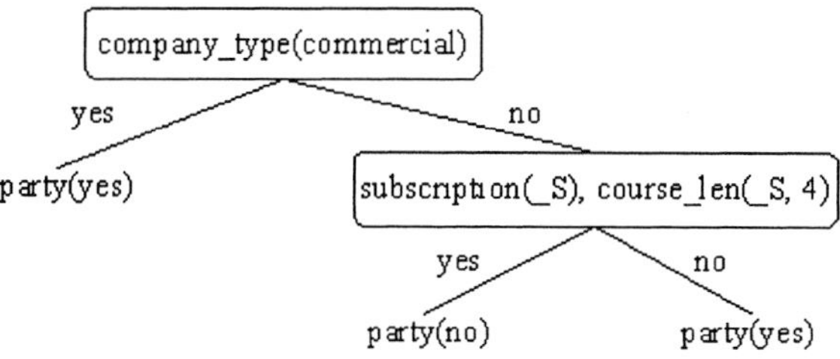

Fig. 5.9. A decision tree for the class *party*.

With association rules, we can induce properties of participants that go the party, such as:

```
job(manager) ; subscription(_S) :- course_len(_S,3) ,
                                    course_type(_S,introductory).
```

This rule says that the considered participants are either managers or they subscribe to all introductory courses of length 3.

Learning from interpretations. The above introduced knowledge representation formalism for hypotheses and examples is quite natural. First, notice that attribute value representations are a straightforward special case. Secondly, notice also that examples roughly correspond to partitions of the database. As a consequence, each example can be considered independently of the other ones. This is important in the light of data mining where one could be given a vast numbers of examples, which excludes the possibility of loading the dataset into main memory (cf. [5.16]). Thus one potentially gains efficiency. Thirdly, we believe this representation is also quite intuitive and elegant (cf. Exercise on Bongard Problem in Section 5.8).

5.3 ICL: Inductive classification logic

The ICL system upgrades the well-known CN2 algorithm by Clark and Niblett [5.10, 5.9] toward inductive logic programming. The CN2 and ICL algorithms will induce a set of production rules such as those listed in Section 2. A simplified algorithm is outlined in Table 5.1.

Table 5.1. A simplified outline of CN2 and ICL.

```
find_hypothesis(P,N)
    H := ∅
    while P is not empty do
        find-rule(P,N,class(positive) :- body)
        add class(positive) :- body to H
        remove from P all examples covered by class(positive) :- body .


find_rule(P,N, class(positive) :- body)
{finds a rule class(positive) :- body that covers some positives in P}
    body := true
    while ?-body succeeds for a negative example in N do
        for all literals lit do
            compute the heuristic value of class(positive) :- body,lit
        body := body, l where l is the literal that scores best
```

The first loop of the algorithm is the so-called "covering" algorithm. It starts off with all positive examples in P and all negatives in N. It will then repeatedly attempt to find a single rule that covers some positives and none of the negatives. Each such rule is added to the hypothesis H and all positives covered by this rule will be removed from P. This process terminates when all positives are covered. To find one such rule, the algorithm performs a top-down search process, whereby it starts from the most general rule class(*positive*) :- true that covers all examples, and repeatedly refines this rule by heuristically choosing the best condition to add. This process terminates when the current rule does not cover any negative example.

Consider the WORN example in Figure 5.10. Suppose we are learning the description of class *positive*. Then ICL would start off by the following rule:

class(*positive*) :- true

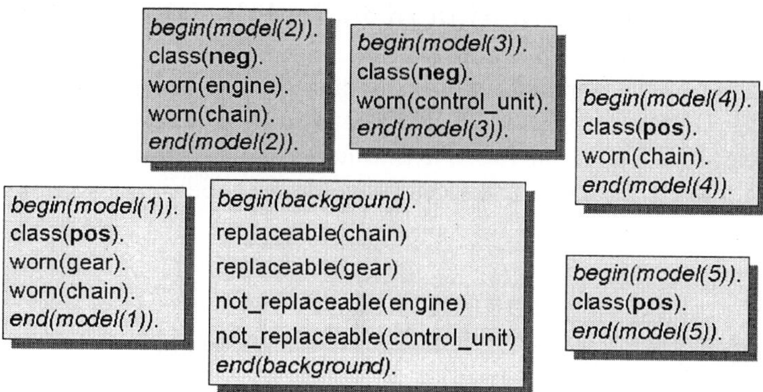

Fig. 5.10. The WORN example with two classes.

It would then discover that this also classifies the negatives as positive and would therefore consider refinements of this rule including:

class(*positive*) :- worn(_X)
class(*positive*) :- not_replaceable(_X)
etc.

Suppose it then chooses the first rule (because it excludes more negatives than the second one). Again, it would notice that it is too general, and start refining. It could then come up with the following correct hypothesis :

class(*positive*) :- worn(_X), not_replaceable(_X).

This rule correctly classifies all examples.

It should be mentioned that the above description of ICL and CN2 is largely simplified. First, it may be the case that the data are noisy in which case a perfect solution may not exist. Under such conditions the algorithms will employ heuristics to find an approximation of the target concept (cf. [5.14]). Second, the find-rule algorithm as outlined above performs hill-climbing : it keeps track of a single current best candidate. ICL and CN2, however, perform a beam-search, where one always keeps track of the m best rules. Finally, the above algorithm does not specify which literals can be added to the hypothesis. This is discussed in Section 5.6.1 on $\mathcal{D}$LAB.

Some additional features of ICL include: it can learn hypotheses in DNF as well as in CNF format, it has some number handling capabilities (using discretization), it can learn multi-classes in case there are more than 2 classes (we use a similar strategy as in CN2, but have also a Bayesian method included), it has some built-in evaluation methods (e.g., 10-fold cross-validation),... (see also [5.26]). Several settings can be used to tune the heuristics and a non-graphical interface is integrated that makes the use of ICL fairly easy.

5.4 Tilde: Top-down induction of logical decision trees

Tilde is a member of the popular top-down induction of decision tree family of algorithms. However, rather than using attribute-value tests in nodes of the tree, it employs logical queries. Furthermore, as in Example WORN in Figure 5.11, a variable name can occur in different nodes. The mean-

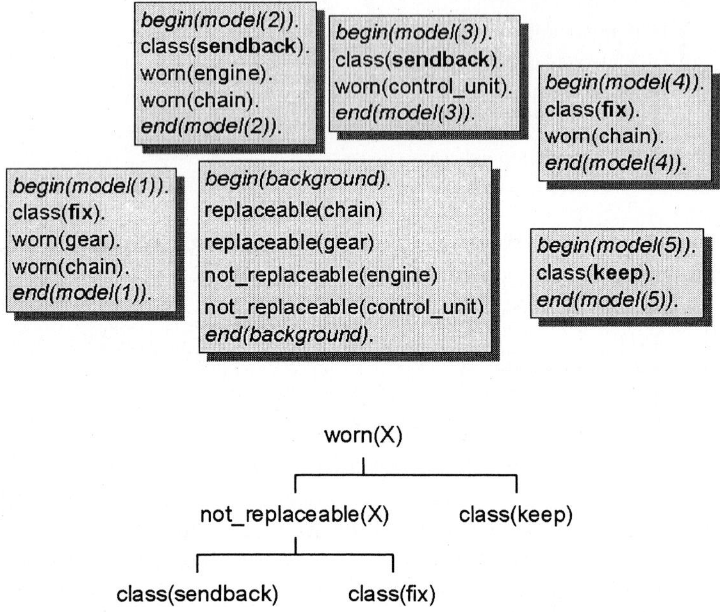

Fig. 5.11. The WORN example with three classes and a corresponding tree.

ing then is that all occurrences denote the same variable. This slightly complicates the classification process with first order decision trees. E.g., to classify an example with the WORN tree, one first tests whether ?-worn(_X) succeeds, if it does not, if belongs to class keep. Otherwise, one should test whether not_replaceable(_X) succeeds. However, since the _X in not_replaceable(_X) already appears higher in the tree, this test corresponds to testing ?-worn(_X),not_replaceable(_X). Only if this query succeeds, the class is sendback; otherwise, it is fix. Hence, an equivalent Prolog program corresponding to the WORN tree is:

class(*sendback*) :- worn(_X), not_replaceable(_X), ! .
class(*fix*) :- worn(_X), ! .
class(*keep*).

The induction algorithm underlying TILDE is similar to C4.5. TILDE will start off with the empty tree. It will then generate all possible tests and compute the heuristic value of these test. Among the possible tests, it will select the one that scores best on the heuristic and place that in the current node. It will then use the partial tree to classify all examples (in the current node) to its sub-nodes. All examples passing the test will be propagated to the left, all the other ones to the right. The procedure will then recursively analyze the left and right subtrees. When a node only contains examples of a single class (or when heuristics indicate it is uninteresting to split a node), it will be turned into a leaf.

Figure 5.12 illustrates the algorithm on Example WORN.

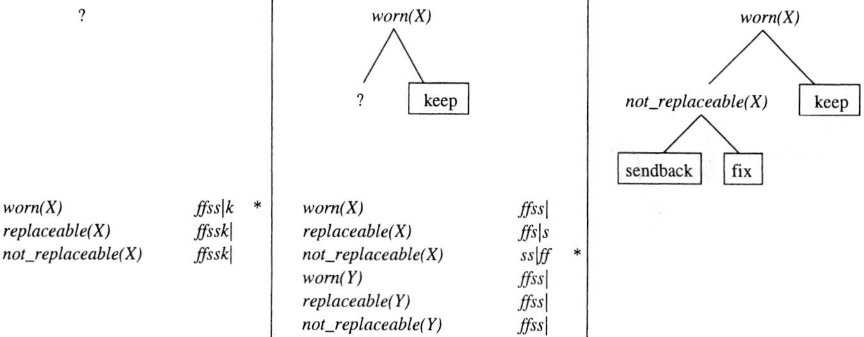

Fig. 5.12. TILDE illustrated on the WORN example. Each step shows the partial tree that has been built, the literals that are considered for addition to the tree, and how each literal would split the set of examples. E.g., *fss*|*k* means that of four examples, one with class `fix` and two with class `sendback` are in the left branch, and one example with class `keep` is in the right branch. The best literal is indicated with an asterisk.

5.5 CLAUDIEN: Clausal discovery

The CLAUDIEN algorithm searches for a set of clausal regularities that hold on the set of examples. Each such clause is of the form

$$h_1; \ldots ; h_n : -b_1, \ldots , b_m$$

A clause of this form expresses that whenever the condition part (the conjunction of the literals b_i holds), that also the conclusion part holds (i.e., at least one of the h_i holds). A clause holds on an example whenever it is the case that if ?-$b_1, \ldots , b_m$ succeeds with a substitution θ then also $? - h\theta$ succeeds.

The clausal discovery algorithm CLAUDIEN does not employ the well-known two-phased algorithm for induction of association rules used for instance in APRIORI [5.1] (but see [5.16] for an upgrade of APRIORI). Instead, it employs an algorithm that directly enumerates all the relevant clauses.

The key idea underlying the CLAUDIEN algorithm is the following: whenever a clause does not hold on an example, it must be that there exists a substitution θ such that ?-$(b_1, ..., b_n)\theta$ holds but $(h_1; ...; h_n)\theta$ does not hold. There are two ways to solve this problem. One can either modify the clause by adding a literal h to its head (for which $h\theta$ might hold), or by adding a literal b to its body (for which $b\theta$ might not hold).

This idea then results in the simple algorithm outlined in Table 5.2. The algorithm starts with a queue containing $false : -true$ as the only clause.

Table 5.2. A simplified outline of CLAUDIEN.

```
Q := { false :- true }
H := ∅
    while Q is not empty do
        delete c from Q
        if c covers all p in P (and H does not entail c)
        then add c to H
        else add all refinements of c to Q
```

This clause does not hold in any example. In the while loop of the algorithm, CLAUDIEN then repeatedly deletes a clause from the queue and tests whether it holds on the data. If it does, it is added to the final hypothesis, otherwise it is refined (by adding a literal to it). The search space of the algorithm for a simple clausal discovery problem is illustrated in Figure 5.13, where m, f, and h denote male, female, and human, respectively.

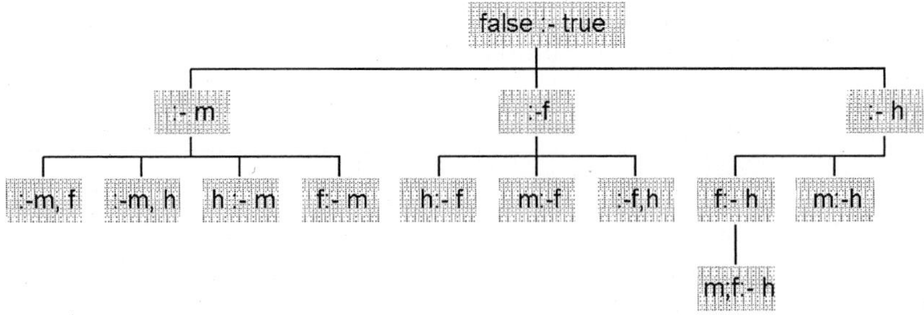

Fig. 5.13. The search space of CLAUDIEN.

5.6 Practical use: Getting started

In this section we use the registration application to guide you through a first session with the three systems CLAUDIEN, ICL and TILDE. Basically, they all require as inputs a knowledge base, a language bias, and some parameter settings.

Once you have constructed the knowledge base as explained in Section 5.2.2, this serves as input to all three systems. Assuming *app* is the name of the application, this input should be stored in a file called *app*.kb. For the definition of language bias, CLAUDIEN and ICL use $\mathcal{D}$LAB; this definition is put in an input file *app*.l. Parameter settings are put in *app*.s. Most of these settings are system specific. Some settings apply to more than one of the companion systems. These are:

- talking(n) specifies how much information a system writes to the screen while inducing hypotheses. n ranges from 0 (no information) to 4 (very detailed). In the experiments described in this text talking(4) is used, so that the induction process can be followed closely. The default setting is system dependent.
- classes(L) defines the classes (only for the classification systems ICL and TILDE). L is a list of atoms C_i; an example belongs to class C_i iff a call to C_i succeeds in the example. The default setting is classes([pos,neg]).

Note that for TILDE, the language bias (specified using rmode settings) is included in the file *app*.s.

All systems write their output to a file called *app*.out. On-line help can be obtained in each system by typing help.

5.6.1 $\mathcal{D}$LAB: A language bias formalism for CLAUDIEN and ICL

The language bias in CLAUDIEN and ICLis defined by means of one or more so-called $\mathcal{D}$LAB templates, written as Prolog facts with a single argument of type string. Several $\mathcal{D}$LAB templates are shown in Figure 5.14(a). Such a template is a compact notation for a finite, but potentially very large, set of clauses. Figure 5.14(a) represents a single clause in a trivial way. We now make the definition of possible clause bodies intensional by combining more literals in a list, and by labeling this list with a discrete interval $Min - Max$ that constrains the minimum and maximum number of elements that can be taken from the list. For instance, the template in Figure 5.14(b) represents the three clauses with the literal party(yes) in the head and 1 or 2 literals from the list [job(researcher),job(manager)] in the body. In the interval specification "len" indicates the length of the associated list, as in the template in Figure 5.14(c), which represents 9 clauses, with three different heads, and, as before, three different bodies. Exactly the same set of clauses is obtained with the template in Figure 5.14(d), in which the interval specifications have moved inside to the level of terms. To exclude the apriori

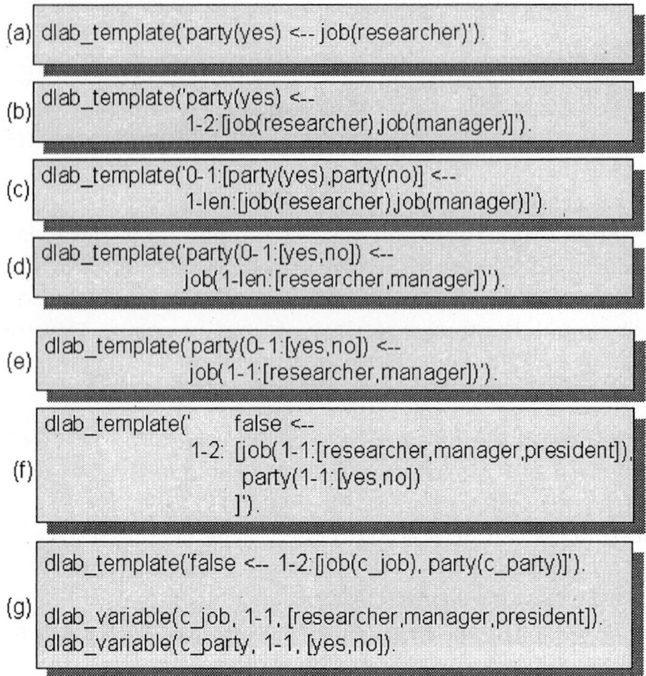

(a) dlab_template('party(yes) <-- job(researcher)').
(b) dlab_template('party(yes) <--
1-2:[job(researcher),job(manager)]').
(c) dlab_template('0-1:[party(yes),party(no)] <--
1-len:[job(researcher),job(manager)]').
(d) dlab_template('party(0-1:[yes,no]) <--
job(1-len:[researcher,manager])').
(e) dlab_template('party(0-1:[yes,no]) <--
job(1-1:[researcher,manager])').
(f) dlab_template(' false <--
1-2: [job(1-1:[researcher,manager,president]),
party(1-1:[yes,no])
]').
(g) dlab_template('false <-- 1-2:[job(c_job), party(c_party)]').
dlab_variable(c_job, 1-1, [researcher,manager,president]).
dlab_variable(c_party, 1-1, [yes,no]).

Fig. 5.14. Examples of $\mathcal{D}$LAB templates.

inconsistent conjunction `job(researcher) and job(manager)` we should alter the template as shown in Figure 5.14(e). The examples above illustrate the basics of $\mathcal{D}$LAB. Two more advanced examples show the use of nesting and dlab_variables. Nesting contributes to the expressiveness, as in Figure 5.14(f), where the template represents 11 clauses, each with 1 or 2 literals in the body. No literal occurs twice, due to the "exclusive or" enforced by the nested 1-1 interval. As a second advanced example, the template in Figure 5.14(g), represents the same set of clauses, and shows how dlab_variables (not to be confused with logical variables that can occur in the template) contribute to readability.

Figure 5.15 further shows the search space that corresponds to a dlab_template presented to CLAUDIEN and ICL. In this example, p stands for party, j for job, r for researcher, m for manager, p for president, ct for company_type, c for commercial and u for university. The dotted lines indicate paths followed by ICL and not by CLAUDIEN. Note that CLAUDIEN's search is optimal in that it never visits the same node twice.

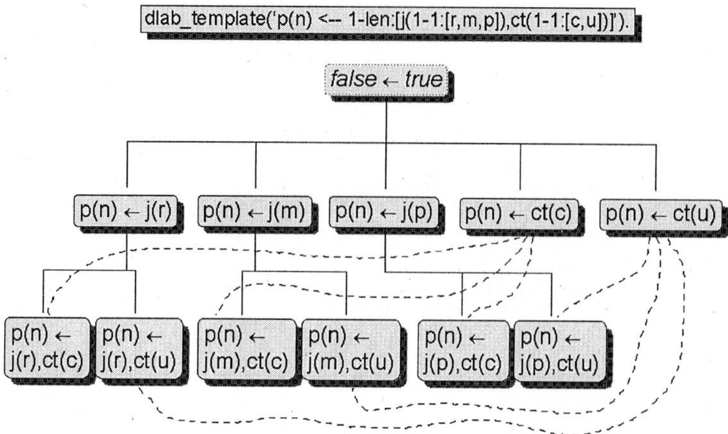

Fig. 5.15. The search space corresponding to a $\mathcal{D}$LAB template.

5.6.2 Running CLAUDIEN

CLAUDIEN requires a knowledge base, in file *registration*.kb and a language, in file *registration*.l. The CLAUDIEN manual [5.17] contains full details on additional optional components such as *registration*.s with non-default settings. We briefly mention some of the more important settings.

- *min_accuracy*(R). If this is set to 1, CLAUDIEN will only accept solutions that are valid on all data. By lowering the accuracy threshold this strong requirement can be relaxed.
- *min_coverage*(N) specifies the required minimum number of examples covered by the clause.
- *search*(S) specifies the search strategy: best-first, beam, breadth-first or depth-first.

If you want CLAUDIEN to discover clauses with at most two literals in the head that express dependencies between job, company type, party, course length, course type you might for instance create a file `registration.l` as shown in Figure 5.16(a). CLAUDIEN then starts up as shown in Figure 5.16(b). Note that the number of clauses in the search space is 71712. The start up phase ends with the display of the prompt `claudien-registration>` . Since no settings file is present, default settings apply.

To initiate the discovery process, issue the command `c` or `claudien`, and CLAUDIEN will respond as shown in Figure 5.17(a). The initial, most general clause in the search space is in this case `false if true`, which is not accepted as a solution but is added to the queue for further refinement. In the second step the first descendant `job(researcher) if true` is evaluated. As this clause is true in half of the examples (there are 3 researchers), its accuracy is 50% (`a(0.5)`), the body succeeds in 6 examples (`t(6)`), the clause

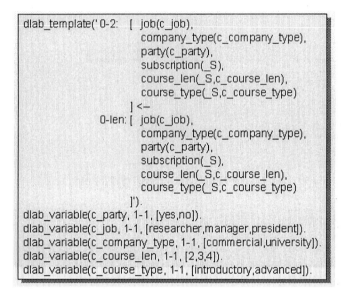

```
dlab_template(' 0-2:  [ job(c_job),
                        company_type(c_company_type),
                        party(c_party),
                        subscription(_S),
                        course_len(_S,c_course_len),
                        course_type(_S,c_course_type)
                      ] <--
              0-len:  [ job(c_job),
                        company_type(c_company_type),
                        party(c_party),
                        subscription(_S),
                        course_len(_S,c_course_len),
                        course_type(_S,c_course_type)
                      ]').
dlab_variable(c_party, 1-1, [yes,no]).
dlab_variable(c_job, 1-1, [researcher,manager,president]).
dlab_variable(c_company_type, 1-1, [commercial,university]).
dlab_variable(c_course_len, 1-1, [2,3,4]).
dlab_variable(c_course_type, 1-1, [introductory,advanced]).
```

(a)

```
% claudien registration
<...>
Now loading files for application registration with configuration
<...>
**Initialising language**
  loading registration.l
Initializing DLAB
dlab_template 1 successfully parsed by DLAB
Size language : 71712
**Initialising knowledge base**
  loading registration.kb
<...>
  -> 6 models have been loaded
<...>
claudien-registration> help
```

(b)

Fig. 5.16. A session with CLAUDIEN: input and startup.

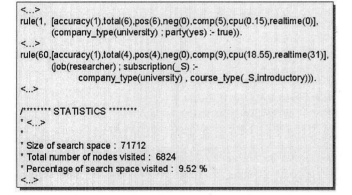

```
claudien-registration> c
<...>
  false if true              [a(0),t(6),p(0),n(6),c(2),val(add to queue)]
New nodes: 1      Remaining nodes: 1
Number of solutions : 0       Consumed cpu : 0.01

Clause being refined : false if true
  job(researcher) if true                 [a(0.5),t(6),p(3),n(3),c(3),val(add to queue)]
  job(manager) if true                    [a(0.33),t(6),p(2),n(4),c(3),val(add to queue)]
  <...>
New nodes: 20      Remaining nodes: 20
Number of solutions : 0       Consumed cpu : 0.07
<...>
  company_type(university);party(yes) if true       +*+*+ S O L U T I O N +*+*+
<...>
Clause being refined : false if job(researcher)
  job(researcher) if job(researcher)         pruned : tautology
  <...>
  company_type(university) if job(researcher)     +*+*+ S O L U T I O N +*+*+
<...>
New nodes: 0      Remaining nodes: 0
Number of solutions : 61       Consumed cpu : 35.49
```

(a)

```
<...>
rule(1, [accuracy(1),total(6),pos(6),neg(0),comp(5),cpu(0.15),realtime(0)],
        (company_type(university) ; party(yes) :- true)).
<...>
rule(60,[accuracy(1),total(4),pos(4),neg(0),comp(9),cpu(18.55),realtime(31)],
        (job(researcher) ; subscription(_S) :-
                company_type(university) , course_type(_S,introductory))).
<...>

/******** STATISTICS ********
* <...>
*
* Size of search space : 71712
* Total number of nodes visited : 6824
* Percentage of search space visited : 9.52 %
<...>
```

(b)

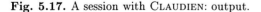

Fig. 5.17. A session with CLAUDIEN: output.

succeeds in 3 examples (`p(3)`) and there are 3 cases where the clause fails (`n(3)`). Accuracy `a(0.5)` and coverage `t(6)` (cf. confidence and support in association rule discovery) have to be higher than user-specified thresholds. By default these thresholds are both set to one. Since the accuracy is not sufficient, this clause is not accepted as a solution, but added back to the queue for further refinement. After each level in the breadth first search (alternative (heuristic) search strategies are available) the number of candidates in the queue is shown, in this case there are 20 nodes in the queue after level 2. The first solution is `company_type(university);party(yes) if true` which says that every participant either works at a university or goes to the party. Solutions are not further refined. In general, CLAUDIEN will prune away nodes that for some reason cannot lead to "interesting" clauses. For instance, `job(researcher) if job(researcher)` is pruned because it is a tautology and will remain so when it is refined. In total 61 (out of 71712) clauses are selected for being "interesting".

All solutions are written to `registration.out`. Figure 5.17(b) shows an extract from this file. For instance, rule 60 says that university participants either do research or take all introductory courses. The body of this rule succeeds in 4 examples (`total(4)`, and in all of these also the entire clause succeeds (`pos(4)`), hence an accuracy of 100%. Apart from the solutions, `registration.out` also contains some search statistics, such as the percentage of search space visited, that is the share of nodes not pruned away, in this case 9.52%.

5.6.3 Running ICL

The input files for ICL are similar to those of CLAUDIEN:

- The **knowledge** base (file *registration*.`kb`) contains *classified* examples.
- The **settings** file *registration*.`s` can be used to change the value of some parameters (settings) by adding a fact *name(value)*, with *name* the name (identifier) of the setting and *value* the new value. For each setting there is a default value.
  Some useful settings are:
  - `significance_level`(*SL*) specifies the confidence level (as percentage) for the significance test; possible values include 0 (no significance testing), 0.80, 0.90 (default), 0.95 and 0.99; a higher percentage will prune more rules.
  - `beam_size`(*N*) with *N* the maximum number of rules to be kept in the beam (default is 5).
- The **language** bias is specified with DLAB in *registration*.`l` (see Section 5.6.1 for an introduction to DLAB). An important difference to CLAUDIEN is that the DLAB bias should not contain the class predicates and the templates should be read as `dlab_template`('[negative literals] <-- [positive literals]').

ICL allows the user to generate `dlab_variable/3` facts automatically by using `dlab_query/3` (a feature of $\mathcal{D}$LAB added by ICL). The first 2 arguments of `dlab_query` are the same as in `dlab_variable`, the third is either `constants`(Q, V) or `discretize`(Q, V) (Q can be any predicate containing the variable V and is used to generate values for V that occur in the examples). The predicate `constants` returns the list of all possible values for V, while `discretize` returns a small subset of all numeric values by using a discretization algorithm (details in [5.26]). The returned list is used as the third argument in the corresponding `dlab_variable`.

For the *registration* application, we would like to classify participants (the examples) into people who attend the party and people who don't. So in the file `registration.`*s* we set `classes`([*party*(*yes*), *party*(*no*)]). A possible $\mathcal{D}$LAB for `registration.`*l* can be found in Figure 5.18. This is very similar to the one used in CLAUDIEN, with the exception that the `party` predicates have been removed. The file `registration.`*kb* is the same as for CLAUDIEN and TILDE.

```
dlab_template(' 0-2:  [ job(c_job),
                        company_type(c_company_type),
                        subscription(_S),
                        course_len(_S,c_course_len),
                        course_type(_S,c_course_type)
                      ] <--
               0-len: [ job(c_job),
                        company_type(c_company_type),
                        subscription(_S),
                        course_len(_S,c_course_len),
                        course_type(_S,c_course_type)
                      ]').
dlab_variable(c_job, 1-1, [researcher,manager,president]).
dlab_variable(c_company_type, 1-1, [commercial,university]).
dlab_variable(c_course_len, 1-1, [2,3,4]).
dlab_variable(c_course_type, 1-1, [introductory,advanced]).
```

Fig. 5.18. The language bias file `registration.`*l* for ICL.

ICL is started as displayed in Figure 5.19 (a)[3]. When everything is loaded successfully, the `icl>` prompt will be displayed from which you can type commands. Type `help` to get a list of available commands (see Figure 5.19 (b)). Quick help on a command can be obtained with the command `help(command)`.

The ICL learning algorithm is started with the command `icl` (or just `i`) as shown in Figure 5.20(a).

[3] ICL will ask for a *configuration* name at start up. We will not discuss this feature in this text, so just type enter. More information can be found on the ICL web pages.

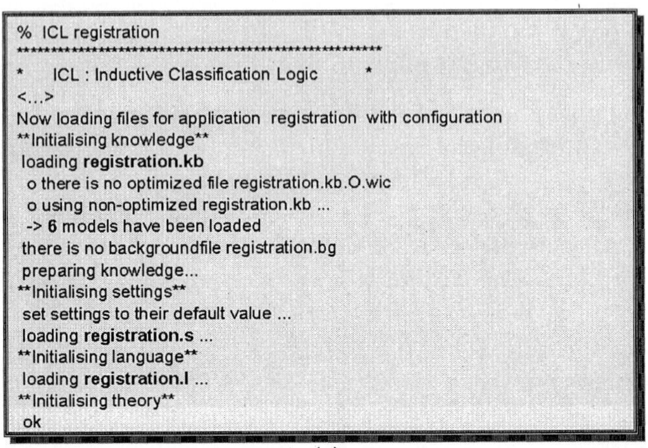

```
% ICL registration
*******************************************************
*    ICL : Inductive Classification Logic          *
<...>
Now loading files for application  registration  with configuration
**Initialising knowledge**
 loading registration.kb
  o there is no optimized file registration.kb.O.wic
  o using non-optimized registration.kb ...
  -> 6 models have been loaded
 there is no backgroundfile registration.bg
 preparing knowledge...
**Initialising settings**
 set settings to their default value ...
 loading registration.s ...
**Initialising language**
 loading registration.l ...
**Initialising theory**
 ok
```

(a)

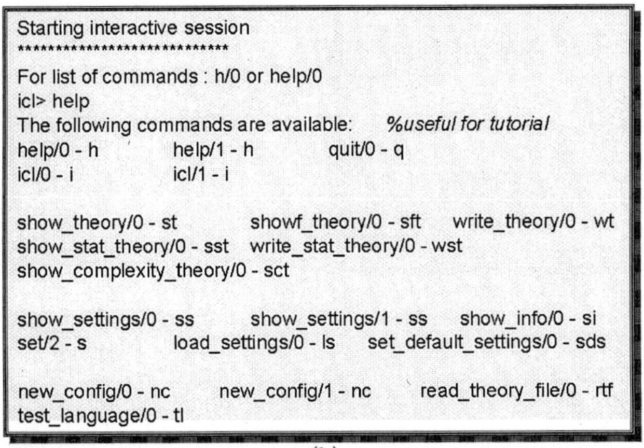

```
Starting interactive session
******************************
For list of commands : h/0 or help/0
icl> help
The following commands are available:    %useful for tutorial
help/0 - h          help/1 - h          quit/0 - q
icl/0 - i           icl/1 - i

show_theory/0 - st          showf_theory/0 - sft     write_theory/0 - wt
show_stat_theory/0 - sst    write_stat_theory/0 - wst
show_complexity_theory/0 - sct

show_settings/0 - ss        show_settings/1 - ss     show_info/0 - si
set/2 - s           load_settings/0 - ls     set_default_settings/0 - sds

new_config/0 - nc       new_config/1 - nc        read_theory_file/0 - rtf
test_language/0 - tl
```

(b)

Fig. 5.19. Starting ICL for the *registration* application.

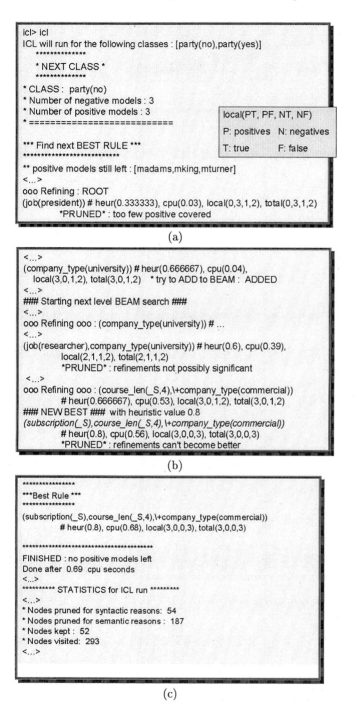

```
icl> icl
ICL will run for the following classes : [party(no),party(yes)]
    **************
    * NEXT CLASS *
    **************
* CLASS :  party(no)
* Number of negative models : 3
* Number of positive models : 3
* ==========================

*** Find next BEST RULE ***
****************************
** positive models still left : [madams,mking,mturner]
<...>
ooo Refining : ROOT
(job(president)) # heur(0.333333), cpu(0.03), local(0,3,1,2), total(0,3,1,2)
          *PRUNED* : too few positive covered
```

```
local(PT, PF, NT, NF)
P: positives   N: negatives
T: true        F: false
```

(a)

```
<...>
(company_type(university)) # heur(0.666667), cpu(0.04),
   local(3,0,1,2), total(3,0,1,2)   * try to ADD to BEAM :  ADDED
<...>
### Starting next level BEAM search ###
<...>
ooo Refining ooo : (company_type(university)) # ...
<...>
(job(researcher),company_type(university)) # heur(0.6), cpu(0.39),
          local(2,1,1,2), total(2,1,1,2)
          *PRUNED* : refinements not possibly significant
 <...>
ooo Refining ooo : (course_len(_S,4),\+company_type(commercial))
          # heur(0.666667), cpu(0.53), local(3,0,1,2), total(3,0,1,2)
### NEW BEST ### with heuristic value 0.8
(subscription(_S),course_len(_S,4),\+company_type(commercial))
          # heur(0.8), cpu(0.56), local(3,0,0,3), total(3,0,0,3)
          *PRUNED* : refinements can't become better
```

(b)

```
***************
***Best Rule ***
***************
(subscription(_S),course_len(_S,4),\+company_type(commercial))
          # heur(0.8), cpu(0.68), local(3,0,0,3), total(3,0,0,3)

******************************************
FINISHED : no positive models left
Done after  0.69  cpu seconds
<...>
********** STATISTICS for ICL run *********
<...>
* Nodes pruned for syntactic reasons:  54
* Nodes pruned for semantic reasons :  187
* Nodes kept :  52
* Nodes visited:  293
<...>
```

(c)

Fig. 5.20. A session with ICL on the *registration* database.

```
    * NEXT CLASS *
* CLASS : party(yes)
<...>
finished : there are still positive models left...
<...>
For this run, the following theory has been found:
*********
*Class  : party(no)
*Status : complete
*Type   : dnf
rule((company_type(university),subscription(_S),course_len(_S,4)),
         [type(dnf),cpu(1.2),heur(0.8),local(3,0,0,3),total(3,0,0,3)])
*********
*Class  : party(yes)
*Status : complete
*Type   : dnf
*********
icl>
```

(a)

```
icl> set(significance_level,0.8)
icl> icl
<...>
*********
*Class  : party(no)
*Status : complete
*Type   : dnf
*********
rule((company_type(university),subscription(_S),course_len(_S,4)),
         [type(dnf),cpu(0.51),heur(0.8),local(3,0,0,3),total(3,0,0,3)])
*********
*Class  : party(yes)
*Status : complete
*Type   : dnf
*********
rule((company_type(commercial)),
         [type(dnf),cpu(0.45),heur(0.75),local(2,1,0,3),total(2,1,0,3)])
rule((course_len(_S,4),\+subscription(_S)),
         [type(dnf),cpu(0.82),heur(0.666667),local(1,0,0,3),total(2,1,0,3)])
```

(b)

Fig. 5.21. A session with ICL on the *registration* database (continued).

The system will try to find a hypothesis for each class. During search, information is written on the screen. Some parts of this are shown in Figure 5.20(a)-(c) and 5.21(a). It seems that a correct theory has been found for class *party(no)*:[4]

H = (party(*no*) :- company_type(*university*), subscription(_S), course_len(_S,4)).

For *party(yes)* however, no theory has been found. When lowering the significance level however, ICL will find a theory for class *party(yes)*. The result is shown in Figure 5.21(b).

During discovery, results and information on the learning process are written to *registration*.`out`. With the command `write_theory`, the learned theories are written in the file *registration*.`theory` so that it can be (re)loaded later on. `write_statistics` will write statistics (like accuracy) of the learned theories w.r.t. the training examples and the test examples (if any). Test examples are those examples that are left out during learning (using the setting `leave_out`).

5.6.4 Running TILDE

The TILDE system uses exactly the same format for the knowledge base as the other systems, but it does not need the *app*.`l` file. The language bias is not specified using the $\mathcal{D}$LAB formalism, but by means of settings in *app*.`s`.

Most of TILDE's settings have default values that need not be changed. We only discuss the most important settings here, and refer to TILDE's user's manual [5.3] for a more detailed discussion.

- `minimal_cases(`n`)` specifies how many examples each leaf must cover. Leaves covering fewer examples are not allowed. The default for n is 2; for very small data sets it may be appropriate to set this to 1.
- `rmode` settings tell TILDE what kind of tests can be put in the nodes of the tree. They partially define the language bias. We discuss them in more detail below.
- `lookahead` settings specify in what way TILDE can perform a local search to generate tests for a node. This is also discussed in detail below.

Rmode settings For TILDE, the hypotheses space is defined mostly by what kind of tests can be put in the nodes of the tree. A set of `rmode` facts is used to specify this. Basically, `rmode(`C`)` tells TILDE that C is allowed as a test. C can be a literal, a conjunction of literals or a more complicated construct.

Recalling the *WORN* example in Figure 5.11, tests that could be allowed in the nodes of the tree are:

[4] Note that the actual output of ICL omits the head of the rules

`worn(wheel)`	does the machine contain a wheel that is worn?
`worn(X)`	(if X a new variable) does the machine contain some worn part X? (if X already exists) is X worn?

Some predicates may require an argument to be a variable that already occurs, others need a new variable, and some may allow for both. So we distinguish three different *modes* for variables: *input* if the variable must already exist in order for it to be used here, *output* if it should not exist, or *any* if it can be either an existing or a new variable.

`Rmode` facts tell TILDE which tests can be used, and if the tests contain variables, indicate the modes of these variables. *Input*, *any* and *output* variables are indicated by writing a +, − or nothing before them, respectively.

The following example shows what kind of tests are generated by certain `rmode` specifications. A represents a variable that already occurs in the tree, B is a new variable.

`rmode(1:worn(gear)).`	worn(gear)
`rmode(5:worn(X)).`	worn(B)
`rmode(5:worn(+X)).`	worn(A)
`rmode(5:worn(-X)).`	both worn(A) and worn(B)

The numbers inside `rmode` indicate how many times it is allowed to occur in any path from the root of the tree to a leaf (i.e., how many times at most it can be tested for one example).

If there are many constants, it is tedious to write similar tests for each of these constants (e.g., `worn(gear)`, `worn(engine)`, ...). Fortunately, TILDE can generate suitable constants itself. It does this in the following way: for each example in the set of examples covered by the node, a query is made that generates one or more constants. Each of these constants is then filled in in the test that is to be put in this node. In order to keep the branching factor of the search space limited, maxima can be given for the number of examples TILDE should look at, as well as for the number of constants that can be generated from each example.

To tell TILDE that a test `worn(c)` can be used, with c being any constant for which `worn` could possibly succeed, the following `rmode` fact could be used:

```
rmode(5: #(15*5*X: worn(X), worn(X))).
           a d c    b        e
```

This specification means that in at most 15 (a) examples, TILDE should run `worn(X)` (b) and see which values X (c) can take; it should return at most 5 (d) values per example. Finally, the test `worn(X)` (e) will be put in the node, but with X changed into one of the constants: `worn(gear)`, `worn(engine)`, ...

In the above example, the constant generating predicate is the same as the test predicate, but this need not always be the case. Another example of the use of constant generation is:

```
rmode(10: #(100*1*C: boundary(C), +X < C)).
```

In at most 100 models one numeric boundary will be computed, and a test should consist of comparing an already bound variable X with this boundary. The computation of a suitable boundary can be defined in background knowledge. It might be done by, e.g., a discretization algorithm.

While the above syntax is a bit awkward, it is very general and allows the generation of constants in many different settings.

Lookahead. In some cases it is possible that a literal will never be chosen by TILDE, because it is not useful as a test. However, such a literal, while not useful in itself, might introduce new variables that make better tests possible later on.

As an example, let us look at the registrations database. In order to allow TILDE to use information about subscriptions, we could provide the following `rmode` facts:

```
rmode(5: subscription(X)).
rmode(5: #(1*3*L: member(L, [2,3,4]), course_len(+X,L)).
```

With only these specifications TILDE will not find a tree where the first test is whether the participant has subscribed to something, and the second test is whether she has subscribed to a course of some specific length. The reason for this is that `subscription(X)` in itself, as a *test*, is never informative: every person has subscribed to at least one course. Thus, this test will never be chosen. Consequently, the test `course_len(X,2)` cannot be chosen either, because it can only test a variable that already occurs.

This problem can be solved by allowing TILDE to *look ahead* in the search space, i.e., to check, immediately after putting `subscription(X)` in a node, whether any tests making use of X are interesting.

In the above case one can add to the `rmode` facts one `lookahead` fact:

```
lookahead(subscription(X),
          #(1*3*L: member(L, [2,3,4]), course_len(X,L))).
```

TILDE will then build a test for one node by first considering `subscription(X)`, then `subscription(X), course_len(X,2)` and so on for 3 and 4. All these tests will be evaluated separately, and the best test is chosen. Note that both literals are put in one single node of the tree.

Sample run. We have now seen all the settings we need to get TILDE running on the registrations data set. Figure 5.22 shows what a suitable `registration.s` would look like. Once this file has been prepared, TILDE can be started by just typing `tilde` at the prompt. The induction is started by typing `go`. Upon finishing, TILDE creates a file `registration.out` containing among other things the tree that has been built, as is shown in Figure 5.22.

registration.s

```
minimal_cases(1).
classes([party(yes),party(no)]).
talking(4).

rmode(1:#(1*3*J:member(J,[researcher,president,manager]),job(J))).
rmode(1:#(1*2*C:member(C,[commercial,university]),company_type(C))).
rmode(5:subscription(X)).
rmode(5:#(1*3*L:member(L,[2,3,4],course_len(+X,L)))).
rmode(5:#(1*2*T:member(T,[introductory,advanced]),course_type(+X,T))).

lookahead(subscription(X),#(1*3*L:member(L,[2,3,4]),course_len(X,L))).
lookahead(subscription(X),#(1*2*T:member(T,[introductory,advanced]),
          course_type(X,T))).
```

Tilde running

```
Tilde 1.34 ready.

Your bidding? go
discretization busy...
building tree...
true,job(manager)
[gain=0,gainratio=0]
true,job(president)
[gain=0.190875,gainratio=0.293643]
(...)

Best test: true, company_type(university)

(true,company_type(university)),job(manager)
[gain=0.122556,gainratio=0.151066]
(...)

Best test: (true,company_type(university)),
            subscription(_2192),course_len(_2192,4)

Output written to file 'registration.out'.

Your bidding?
```

registration.out

```
(...)

Compact notation of pruned tree:

company_type(university) ?
+--yes: subscription(A), course_len(A,4) ?
|    +--yes: party(no)[3/3]
|    +--no:  party(yes)[1/1]
+--no: party(yes)[2/2]

(...)
```

Fig. 5.22. Settings file, screendump and output file of TILDE on the registration database.

5.7 Sample application: Mutagenesis

The data in the mutagenesis domain (see [5.23]) consists of 188 molecules, of which 125 are active (thus mutagenic) and 63 are inactive. A molecule is described by listing its atoms `atom(AtomID,Element,Type,Charge)` (the number of atoms differs between molecules, ranging from 15 to 35) and the bonds `bond(Atom1,Atom2,BondType)` between atoms.

In [5.23], four different backgrounds have been defined. We will illustrate our three systems on the mutagenesis data with the simplest of these (BG1 in [5.23]). Figure 5.23 shows a snapshot of the knowledge base file *muta*.kb.

muta.kb

```
begin(model(1)).
pos.
atom(d1_1,c,22,-0.117).      bond(d1_1,d1_2,7).
atom(d1_2,c,22,-0.117).      bond(d1_2,d1_3,7).
atom(d1_3,c,22,-0.117).      bond(d1_3,d1_4,7).
atom(d1_4,c,195,-0.087).     ...
...
end(model(1)).
```

Fig. 5.23. Snapshot of the knowledge base file for the mutagenesis data.

Examples of settings, language and result files can be found in Figures 5.24, 5.25 and 5.26.

To experiment on this data yourself, you can retrieve the necessary input files from our website (see Section 5.9).

Figure 5.24 illustrates a setup for CLAUDIEN. The size of the language shown in Figure 5.24(a) is in the order of 10^8. The settings file in Figure 5.24(b) is added to enforce a heuristic, best first search and accuracy and coverage thresholds of respectively 90% and 20 molecules. One of the first discovered rules, after four cpu minutes, is shown in Figure 5.24(c). This rule has an accuracy of 90% and the body of the rule holds for 62 of the 188 molecules.

For ICL, we use the same language file `muta.l` as the one for CLAUDIEN in Figure 5.24(a), except for `1-1:[pos,neg]` in the `dlab_template` (which must be replaced by `false`). The settings file `muta.s` contains `classes([pos,neg])`. ICL learns a DNF theory for both classes *pos* and *neg*. For class *neg*, the theory in Figure 5.25 has been found in about 81 seconds. The accuracy of ICL (estimated by 10-fold cross-validation) ranges from 80.9% (for BG1) up to 88.3% (for BG4) for multi-class theories.

TILDE's language bias definition is quite different from that of the other systems, as Figure 5.26 shows. With this language[5], which allows TILDE to

[5] No other settings need to be given here.

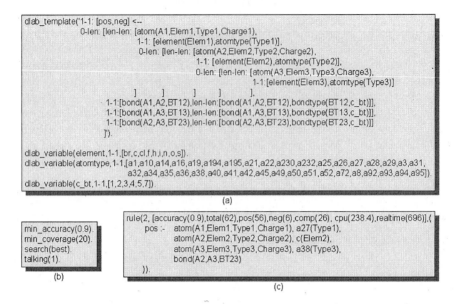

```
dlab_template('1-1: [pos,neg] <--
            0-len: [len-len: [atom(A1,Elem1,Type1,Charge1),
                              1-1: [element(Elem1),atomtype(Type1)],
                              0-len: [len-len: [atom(A2,Elem2,Type2,Charge2),
                                                1-1: [element(Elem2),atomtype(Type2)],
                                                0-len: [len-len: [atom(A3,Elem3,Type3,Charge3),
                                                                  1-1:[element(Elem3),atomtype(Type3)]
                             ]         ]         ]         ]         ],
                  1-1:[bond(A1,A2,BT12),len-len:[bond(A1,A2,BT12),bondtype(BT12,c_bt)]],
                  1-1:[bond(A1,A3,BT13),len-len:[bond(A1,A3,BT13),bondtype(BT13,c_bt)]],
                  1-1:[bond(A2,A3,BT23),len-len:[bond(A2,A3,BT23),bondtype(BT23,c_bt)]]
                  ]').

dlab_variable(element,1-1,[br,c,cl,f,h,i,n,o,s]).
dlab_variable(atomtype,1-1,[a1,a10,a14,a16,a19,a194,a195,a21,a22,a230,a232,a25,a26,a27,a28,a29,a3,a31,
                          a32,a34,a35,a36,a38,a40,a41,a42,a45,a49,a50,a51,a52,a72,a8,a92,a93,a94,a95]).
dlab_variable(c_bt,1-1,[1,2,3,4,5,7]).
```

(a)

```
min_accuracy(0.9).
min_coverage(20).
search(best).
talking(1).
```

(b)

```
rule(2, [accuracy(0.9),total(62),pos(56),neg(6),comp(26), cpu(238.4),realtime(696)],(
      pos :-   atom(A1,Elem1,Type1,Charge1), a27(Type1),
               atom(A2,Elem2,Type2,Charge2), c(Elem2),
               atom(A3,Elem3,Type3,Charge3), a38(Type3),
               bond(A2,A3,BT23)
      )).
```

(c)

Fig. 5.24. CLAUDIEN applied to the mutagenesis data set. (a) language bias; (b) settings file; (c) extract from the results file.

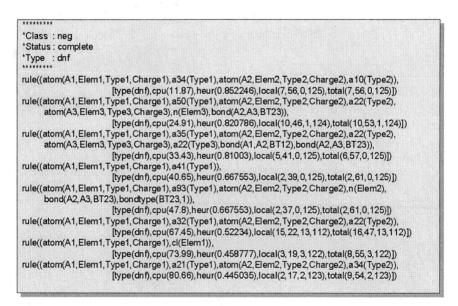

```
*********
*Class  : neg
*Status : complete
*Type   : dnf
*********
rule((atom(A1,Elem1,Type1,Charge1),a34(Type1),atom(A2,Elem2,Type2,Charge2),a10(Type2)),
                          [type(dnf),cpu(11.87),heur(0.852246),local(7,56,0,125),total(7,56,0,125)])
rule((atom(A1,Elem1,Type1,Charge1),a50(Type1),atom(A2,Elem2,Type2,Charge2),a22(Type2),
      atom(A3,Elem3,Type3,Charge3),n(Elem3),bond(A2,A3,BT23)),
                          [type(dnf),cpu(24.91),heur(0.820786),local(10,46,1,124),total(10,53,1,124)])
rule((atom(A1,Elem1,Type1,Charge1),a35(Type1),atom(A2,Elem2,Type2,Charge2),a22(Type2),
      atom(A3,Elem3,Type3,Charge3),a22(Type3),bond(A1,A2,BT12),bond(A2,A3,BT23)),
                          [type(dnf),cpu(33.43),heur(0.81003),local(5,41,0,125),total(6,57,0,125)])
rule((atom(A1,Elem1,Type1,Charge1),a41(Type1)),
                          [type(dnf),cpu(40.65),heur(0.667553),local(2,39,0,125),total(2,61,0,125)])
rule((atom(A1,Elem1,Type1,Charge1),a93(Type1),atom(A2,Elem2,Type2,Charge2),n(Elem2),
      bond(A2,A3,BT23),bondtype(BT23,1)),
                          [type(dnf),cpu(47.8),heur(0.667553),local(2,37,0,125),total(2,61,0,125)])
rule((atom(A1,Elem1,Type1,Charge1),a32(Type1),atom(A2,Elem2,Type2,Charge2),a22(Type2)),
                          [type(dnf),cpu(67.45),heur(0.52234),local(15,22,13,112),total(16,47,13,112)])
rule((atom(A1,Elem1,Type1,Charge1),cl(Elem1)),
                          [type(dnf),cpu(73.99),heur(0.458777),local(3,19,3,122),total(8,55,3,122)])
rule((atom(A1,Elem1,Type1,Charge1),a21(Type1),atom(A2,Elem2,Type2,Charge2),a34(Type2)),
                          [type(dnf),cpu(80.66),heur(0.445035),local(2,17,2,123),total(9,54,2,123)])
```

Fig. 5.25. ICL applied to the mutagenesis data set.

muta.s

```
lookahead(bond(A1, A2, BT),
         #(1*7*C: member(C, [1,2,3,4,5,7]), BT=C)).
lookahead(bond(A1, A2, BT),
         #(230*9*E: atom(A2, E, _, _), atom(A2, E, _, _))).
lookahead(bond(A1, A2, BT),
         #(230*37*T: atom(A2, _, T, _), atom(A2, _, T, _))).

rmode(20: #(230*37*T: atom(_, _, T, _), atom(A, E, T, Ch))).
rmode(20: #(230*9*E: atom(_, E, _, _), atom(A, E, T, Ch))).
rmode(20: bond(+A1, -A2, BT)).
```

muta.out

```
atom(A,B,41,C) ?
+--yes: neg [2 / 2]
+--no:  atom(D,E,50,F) ?
     +--yes: neg [10 / 12]
     +--no:  atom(G,H,35,I) ?
          +--yes: neg [5 / 6]
          +--no:  atom(J,K,34,L) ?
               +--yes: bond(J,M,N) , atom(M,c,O,P) ?
               |    +--yes: neg [9 / 10]
               |    +--no:  pos [2 / 2]
               +--no:  atom(Q,R,93,S) ?
                    +--yes: neg [5 / 7]
                    +--no:  atom(T,U,1,V) ?
                         +--yes: atom(W,X,14,Y) ?
                         |    +--yes: pos [4 / 5]
                         |    +--no:  neg [14 / 24]
                         +--no:  atom(Z,A_1,92,B_1) ?
                              +--yes: atom(C_1,D_1,29,E_1) ?
                              |    +--yes: pos [2 / 2]
                              |    +--no:  neg [4 / 5]
                              +--no:  pos [100 / 113]
```

Fig. 5.26. TILDE applied to the mutagenesis data set.

use only the background information in BG1, TILDE needs approximately 18 seconds (on a SUN Sparc Ultra-2 machine) to output the tree shown in Figure 5.26. A 10-fold cross-validation using the same language yields 75% accuracy.

5.8 An exercise

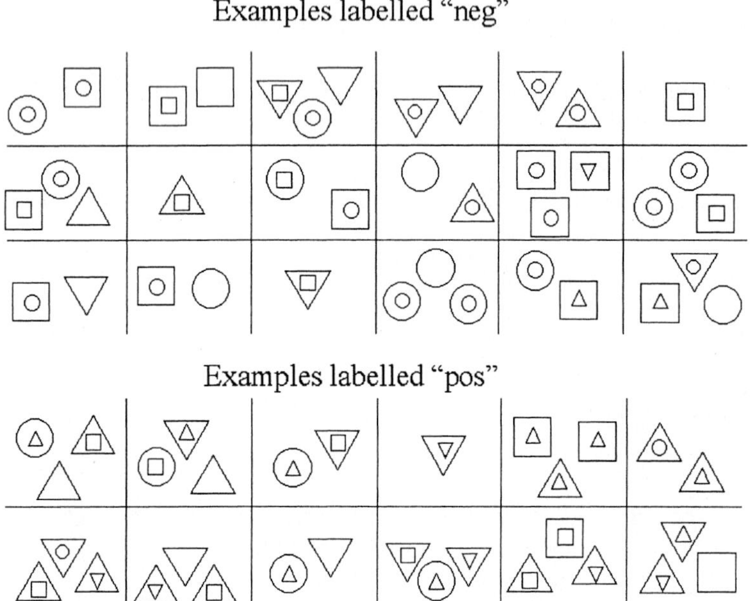

Fig. 5.27. Bongard problems.

This exercise is based on some pattern recognition tasks, invented by the Russian scientist M. Bongard [5.7]. A number of drawings are given, each labeled with $\oplus$ or $\ominus$. The task is to find a theory that allows to discriminate between both classes based on the objects in the drawings.

Figure 5.27 shows a number of these drawings. Drawings can contain circles, squares and triangles. For triangles the direction in which they point (up or down) may be relevant. Finally, whether an object is inside another object may also be relevant. So, the aim is to find a theory that discriminates positive from negative drawings by looking at the kind of objects they consist of, their configuration (for triangles, up or down) and their relative positions (objects inside one another).

You might want to take a look at Figure 5.27 to see whether you can find a good rule yourself. There exists a rule predicting examples to be positive that is correct and almost complete (i.e., the condition of the rule holds for nearly all positives and for no negatives) for the examples in this drawing. If you find a good rule, then you can run the inductive systems and see whether they yield the same rule, better rules, or worse ones.

With respect to using the inductive systems on these data, we suggest the following actions:

- First think about a representation for these drawings. Write down some examples from the drawing.
- Download the whole data set (`bongard.kb`, > 300 examples) from the URL given below. Compare the representation used in this file with yours.
- Try to run CLAUDIEN, ICL and/or TILDE on these data sets. You will need to construct the `bongard.l` and `bongard.s` files yourself now. See what results you get, try to improve them.
- If you are satisfied with your results or get stuck, download the example `bongard.s` and `bongard.l` files from the above site. Compare them with your files.
- Finally, run the systems with the settings and language files you downloaded. See whether you get better or worse results than with your files. Also compare these results with the ones you have found manually.

5.9 Conclusions and practical info

We have presented three companion systems for first order data mining. They all learn from interpretations and use the same or similar inputs. This tutorial is meant as a gentle introduction to the three systems and their use. However, we wish to stress that all three systems have a number of additional features that were not mentioned here. These are described in the manuals [5.17, 5.3, 5.25].

CLAUDIEN, ICL and TILDE are freely available (for academic use) via the Internet. Information on how to obtain them can be found at the following URL:

`http://www.cs.kuleuven.ac.be/~ml/Tutorial/`

This web page also contains links to the data files mentioned in this text.

Acknowledgments

At the time of writing this chapter, Luc De Raedt and Wim Van Laer were supported by the Fund for Scientific Research, Flanders and Hendrik Blockeel was supported by the Flemish Institute for the Promotion of Scientific and Technological Research in the Industry (IWT). This work was also supported by the European Community Esprit project no. 20237, Inductive Logic Programming 2.

References

5.1 R. Agrawal, H. Mannila, R. Srikant, H. Toivonen, and A. Verkamo. Fast discovery of association rules. In U. Fayyad, G. Piatetsky-Shapiro, P. Smyth, and R. Uthurusamy, editors, *Advances in Knowledge Discovery and Data Mining*, pages 307–328. MIT Press, Cambridge, MA, 1996.

5.2 R. Agrawal, T. Imielinski, and A. Swami. Mining association rules between sets of items in large databases. In *Proceedings of the ACM SIGMOD Conference on Management of Data*, pages 207–216. ACM Press, New York, 1993.

5.3 H. Blockeel. TILDE 1.3 User's manual, 1997. `http://www.cs.kuleuven.ac.be/ ~ml/Tilde/`.

5.4 H. Blockeel. *Top-down induction of first order logical decision trees.* PhD thesis, Department of Computer Science, Katholieke Universiteit, Leuven, 1998. `http://www.cs.kuleuven.ac.be/~ml/PS/blockeel98:phd.ps.gz`.

5.5 H. Blockeel and L. De Raedt. Lookahead and discretization in ILP. In *Proceedings of the Seventh International Workshop on Inductive Logic Programming*, pages 77–85. Springer, Berlin, 1997.

5.6 H. Blockeel and L. De Raedt. Top-down induction of first order logical decision trees. *Artificial Intelligence*, 101(1-2): 285–297, 1998.

5.7 M. Bongard. *Pattern Recognition.* Spartan Books, 1970.

5.8 I. Bratko and S. Muggleton. Applications of inductive logic programming. *Communications of the ACM*, 38(11): 65–70, 1995.

5.9 P. Clark and R. Boswell. Rule induction with CN2: Some recent improvements. In *Proceedings of the Fifth European Working Session on Learning*, pages 151–163. Springer, Berlin, 1991.

5.10 P. Clark and T. Niblett. The CN2 algorithm. *Machine Learning*, 3(4): 261–284, 1989.

5.11 L. De Raedt. Logical settings for concept learning. *Artificial Intelligence*, 95: 187–201, 1997.

5.12 L. De Raedt and M. Bruynooghe. A theory of clausal discovery. In *Proceedings of the Thirteenth International Joint Conference on Artificial Intelligence*, pages 1058–1063. Morgan Kaufmann, San Mateo, CA, 1993.

5.13 L. De Raedt and L. Dehaspe. Clausal discovery. *Machine Learning*, 26: 99–146, 1997.

5.14 L. De Raedt and W. Van Laer. Inductive constraint logic. In *Proceedings of the Sixth International Workshop on Algorithmic Learning Theory*, pages 80–94. Springer, Berlin, 1995.

5.15 L. Dehaspe and L. De Raedt. Mining a natural language corpus for multi-relational association rules. In *ECML'97 - Workshop Notes on Empirical Learning of Natural Language Processing Tasks*, pages 35–48. Laboratory of Intelligent Systems, Faculty of Informatics and Statistics, University of Economics, Prague, Czech Republic, 1997.

5.16 L. Dehaspe and L. De Raedt. Mining association rules in multiple relations. In *Proceedings of the Seventh International Workshop on Inductive Logic Programming*, pages 125–132. Springer, Berlin, 1997.

5.17 L. Dehaspe, L. De Raedt, and W. Van Laer. Claudien – a clausal discovery engine: A user manual. Technical Report CW-239, Department of Computer Science, Katholieke Universiteit Leuven, Belgium, 1996.

5.18 S. Džeroski, N. Jacobs, M. Molina, C. Moure, S. Muggleton, and W. Van Laer. Detecting traffic problems with ILP. In *Proceedings of the Eighth International Conference on Inductive Logic Programming*, pages 281–290. Springer, Berlin, 1998.

5.19 R. S. Michalski. A theory and methodology of inductive learning. In R. S. Michalski, J. Carbonell, and T. Mitchell, editors, *Machine Learning: An Artificial Intelligence Approach*, pages 83–134. Morgan Kaufmann, San Mateo, CA, 1983.
5.20 S. Muggleton. Inverse entailment and Progol. *New Generation Computing, Special issue on Inductive Logic Programming*, 13(3-4): 245–286, 1995.
5.21 J. R. Quinlan. *C4.5: Programs for Machine Learning*. Morgan Kaufmann, San Mateo, CA, 1993.
5.22 J. R. Quinlan. Learning logical definitions from relations. *Machine Learning*, 5: 239–266, 1990.
5.23 A. Srinivasan, S. Muggleton, M. J. E. Sternberg, and R. D. King. Theories for mutagenicity: A study in first-order and feature-based induction. *Artificial Intelligence*, 85(1,2): 277–299, 1996.
5.24 H. Toivonen, M. Klemettinen, P. Ronkainen, K. Hätönen, and H. Mannila. Pruning and grouping discovered association rules. In *Proceedings of the ML-net Familiarization Workshop on Statistics, Machine Learning and Knowledge Discovery in Databases*, pages 47–52. FORTH, Heraklion, Greece, 1995.
5.25 W. Van Laer. Web pages for the ICL system, 1997. `http://www.cs.kuleuven.ac.be/~ml/ICL/`.
5.26 W. Van Laer, L. De Raedt, and S. Džeroski. On multi-class problems and discretization in inductive logic programming. In *Proceedings of the Tenth International Symposium on Methodologies for Intelligent Systems*, pages 277–286. Springer, Berlin, 1997.

6. Inducing Classification and Regression Trees in First Order Logic

Stefan Kramer[1] and Gerhard Widmer[2,3]

[1] Institut für Informatik, Albert-Ludwigs-Universität Freiburg
Am Flughafen 17, D-79110 Freiburg, Germany

[2] Department of Medical Cybernetics and Artificial Intelligence,
University of Vienna, Freyung 6/2, A-1010 Vienna, Austria

[3] Austrian Research Institute for Artificial Intelligence
Schottengasse 3, A-1010 Vienna, Austria

Abstract

In this chapter, we present a system that enhances the representational capabilities of decision and regression tree learning by extending it to first-order logic, i.e., relational representations as commonly used in Inductive Logic Programming. We describe an algorithm named *Structural Classification and Regression Trees* (S-CART), which is capable of inducing first-order trees for both *classification* and *regression problems*, i.e., for the prediction of either discrete classes or numerical values. We arrive at this algorithm by a strategy called *upgrading* – we start from a propositional induction algorithm and turn it into a relational learner by devising suitable extensions of the representation language and the associated algorithms. In particular, we have upgraded CART, the classical method for learning classification and regression trees, to handle relational examples and background knowledge. The system constructs a tree containing a literal (an atomic formula or its negation) or a conjunction of literals in each node, and assigns either a discrete class or a numerical value to each leaf. In addition, we have extended the CART methodology by adding linear regression models to the leaves of the trees; this does not have a counterpart in CART, but was inspired by its approach to pruning. The regression variant of S-CART is one of the few systems applicable to Relational Regression problems. Experiments in several real-world domains demonstrate that the approach is useful and competitive with existing methods, indicating that the advantage of relatively small and comprehensible models does not come at the expense of predictive accuracy.

6.1 Introduction

Decision trees and regression trees (or model trees) are among the most popular types of models used in data mining and knowledge discovery, and many algorithms for inducing them from data have been developed in the past twenty years [6.4, 6.28]. Virtually every data mining tool includes a decision tree learning algorithm in its repertoire of methods, and numerous successful data mining applications of decision and regression tree algorithms have been reported in recent years. The popularity and success of these methods is not surprising, as decision tree induction algorithms possess a number of nice properties:

Low computational complexity: The induction of decision trees has a time complexity linear in the number of examples. It is therefore applicable to large real-world datasets.

Good acceptance by users: Decision and regression trees are known to be well accepted by users. They are easily understood and can readily be interpreted by domain experts.

Effective handling of noise and uncertainty: Sophisticated pruning techniques have been developed for dealing with noise in the data [6.28]. Class probability trees [6.5] can predict class distributions instead of categorical values.

Well-understood theoretical basis: There is a growing body of literature on the theoretical foundations of decision tree learning [6.5, 6.1, 6.18, 6.11, 6.19] and the basic characteristics of decision tree induction algorithms are well understood.

But the decision tree family of models also has its limitations. Some of these are primarily linked to the representation language, while others are of an algorithmic origin. The central problem is the restriction of the knowledge representation to so-called *propositional representations.* Training examples must be represented as fixed-length vectors of attribute values; a training database is then a simple two-dimensional table of values. Aspects of the internal structure of training examples cannot be described, and the induced trees cannot refer to such structural properties. This seriously compromises the applicability of decision trees in domains where the internal structure of the objects of study is of central importance, such as chemistry, biology, language, or other complex systems. In the remainder of this chapter, we will present an algorithm that removes this limitation.

Other problems of decision trees that are a consequence of the model representation language are the so-called *replication and fragmentation problem* [6.25], which can be overcome by feature construction techniques, and the restriction to *axis-parallel splits* in the instance space (exceptions are the so-called oblique decision trees by Heath *et al.* [6.15]). A property that was considered a problem until recently is the fact that tree induction algorithms are quite *unstable* learners, i.e., their predictions are very sensitive to small perturbations in the training data. While this may negatively affect learning in general, recent developments in machine learning research have shown that it is in fact beneficial in the context of various so-called ensemble methods like bagging or boosting [6.29] (see also Chapter 12).

In this chapter, we present a system that enhances the representational capabilities of decision and regression tree learning by extending it to first-order logic, i.e., relational representations as commonly used in Inductive Logic Programming. We describe an algorithm named *Structural Classification and Regression Trees* (S-CART), which is capable of inducing first-order trees for both *classification* and *regression problems*, i.e., for the prediction of either discrete classes or numerical values. We arrive at this algorithm by a strategy called *upgrading* – we start from a propositional induction algorithm and turn it into a relational learner by devising suitable extensions of the representation language and the associated algorithms. In particular, we have upgraded CART [6.4], the classical method for learning classification and regression trees, to handle relational examples and background knowledge. Similar approaches to upgrading can be found in Chapters 5, 9, and 10. In addition, we have extended the CART methodology by adding linear regression models to the leaves of the trees; this does not have a counterpart in CART, but was inspired by its approach to pruning.

The chapter is organized as follows: In Section 6.2, we present basic ideas of learning trees in first-order logic. Section 6.3 provides a top-level view of S-CART, our upgraded CART variant. Subsequently, the individual components of the algorithm are described in detail: Section 6.4 deals with growing a tree in first-order logic, Section 6.5 recalls the important issue of model selection by error/cost-complexity pruning as originally introduced in CART and as implemented in S-CART, and Section 6.6 extends the CART methodology with multiple linear regression models in the leaves of the trees. Section 6.7 illustrates the applicability of S-CART by briefly reviewing some recent applications of the system to different domains. In Section 6.8 we review related work and in Section 6.9 we summarize the main points of this chapter.

6.2 Tree induction in logic

Let us first turn to the notion of decision and regression trees and see what form and interpretation they can take in first-order logic, and how they relate to logic programs as commonly used in Inductive Logic Programming.

Figure 6.1 shows a graphical example of what we call a *structural regression tree.* It is said to be "structural" because it can consider structural properties of examples in its decision nodes. We have adopted this notion from Watanabe and Rendell [6.33] and their "structural decision trees". A *classification* tree looks just like this, except that it has symbolic class labels in the leaves. The tree predicts the biological activity of some chemical compound C from its structure and its characteristics. Depending on the conditions in the nodes, the theory assigns either 7.82, 7.51, 6.08 or 6.73 to every unseen instance. If proving $atom(C, A1, cl), bond(C, A1, A2, BT), atom(C, A2, n)$ succeeds, the value 7.82 is predicted for a compound C. If this cannot be proven, but still $atom(C, A1, cl)$ alone succeeds, 7.51 is predicted. If $atom(C, A1, cl)$ fails, but $atom(C, A3, o)$ succeeds, 6.08 is predicted for C, and if $atom(C, A3, o)$ fails as well, 6.73 is predicted.

This example suggests how such a tree can be turned into a logic program [6.2] (see Table 6.1). We assume that for the application of such a theory, variable C is bound to an example identifier, and the value of variable A is

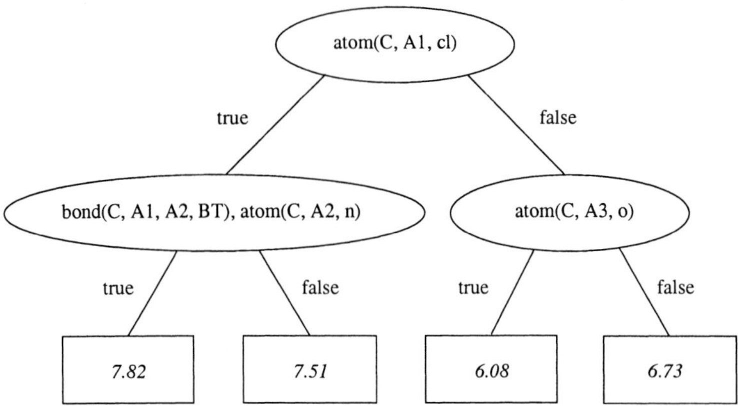

Fig. 6.1. A structural regression tree for predicting the biodegradability of a compound from its structure. The quantity to be predicted is the logarithm of the half-life time of biodegradation of the compound in water. Variable `C` is a compound identifier, `A1`, `A2` and `A3` are atom identifiers and `BT` represents a bond type.

Table 6.1. A Prolog representation of the structural regression tree in Figure 6.1.

```
activity(C, A) :- atom(C, A1, cl),
                  bond(C, A1, A2, BT),
                  atom(C, A2, n),
                  A is 7.82,
                  !.
activity(C, A) :- atom(C, A1, cl),
                  A is 7.51,
                  !.
activity(C, A) :- atom(C, A3, o),
                  A is 6.08,
                  !.
activity(C, A) :- A is 6.73.
```

to be determined. The transformation makes use of the cut symbol. The cut symbol is an important facility for specifying control information in Prolog. The use of the cut makes the logic program in Table 6.1 a decision list: the first applicable clause is taken to predict the value of the activity of a compound.

Generalizing the example above, the transformation of a logical classification tree or regression tree is straightforward. We have to traverse the tree depth-first, and output the positive tests of the paths along the way to a respective leaf, adding a cut to the resulting rule to make sure that only the solution of the first succeeding clause is used for prediction.

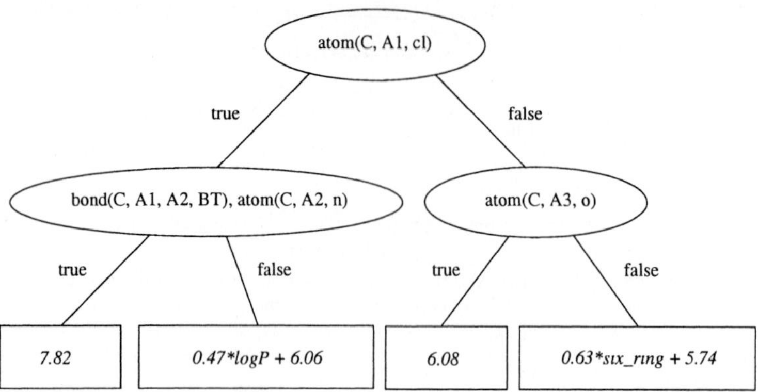

Fig. 6.2. A structural model tree for predicting the biodegradability of a compound from its structure. The variables are named as in Figure 6.1. In addition, *logP* denotes the logP value of a compound C, and *six_ring* denotes the number of rings of size six in the compound.

Analogously, structural regression trees having linear regression models in the leaves are called *structural model trees* (see Figure 6.2). In the example, an increase in *logP* or in the number of rings of size six also increases the activity predicted. The Prolog representation simply contains additional literals calculating the numerical value to be predicted (see Table 6.2).

Table 6.2. A Prolog representation of the structural model tree in Figure 6.2.

```
activity(C, A) :- atom(C, A1, cl),
                  bond(C, A1, A2, BT),
                  atom(C, A2, n),
                  A is 7.82,
                  !.
activity(C, A) :- atom(C, A1, cl),
                  logP(C, L),
                  A is 0.47*L + 6.06,
                  !.
activity(C, A) :- atom(C, A3, o),
                  A is 6.08,
                  !.
activity(C, A) :- six_ring(C, S),
                  A is 0.63*S + 5.74.
```

6.3 Structural classification and regression trees (S-CART): The top level algorithm

S-CART is an algorithm that learns a theory for the prediction of either discrete classes or numerical values from examples and relational background knowledge. The algorithm constructs a tree containing a literal (an atomic formula or its negation) or a conjunction of literals in each node, and assigns a class value or a numerical value to each leaf.

At the top-level, the main stages of S-CART are the following:

Growing the tree: An initial (usually very complex) "main tree" is grown based on the training set.

Pruning the tree: In the pruning phase, the tree is cut back to appropriate size, in order to avoid overfitting. In S-CART, pruning consists of estimating the optimal value of α, the so-called *complexity parameter*. This is done based on a separate prune set or cross-validation (see below).

Adding linear regression models: Optionally, one can perform a step that adds linear regression models to the leaves of the pruned tree. This is a "conservative" extension of the CART methodology, since the "right" value of a significance parameter used for forward selection in step-wise multiple linear regression is estimated based on a prune set or cross-validation as well.

In the following sections, we describe these steps in some detail.

6.4 Growing a tree in first-order logic

For the construction of a single tree, S-CART employs the so-called divide-and-conquer strategy, much as other well-known approaches to the top-down induction of decision trees [6.28].

For the propositional case, divide-and-conquer proceeds as follows (see Table 6.3): As long as the termination condition does not hold, the algorithm searches for the test that best reduces some error measure if examples were split according to that test. Next, the examples are split up into n subsets $Split_i$ according to the best test. For each split, the procedure is recursively applied, obtaining subtrees for the respective splits. The recursive partitioning stops when the set of training instances belonging to a leaf in a partial tree is "pure", i.e., the values of the dependent variable are the same, or, in the case of regression problems, when further splitting would make the numbers of instances drop below a given threshold. For impure leaves, the majority class is predicted for classification, and the average value is predicted for regression.

In the propositional case, tests are simple conditions of the form $f_i = c$ (a nominal feature f_i of an example takes a value c) or $f_j > t$, $f_j \geq t$, etc.

Table 6.3. A divide-and-conquer tree learning algorithm for propositional logic.

```
procedure DivideAndConquer(Examples)

if TerminationCondition(Examples)
then
   NewLeaf = CreateNewLeaf(Examples)
   return NewLeaf
else
   BestTest = FindBestTest(Examples)
   Splits = SplitExamples(Examples, BestTest)
   Subtrees = []
   for each Split_i ∈ Splits do
      Subtree_i = DivideAndConquer(Split_i)
      Subtrees = [Subtree_i|Subtrees]
   return [BestTest|Subtrees]
```

(an integer, real or ordinal feature f_j is greater than, greater than or equal, etc. a threshold t). Another possibility is to test for intervals (f_k *in* (a, b)).

For the first-order case, the divide-and-conquer strategy has to be extended (see Table 6.4). The reason for this is that the tests are no longer simple propositions, but literals or conjunctions of literals containing variables. Since two tests in a tree might share some variables, tests cannot be viewed as independent anymore. At each point, the set of possible tests depends on previous tests in the tree above. Consider for instance the tree in Figure 6.1: The conjunction of literals `bond(C, A1, A2, BT), atom(C, A2, n)` shares the two variables `C` (the compound identifier) and `A1` (an atom identifier) with the test in the root `atom(C, A1, cl)`. Thus, we not only have to keep track of the examples during tree construction, but also of the tests chosen so far. More precisely, we only have to keep track of the *positive tests* along the path, since in the case of a negative outcome no variables can be shared with that test.

Whereas in general a propositional decision or regression tree may contain multi-valued splits (see Table 6.3), a first-order tree only contains binary decisions: either proving $PosTestsSofar \wedge BestTest$ succeeds, or not.

The selection of the literal or conjunction is performed as follows. Let *Examples* be the set of training instances covered by the current node, and $PosTestsSofar$ be the conjunction of all positive tests in the path from the root of the tree to this node. (For a definition of all terms used see Table 6.6.) Assume that we have computed a set of candidate tests or "refinements" (literals or conjunctions of literals) $refs(PosTestsSofar)$ that can be used to further split the node – how this set is computed is described below. Each possible refinement $Ref \in refs(PosTestsSofar)$ is evaluated according to the resulting partitioning of the training instances. The instances *Examples* are partitioned into the instances $Split_1 \subseteq Examples$ for which proving

Table 6.4. A divide-and-conquer tree learning algorithm for first-order logic.

---

procedure DIVIDEANDCONQUER(*PosTestsSofar, Examples*)

if TERMINATIONCONDITION(*Examples*)
then
  $NewLeaf$ = CREATENEWLEAF($Examples$)
  return $NewLeaf$
else
  $BestTest$ = FINDBESTTEST(*PosTestsSofar, Examples*)
  $(Split_1, Split_2)$ = SPLITEXAMPLES(*Examples, PosTestsSofar, BestTest*)
  $LeftSubtree$ = DIVIDEANDCONQUER($PosTestsSofar \wedge BestTest, Split_1$)
  $RightSubtree$ = DIVIDEANDCONQUER($PosTestsSofar, Split_2$)
  return $[BestTest, LeftSubtree, RightSubtree]$

$PosTestsSofar \wedge Ref$ succeeds, and into the instances $Split_2 \subseteq Examples$ for which proving $PosTestsSofar \wedge Ref$ fails. For every possible refinement $Ref \in refs(PosTestsSofar)$ we calculate a quality measure (the Gini Index based on the class frequencies $f_i(c)$ in $Split_i$ for classification problems, and the mean squared error of the values of the dependent variable $y_{i,j}$ for regression problems (see Table 6.5)). The Gini Index is a measure that weights the impurity of splits by the proportion of examples in the respective split. From all possible refinements, S-CART selects $BestTest \in refs(PosTestsSofar)$ which minimizes the respective error measure.

The set of possible refinements is specified by the user by means of so-called *schemata*. Using schemata, one can specify the form of possible hypotheses declaratively. So, schemata are a form of what is called a *declarative language bias* in ILP. They are similar to relational clichés [6.30] or the declarative language bias of related ILP systems such as TILDE [6.2]. The idea of schemata can easily be explained using an example:

```
schema((bond(V, W, X, Y), atom(V, X, Z)),
       [V:chemical:'+',  W:atomid:'+',   X:atomid:'-',
        Y:bondtype:'-',  Z:element: =]).
```

This expression specifies an admissible refinement of a given clause or tree by the addition of a conjunction of literals (the conjunction `(bond(V, W, X, Y), atom(V, X, Z))`, in this case). It defines the conditions under which this conjunction of literals can be added to a clause or tree under construction. Each refinement step requires such a schema. The first subexpression may also consist of a single condition only. The refinement is constrained by the specifications in the second subexpression, a list containing argument descriptions. Argument descriptions are triples consisting of a variable, a variable type and a so-called mode. In the example, `chemical` or `atomid` are variable types. The variables refer to argument positions in the conjunction of

Table 6.5. Definition of the resubstitution estimate, the standard error, the evaluation of splits and the predicted value for classification and regression.

Classification:

Resubstitution Estimate:

$$R = \frac{1}{N}\sum_{i=1}^{N} 0/1\text{-loss}(y_i, d(x_i))$$

Standard Error:

$$SE = \sqrt{\frac{R^{CV}(1 - R^{CV})}{N}}$$

Evaluation of Splits:

Let *Classes* be the set of given classes $\{c_1, c_2, ..., c_m\}$ and $f_i(c)$ denote the relative frequency of class $c \in Classes$ in split i of a leaf in a partial tree. Then the Gini Index is defined as:

$$Gini\ Index = \frac{1}{N'}\sum_{i=1}^{2} N_i(1 - \sum_{c \in Classes} f_i(c)^2)$$

Predicted Value:

For the examples in $Split_i$ of a leaf in a (partial) tree, the prediction is the majority class:

$$Prediction = argmax_{c \in Classes}(f_i(c))$$

Regression:

Resubstitution Estimate:

$$R = \frac{1}{N}\sum_{i=1}^{N}(y_i - d(x_i))^2$$

Standard Error:

$$SE = \frac{1}{\sqrt{N}}\sqrt{\frac{1}{N}\sum_{i=1}^{N}(y_i - d(x_i))^4 - {R^{CV}}^2}$$

Evaluation of Splits:

$$Mean\ Squared\ Error = \frac{1}{N'}\sum_{i=1}^{2}\sum_{j=1}^{N_i}(y_{i,j} - \overline{y_{i,j}})^2$$

Predicted Value:

For the examples in split i belonging to a leaf in a (partial) tree, the prediction is the average value of the dependent variable:

$$Prediction = \overline{y_{i,j}}\ (see\ Table\ 6.6)$$

Table 6.6. Definition of terms used in description of S-CART.

T_{max} – the fully grown tree

$T_i \prec T_j$ – T_i is a *subtree* of T_j, i.e., T_i is a result of pruning T_j in some way

R – resubstitution estimate (training set error)

N – number of training instances

α – complexity parameter

$\tilde{T}$ – the number of leaves in tree T

m – number of folds used for cross-validating the model parameters α (the complexity parameter) and SL (the significance parameter for the multiple linear regression step)

0/1-*loss* – function returning 1 in case of misclassifying an instance, 0 otherwise

y_i – actual value of dependent variable for example x_i

$d(x_i)$ – predicted value for instance x_i

R^{CV} – cross-validation estimate of error R

Examples – set of training instances covered by a leaf in a partial tree

N' – number of examples in a leaf in a partial tree

$PosTestsSofar$ – the conjunction of all positive tests in the path from the root of the tree to a leaf in a partial tree

$refs(PosTestsSofar)$ – set of all possible refinements of $PosTestsSofar$

$Split_1$ – subset of *Examples* for which proving $PosTestsSofar \wedge Ref$ succeeds, $Ref \in refs(PosTestsSofar)$

$Split_2$ – subset of *Examples* for which proving $PosTestsSofar \wedge Ref$ fails ($Examples = Split_1 \cup Split_2,\ Split_1 \cap Split_2 = \emptyset$)

N_i – number of examples in $Split_i$ of a leaf in a partial tree, $i \in \{1, 2\}$

$f_i(c)$ – relative class-frequency of class c in $Split_i$ of a leaf in a partial tree

$y_{i,j}$ – value of dependent variable of example j in $Split_i$ of a leaf in a partial tree

$\overline{y_{i,j}}$ – average value of dependent variable in $Split_i$ of a leaf in a partial tree

literals. All variables occurring in the literals have to occur in the second subexpression, and vice versa.

If a given branch is to be refined by means of such a schema, the variables labeled as '+' are unified with variables already bound, provided that the types are matching as well. If a variable is labeled as '–', it means that this variable must not be bound before the application of the schema. A new variable can be used in subsequent refinement steps, given that the mode is '+' there and the types are matching. So, in the example, variable `X` can be used in any subsequent literal, if all constraints are fulfilled. The mode declaration '=' means that a constant is inserted at the respective argument position.

Example 6.4.1. Consider the tree in Figure 6.1. To illustrate the above notions, suppose we currently only have the root (`atom(C, A1, cl)`) and we want to generate tests for refining it, given the above and the following schema:

```
schema(bond(W, X, Y, Z),
 [W:chemical:(+), X:atomid:(+), Y:atomid:(+), Z:bondtype: =]).
```

Given these schemata, the refinements of $PosTestsSofar$ = `atom(C, A1, cl)` would simply be:

```
bond(C, A1, A2, BT), atom(C, A2, s)
bond(C, A1, A2, BT), atom(C, A2, o)
bond(C, A1, A2, BT), atom(C, A2, n)
bond(C, A1, A2, BT), atom(C, A2, cl)
bond(C, A1, A2, BT), atom(C, A2, c)
...
bond(C, A1, A2, 1)
bond(C, A1, A2, 2)
bond(C, A1, A2, 3)
bond(C, A1, A2, 7)
```

where `s`, `o`, `n`, `cl`, `c` etc. are valid constants of type `element`, and `1`, `2`, `3`, and `7` are valid `bondtypes`. In the example, both variables `C` (the compound identifier) and `A1` (an atom identifier) are "re-used" in the generated tests. After testing all such candidates, `bond(C, A1, A2, BT), atom(C, A2, n)` is chosen as the best refinement and added to the tree.
□

6.5 Model selection by error/cost complexity pruning

S-CART's criterion for deciding when to stop refining a tree is a very simple one: it stops if all examples in a leaf of a partial tree share the same value of the dependent variable, or if further splitting would make the number of instances drop below a given threshold. The resulting trees will thus usually be very complex and will tend to overfit the data. It is the task of the subsequent *pruning phase* to empirically determine the appropriate size and complexity of the final tree.

The pruning method is based directly on the strategy used in CART. CART's solution to this problem is based on either a separate prune set or cross-validation. This choice depends on the size of the data-set: If a sufficient number of examples is available, a prune set is used, otherwise cross-validation is performed. Both methods are used to tune the so-called *complexity parameter* α.

This parameter is defined as follows (all relevant terms used here are summarized in Table 6.6.):[1] For any subtree $T \prec T_{max}$ (T_{max} being the fully grown tree), define its complexity as $|\tilde{T}|$, the number of leaves in T. Let the cost-complexity measure $R_\alpha(T)$ be defined as a linear combination of the cost of the tree $R(T)$[2] and its complexity:

$$R_\alpha(T) = R(T) + \alpha|\tilde{T}|$$

So, the complexity parameter α is defined as a real number weighting the complexity of the tree relative to the misclassification cost. The cost-complexity measure $R_\alpha(T)$ adds a cost penalty for complexity to the misclassification cost.

For each value of α, the task is to find a subtree $T(\alpha) \prec T_{max}$ which minimizes $R_\alpha(T)$:

$$R_\alpha(T(\alpha)) = \min_{T \prec T_{max}} R_\alpha(T)$$

If α is small, the penalty for having a large number of leaves is small and $T(\alpha)$ will be large. As the penalty α increases, the minimizing subtrees $T(\alpha)$ will have fewer leaves. Finally, for α sufficiently large, the minimizing subtree T_α will consist of the root node only, and the tree T_{max} will be completely pruned.

The procedure for error/cost-complexity pruning determines the sequence of trees which is optimal with respect to α, i.e., each tree in the sequence is optimal for some interval of α. A separate prune-set or cross-validation is used to evaluate and select from this sequence based on the error of the trees.

For the case of a prune set, we divide the training set into a *grow* and *prune set*. Usually, 2/3 of the examples in the training set are used for the grow set, and 1/3 for the prune set. Subsequently, a tree is constructed from the grow set (for details see the previous section). Next, the fully grown tree is pruned by a process that is called *weakest link cutting* [6.4].

In each iteration of weakest link cutting, we determine the internal node which is the least beneficial to keep in the tree. The benefit of an internal node is expressed as the increase in the error rate when removing the subtree underneath and turning it into a leaf node, divided by the number of nodes in the subtree. Then, the node with the lowest benefit is turned into a leaf node. This process iterates until only the root node is left.

---

[1] The following description of the CART methodology for tree growing and pruning is partly quoted from Breiman *et al.* [6.4], partly paraphrased. We describe this method in some detail because the algorithmic details of CART are not as widely known as the system itself.

[2] Here, the notion of "costs" subsumes both error rates for classification and numerical error measures for regression. If actual costs for misclassifications were known in application domains, they can easily be considered in S-CART.

For each tree in the resulting sequence, S-CART determines the error rate on the held-out prune set. Subsequently, a tree is selected from the sequence of trees. Since the selection of the tree with the lowest estimated error (estimated on the prune set) would be somewhat arbitrary, S-CART, like CART, selects the smallest tree within one standard error of the best tree in the sequence (for a definition of the standard error see Table 6.5).

If the size of the dataset is relatively small, one cannot afford to waste data for a separate prune set, and needs to make more efficient use of the data. In this case, the variant of error/cost-complexity pruning with cross-validation is recommended. A tree is grown and pruned on the complete training set. In the following, we refer to the fully grown tree as the "main tree", and to the resulting sequence of trees as the "main sequence". As in the case for the prune set, the selection of the final tree is done based on the error, but this time the error of a tree is estimated not directly using the prune set, but indirectly using cross-validation.

The training set is divided into k folds, and for all j (from 1 to k), on $\cup_{i=1}^{k} fold_i - fold_j$ in turn, we grow and prune trees as for the "main tree". In this way, we obtain k auxiliary sequences of trees pruned by different amounts. The basic idea then is to estimate the errors of each tree t in the main sequence by the errors of those k trees corresponding to t in the k auxiliary sequences (one in each of the k sequences).

This correspondence is established based on the parameter α. The whole pruning process is defined in a way such that each tree in a sequence represents the optimal tree with respect to some interval in the possible values of α. So we find, for each tree in the main sequence that is optimal for some α, the corresponding k trees (one in each of the k sequences) that are also optimal for this α. Since these values are unlikely to be identical, there is a particular way of matching the trees for different values of α.[3]

Once the correspondence between the trees is established, the method estimates the errors made by each tree in the main sequence by the errors made by the corresponding trees in the auxiliary sequences on the respective held-out sets. Finally, the selection of the final tree is performed as for the case with the prune set: we choose the smallest tree in the "main sequence" which is within one standard error of the best one.

6.6 First-order model trees

The above procedure can be employed for the induction of both classification and regression trees (where a regression tree predicts a real number – the mean of the respective training instances – in each leaf). With regard to

[3] A comprehensive description is beyond the scope of this chapter. The interested reader is referred to Breiman *et al.* [6.4] for details.

regression, we have devised a "conservative" extension of the CART methodology that adds multiple linear regression models to the leaves of the trees (see Figure 6.2). Following Quinlan [6.27], we call the resulting structures *model trees*. For each leaf in the pruned tree, multiple linear step-wise regression [6.14] is applied to the examples in the respective leaf, with varying values of the significance parameter that determines whether a variable is included in the regression model or not. Only when a variable contributes significantly to "explaining" the variation of the dependent variable is it included in the linear regression model. The significance parameter is set globally for the tree, since we are only interested in a rough estimate of the appropriate model complexity of the model trees. We vary this parameter following the usual suggestions found in the statistical literature (i.e., setting the significance level SL to the values in $\{0.90, 0.95, 0.99, 0.999\}$). Tuning this parameter is done exactly as tuning α in pruning. For the cross-validation variant, this step re-uses the auxiliary trees that resulted from the pruning procedure as by-products. The significance parameter yielding the lowest error is returned from the procedure and subsequently used for the "main tree" selected in the pruning step.

6.7 Applications

One of the major recent applications of S-CART was the prediction of the biodegradability of organic compounds [6.12]. The prediction of biodegradability is a novel task in the area of quantitative structure-activity relationships (QSARs). The quantity to be predicted is the biodegradability of chemical compounds in water. In particular, the target variable is the mean of the lower bound and the upper bound of the half-life time in water for aerobic aqueous biodegradation. Structural descriptions of the compounds were given in terms of the well-known atom-bonds representation and functional groups. Predicting biodegradability is essentially a regression problem, but we also considered a discretized version of the target variable.

Our most interesting result has been obtained with a reformulation of the learning problem: we predicted the lower bound and upper bound of the half-life time separately, and merged these values into one overall prediction. So, in effect, S-CART (and the other systems we applied) were learning to predict an interval. However, we evaluated the performance with respect to the mean of the lower and the upper bound (see above).

In the experiments, $M5'$ [6.32] (applied to a "propositionalized" version of the learning problem, cf. Chapter 11) and S-CART (applied to the relational background knowledge) gave the best quantitative results (M5': $RMSE = 1.177/r = 0.663$, S-CART: $RMSE = 1.198/r = 0.659$). The main advantage of S-CART was the relatively small size of the models. The regression trees produced by S-CART contained around 50 descriptors, whereas $M5'$ trees contained around 300 descriptors on the average.

We also conducted experiments in the well-known ILP domains of mutagenicity prediction [6.31] and carcinogenicity prediction [6.20]. For mutagenicity, the predictive accuracy turned out to be not as good as the one of strong propositional learners (using default settings). However, the average size of the models is significantly smaller than for other methods. For carcinogenicity, S-CART performs well in terms of predictive accuracy and the models are much smaller than models induced by other methods. Summing up, S-CART performed well quantitatively for carcinogenicity, but not so much for mutagenicity. Still, it may be the method of choice if the user is interested in small, comprehensible models.

In another application, we studied durational phenomena in spoken language, in particular, list-like enumerations in German [6.26]. Due to its highly structured and uniform nature this rather specialized utterance type seemed especially well suited for investigating principles of the rhythmical organization of speech. The learning task was to predict the duration of a syllable in some normalized form. Since the duration heavily depends on the "neighborhood" of a syllable, this is basically a relational learning task. We inspected the results qualitatively in order to check the plausibility in the light of previously performed investigations on durational effects: interestingly, S-CART re-discovered known effects in phonetic research. In summary, this new application indicated the potential of ILP in language applications (see also the recent book on learning language in logic [6.8]).

6.8 Related work

6.8.1 Upgrading propositional algorithms

In recent years, a number of novel Inductive Logic Programming systems [6.10, 6.2, 6.9] have been built using the methodology of "upgrading" propositional learning algorithms. The basic idea is that ILP systems can be obtained by extending propositional systems to handle Herbrand interpretations instead of feature-vectors as examples. TILDE [6.2] upgrades C4.5 [6.28], ICL [6.10] upgrades CN2 [6.6] and WARMR [6.9] upgrades association rules algorithms. The relational learners described in Chapter 9 are upgrades of simple, but effective instance- or distance-based methods from the domain of propositional learning and clustering.

6.8.2 Tree induction in logic

There has been some previous work on classification trees in first-order logic. Manago's KATE [6.24] learns decision trees from examples represented in a frame-based language that is equivalent to first-order predicate calculus. KATE makes extensive use of a given hierarchy and heuristics to generate the branch tests. To our knowledge, KATE was the first system to induce first-order theories in a divide-and-conquer fashion.

Watanabe and Rendell [6.33] also investigated the use of divide-and-conquer for learning first-order theories. Their system STRUCT induces so-called structural decision trees and differs from S-CART mainly in that it can only be used for the prediction of discrete classes.

The closest work in the literature is TILDE [6.2], an upgrade of C4.5. The main contribution of the paper on TILDE is that it clarifies the theoretical foundations of tree induction in logic. Although the predecessor of S-CART, SRT [6.21], has been developed before TILDE, it fits into the theoretical framework worked out by Blockeel and De Raedt [6.2].

6.8.3 Regression

The classical statistical model for the prediction of numerical values is linear least-squares regression. Refinements and extensions like non-linear models are also well-known and used in many real-world applications. However, such regression models have several limitations: First of all, they are often hard to understand. Secondly, classical statistical methods assume that all features are equally relevant for all parts of the instance space. Thirdly, such regression models do not allow for easy utilization of domain knowledge. The only way to include knowledge is to "engineer" features, and to map these symbolic features to real-valued features.

In order to solve some of these problems, regression tree methods (CART [6.4], RETIS [6.16], M5 [6.27]) have been developed. Regression trees are supposed to be more comprehensible than traditional regression models. Furthermore, regression trees by definition do not treat all features as equally relevant for all regions of the instance space. The basic idea of regression trees according to CART is to minimize the mean squared error for the next split of a node in the tree, and to predict, for unseen instances, the average of the dependent variable of all training instances covered by the matching leaf. RETIS and M5 differ in that they do not assign single values to the leaves, but linear regression models.

So far, only few other methods have been applied to the problem of regression in the context of relational data: DINUS/RETIS [6.13], a combination of DINUS [6.23] and RETIS [6.16], FORS [6.17] and TILDE [6.2].

DINUS/RETIS transforms the learning problem into a propositional language, and subsequently applies RETIS, a propositional regression tree algorithm, to the transformed problem. However, this transformation does not work for non-determinate background knowledge, which is a limitation of the approach.[4]. S-CART, in contrast, solves the problem in its original representation, and does not require transforming the problem.

[4] For a recent extension of this method to handle non-determinate literals, see Chapter 11.

FORS is an algorithm that can be applied to the same class of learning problems as S-CART, since it can deal with non-determinate background knowledge as well. It is also the case that TILDE has been extended for regression problems. The regression variant of TILDE employs the RMSE as its splitting criterion and a simple pre-pruning strategy based on the F-test. FORS differs from S-CART and TILDE in its use of separate-and-conquer instead of divide-and-conquer (i.e., it is a covering algorithm, not a tree-based algorithm). Generally, all the advantages and disadvantages known from other algorithms of these types (tree-based vs. covering) apply. Boström [6.3] discusses both strategies in the context of top-down induction of logic programs. On one hand, the hypothesis space for separate-and-conquer is larger than for divide-and-conquer. Thus, more compact hypotheses might be found using separate-and-conquer. On the other hand, building a tree is computationally cheaper than searching for rules. Weiss and Indurkhya [6.34] basically draw the same conclusions in their comparison of tree induction and rule induction in propositional regression. However, Weiss and Indurkhya's approach to rule-based regression [6.34] is different from FORS, since it involves the discretization of the numerical dependent variable.

6.9 Conclusion

In this chapter, we have presented Structural Classification and Regression Trees (S-CART), a learning and data mining algorithm that can be applied to the prediction of either discrete classes or continuous values from examples and relational (and non-determinate) background knowledge. S-CART can be viewed as bringing the statistical method of classification and regression trees [6.4] into the world of Inductive Logic Programming.

Experiments in several real-world domains demonstrated that S-CART is competitive with existing methods in terms of predictive accuracy, indicating that its advantages (the applicability to relational classification and regression given non-determinate background knowledge and the comprehensibility of the rules) do not come at the expense of predictive accuracy.

With respect to computational efficiency, it is clear that relational formalisms (especially when they allow the introduction of new variables) open up a much larger search space. Inductive Logic Programming algorithms are often thought to be too slow for real applications. While the underlying problem cannot simply be made to disappear – there is a price to be paid for a more expressive representation language – it can be largely circumvented by giving the knowledgeable user a means to restrict the search space in a problem-specific way. S-CART offers the data miner an expressive language to specify constraints on the search space. This *declarative bias* language includes schemata to define which literals and conjunctions of literals may be used to refine a partial tree, as well as argument modes and types. In fact, we have developed a *meta-language* in which the declarative bias for whole classes of problems can conveniently be expressed, as well as an algorithm that automatically compiles such high-level specifications into operational schemata etc. for a given specific application [6.22].

That said, we do believe that Inductive Logic Programming methods can play an essential role in data mining, and that upgrading "propositional" algorithms to the world of relational data mining and ILP is a fruitful way to go, as long as one pays careful attention to efficiency issues. The expressiveness of first-order logic simply is too attractive a feature for data mining research to ignore, and we expect to see more of it not only in new algorithms, but also in exciting knowledge discovery applications.

Acknowledgments

This research was partly funded by the Austrian *Fonds zur Förderung der Wissenschaftlichen Forschung (FWF)* under grants no. P10489-MAT and P12645-INF. It was also funded by the Austrian Federal Ministry of Education, Science and Culture (GZ 70.017/1-Pr/4/2000). The Austrian Research Institute for Artificial Intelligence is supported by the Austrian Federal Ministry for Transport, Innovation and Technology.

References

6.1 P. Auer, W. Maass, and R. Holte. Theory and applications of agnostic PAC-learning with small decision trees. In *Proceedings of the Twelfth International Conference on Machine Learning*, pages 21–29. Morgan Kaufmann, San Francisco, CA, 1995.
6.2 H. Blockeel and L. De Raedt. Top-down induction of first order logical decision trees. *Artificial Intelligence*, 101(1-2):285–297, 1998.
6.3 H. Boström. Covering vs. divide-and-conquer for top-down induction of logic programs. In *Proceedings of the Fourteenth International Joint Conference on Artificial Intelligence*, pages 1194–1200. Morgan Kaufmann, San Francisco, CA, 1995.
6.4 L. Breiman, J. Friedman, R. Olshen, and C. Stone. *Classification and Regression Trees*. Wadsworth, Belmont, CA, 1984.
6.5 W. Buntine. *A theory of learning classification rules*. PhD thesis, School of Computing Science, University of Technology, Sydney, Australia, 1992.
6.6 P. Clark and T. Niblett. The CN2 induction algorithm. *Machine Learning*, 3:261–283, 1989.
6.7 W. Cohen. Grammatically biased learning: Learning logic programs using an explicit antecedent description language. *Artificial Intelligence*, 68(2):303–366, 1994.
6.8 J. Cussens and S. Džeroski, editors. *Learning Language in Logic*. Springer, Berlin, 2000.
6.9 L. Dehaspe and H. Toivonen. Discovery of frequent datalog patterns. *Data Mining and Knowledge Discovery*, 3(1):7–36, 1999.
6.10 L. De Raedt and W. Van Laer. Inductive constraint logic. In *Proceedings of the Fifth Workshop on Algorithmic Learning Theory*, pages 80–94. Springer, Berlin, 1995.

6.11 T. Dietterich, M. Kearns, and Y. Mansour. Applying the weak learning framework to understand and improve C4.5. In *Proceedings of the Thirteenth International Conference on Machine Learning*, pages 96–104. Morgan Kaufmann, San Francisco, CA, 1996.
6.12 S. Džeroski, H. Blockeel, B. Kompare, S. Kramer, B. Pfahringer, and W. Van Laer. Experiments in predicting biodegradability. In *Proceedings of the Ninth International Workshop on Inductive Logic Programming*, pages 80–91. Springer, Berlin, 1999.
6.13 S. Džeroski, L. Todorovski, and T. Urbančič. Handling real numbers in inductive logic programming: A step towards better behavioural clones. *Proceedings of the Eighth European Conference on Machine Learning*, pages 283–286. Springer, Berlin, 1995.
6.14 R. Freund and W. Wilson. *Regression analysis*. Academic Press, London, 1998.
6.15 D. Heath, S. Kasif, and S. Salzberg. Induction of oblique decision trees. In *Proceedings of the Thirteenth International Joint Conference on Artificial Intelligence*, pages 1002–1007. Morgan Kaufmann, San Francisco, CA, 1993.
6.16 A. Karalič. Employing linear regression in regression tree leaves. In *Proceedings of the Tenth European Conference on Artificial Intelligence*, pages 440–441. John Wiley and Sons, 1992.
6.17 A. Karalič and I. Bratko. First order regression. *Machine Learning*, 26(2/3): 147–176, 1997.
6.18 M. Kearns. Boosting theory towards practice: Recent developments in decision tree induction and the weak learning framework. In *Proceedings of the Thirteenth National Conference on Artificial Intelligence*, pages 1337–1339. AAAI Press, Menlo Park, CA, 1996.
6.19 M. Kearns and Y. Mansour. A fast, bottom-up decision tree pruning algorithm with near-optimal generalization. In *Proceedings of the Fifteenth International Conference on Machine Learning*, pages 269–277. Morgan Kaufmann, San Francisco, CA, 1998.
6.20 R. King and A. Srinivasan. Prediction of rodent carcinogenicity bioassays from molecular structure using Inductive Logic Programming. *Environmental Health Perspectives*, 104(5):1031–1040, 1996.
6.21 S. Kramer. Structural regression trees. In *Proceedings of the Thirteenth National Conference on Artificial Intelligence*, pages 812–819. AAAI Press, Menlo Park, CA, 1996.
6.22 S. Kramer. *Relational learning vs. propositionalization: investigations in inductive logic programming and propositional machine learning*. PhD thesis, Vienna University of Technology, Vienna, Austria, 1999.
6.23 N. Lavrač and S. Džeroski. *Inductive Logic Programming: Techniques and Applications*. Ellis Horwood, Chichester, 1994. Freely available at `http://www-ai.ijs.si/SasoDzeroski/ILPBook/`.
6.24 M. Manago. Knowledge-intensive induction. In *Proceedings of the Sixth International Workshop on Machine Learning*, pages 151–155. Morgan Kaufman, San Francisco, CA, 1989.
6.25 G. Pagallo and D. Haussler. Boolean feature discovery in empirical learning. *Machine Learning*, 5(1):71–100, 1990.
6.26 H. Pirker and S. Kramer. Listening to lists: Studying durational phenomena in enumerations. In *Proceedings of the Fourteenth International Conference of Phonetic Sciences*, pages 273–276. Department of Linguistics, University of California, Berkeley, CA, 1999.

6.27 J. Quinlan. Learning with continuous classes. *Proceedings of the Second Australian Conference on Artificial Intelligence*, pages 343–348. World Scientific, Singapore, 1992.

6.28 J. Quinlan. *C4.5: programs for machine learning*. Morgan Kaufmann, San Francisco, CA, 1993.

6.29 J. Quinlan. Bagging, boosting and C4.5. In *Proceedings of the Thirteenth National Conference on Artificial Intelligence*, pages 725–730. AAAI Press, Menlo Park, CA, 1996.

6.30 G. Silverstein and M. Pazzani. Relational clichés: Constraining constructive induction during relational learning. In *Proceedings of the Eighth International Workshop on Machine Learning*, pages 203–207. Morgan Kaufmann, San Francisco, CA, 1991.

6.31 A. Srinivasan, S. Muggleton, R. King, and M. Sternberg. Theories for mutagenicity: a study of first-order and feature based induction. *Artificial Intelligence*, 85(1-2):277–299, 1996.

6.32 Y. Wang and I. Witten. Inducing model trees for continuous classes. In *Poster Papers of the Ninth European Conference on Machine Learning*, pages 128–137. Faculty of Informatics and Statistics, University of Economics, Prague, 1997.

6.33 L. Watanabe and L. Rendell. Learning structural decision trees from examples. In *Proceedings of the Twelfth International Joint Conference on Artificial Intelligence*, pages 770–776. Morgan Kaufmann, San Francisco, CA, 1991.

6.34 S. Weiss and N. Indurkhya. Rule-based machine learning methods for functional prediction. *Journal of Artificial Intelligence Research*, 3:383–403, 1995.

7. Relational Rule Induction with CPROGOL4.4: A Tutorial Introduction

Stephen Muggleton and John Firth

Department of Computer Science, University of York
Heslington, York YO1 5DD, UK

Abstract

This chapter describes the theory and use of CPROGOL4.4, a state-of-the-art Inductive Logic Programming (ILP) system. After explaining how to download the source code, the reader is guided through the development of PROGOL input files containing type definitions, mode declarations, background knowledge, examples and integrity constraints. The theory behind the system is then described using a simple example as illustration. The main algorithms in PROGOL are given and methods of pruning the search space of possible hypotheses are discussed. Next the application of built-in procedures for estimating predictive accuracy and statistical significance of PROGOL hypotheses is demonstrated. Lastly, the reader is shown how to use the more advanced features of CPROGOL4.4, including positive-only learning and the use of metalogical predicates for pruning the search space.

7.1 Introduction

The theory and implementation of the Inductive Logic Programming (ILP) system CPROGOL4.1 was first described in [7.8]. Since then a number of advances have been made over the original CPROGOL4.1 in systems such as CPROGOL4.2 [7.9], PPROGOL2.1, PPROGOL2.2, and ALEPH [7.14]. The development of these systems has been informed by feedback from experiments on a variety of real-world applications [7.3, 7.6, 7.5, 7.15, 7.2].

This chapter describes the theory and use of CPROGOL4.4, a publicly distributed version of the PROGOL family of ILP systems. In order to follow the examples in this chapter, it is assumed that the reader is familiar with the aims of ILP, Horn clause logic and Prolog notation for clauses. It is also assumed that the reader is familiar with the concepts of deductive logic in order to understand the theoretical foundations of PROGOL. Furthermore it is assumed that the reader has access to a machine running UNIX with an Internet connection. This will be necessary in order to obtain the software and run the examples described in the chapter.

The chapter has the following structure. Section 7.2 describes how to use anonymous ftp to obtain CPROGOL4.4, together with the associated distribution datasets and documentation. Having obtained the system, Section 7.3 guides the reader through the general approach that should be taken to developing a PROGOL input file containing examples and background knowledge. The theory behind PROGOL is then described in Section 7.4 and illustrated by a simple example. Here the main algorithms inside the system are given in quite a good deal of detail, hopefully without being too technical. The reader is then shown in Section 7.5 how to make use of built-in procedures which support the estimation of predictive accuracy and statistical significance of hypothesised theories. Section 7.6 describes how to use integrity constraints and prune statements to control search in CPROGOL4.4. The setting of resource bounds is discussed in Section 7.7. Section 7.8 describes the facilities available to support the debugging of a CPROGOL4.4 input file. Finally, Section 7.9 summarises the philosophy behind PROGOL and explains why it was designed in the way it was. Mention is made of some of the more important applications of the system to real-world problems.

7.2 How to obtain CPROGOL4.4

It should be noted that CPROGOL4.4 is available free for academic research and teaching. For commercial research, a license should be obtained by writing to the first author. CPROGOL4.4 can be obtained through the web page `http://www.cs.york.ac.uk/mlg/progol.html` or by anonymous ftp from `ftp.cs.york.ac.uk` in the directory `pub/ML_GROUP/progol4.4`.

The ftp site contains a README file with instructions on installing the software. For this you will need to be running a UNIX (or Linux) operating system with a gcc compiler. In fact, CPROGOL4.4 is automatically compiled in the source directory when you obtain all the files from the ftp site and run the *expand* script. This command additionally compiles a program called qsample, which is useful for randomly sampling training and test sets (instructions for running qsample are in the comments at the top of qsample.c).

To test that PROGOL is correctly installed after running expand, move to the source directory and issue the following command from the UNIX prompt.

```
$ progol
```

In response you should see the following.

```
CProgol Version 4.4

|-
```

To see the list of available commands type `help?` at the PROGOL command prompt `|-`.

Note that in interactive mode PROGOL is similar to a Prolog interpreter. In fact, PROGOL uses the Edinburgh Prolog syntax and contains almost all of the built-in predicates used in Clocksin and Mellish's introduction to Prolog [7.1]. One immediate difference however is the use of '?' at the end of queries instead of '.'. Any statement ending with a '.' is asserted into the clause-base.

You can get one-line description of PROGOL's built-in commands using the command help in the following fashion.

```
|- help(tell/1)?
```

To quit PROGOL type either an end-of-file (usually Control-D) or 'quit?'.

For the following, you should include the source directory as part of the path description in your .login file in order to be able to use the 'progol' command in other directories.

7.3 Developing an input file for CPROGOL4.4

As with all ILP systems, PROGOL constructs logic programs from examples and background knowledge. For instance, given various examples of "aunt_of"

```
aunt_of(jane,henry).
aunt_of(sally,jim).
...
```

and background knowledge concerning "parent_of", "father_of", "mother_of" and "sister_of"

```
parent_of(Parent,Child) :- father_of(Parent,Child).
parent_of(Parent,Child) :- mother_of(Parent,Child).
...
father_of(sam,henry).
...
mother_of(sarah,jim).
...
sister_of(jane,sam).
sister_of(sally,sarah).
...
```

together with a list of types and mode declarations, PROGOL can construct a definition for the aunt_of predicate. In order to do this PROGOL needs to be told the following information about the learning task.

7.3.1 Types

These describe the categories of objects (numbers, lists, names, etc.) in the world under consideration. In this example we only need the type *person*, since all objects in the given relations are of this type.

```
person(jane).
person(henry).
person(sally).
...
```

This simply states that all of the elements involved satisfy the unary predicate *person*. In other words they have type *person*.

7.3.2 Modes

These describe the relations (predicates) between objects of given types which can be used either in the head (*modeh* declarations) or body (*modeb* declarations) of hypothesized clauses. Modes also describe the form these atoms can take within a clause. For the head of any general rule defining *aunt_of* we might give the following head mode declarations

```
:- modeh(1,aunt_of(+person,+person))?
```

The declaration states that head atom has predicate symbol *aunt_of* and has 2 variables of type *person* as arguments. The '+' sign indicates that an argument is an input variable. A '-' sign would indicate an output variable, and a '#' would indicate that a constant should be placed at the position in the hypothesis. Thus the declaration above indicates that

```
aunt_of(X, Y)
```

is allowed as the head atom of a general definite clause, and that goals calling this clause should bind both variables.

All modeh (and modeb) declarations contain a number called the recall. This is used to bound the number of alternative solutions for instantiating the atom. Here it is 1 because the aunt_of predicate gives a unique answer (yes or no) when given two input arguments. For a predicate such as *square_root* the recall would be 2 since a number has at most two square roots. If in doubt use a large number or '*' as the recall. The latter means no limit to the number of instantiations that might be made but, in practice, Progol will set its own upper limit of 100.

For atoms in the body of a general rule, body mode declarations must be given as follows.

```
:- modeb(*,parent\_of(-person,+person))?
:- modeb(*,parent\_of(+person,-person))?
:- modeb(*,sister\_of(+person,-person))?
```

The first of these declarations can be used to add *parent_of* atoms to the body of a hypothesis which introduce one or more parents of a given child. Similarly, the second allows *parent_of* to be used in the body to find one or more children of a given parent and the last can be used in the body to find one or more sisters of a given individual.

7.3.3 Settings

These describe some of the runtime parameter settings that Progol uses. For instance, in the following

```
:- set(posonly)?
```

posonly turns on the positive-only evaluation mechanism [7.9]. This allows the user to avoid incorporating negative examples, which are often unnatural to define, and also often unavailable in real-world domains.

All the other settings in PROGOL are described in the manual that is part of the source distribution.

Table 7.1. The PROGOL input file for learning the relation *aunt_of*.

```
% Learning aunt_of from parent_of and sister_of.

% Settings

:- set(posonly)?

% Mode declarations

:- modeh(1,aunt_of(+person,+person))?
:- modeb(*,parent_of(-person,+person))?
:- modeb(*,parent_of(+person,-person))?
:- modeb(*,sister_of(+person,-person))?

% Types

person(jane).
person(henry).
person(sally).
person(jim).
person(sam).
person(sarah).
person(judy).

% Background knowledge

parent_of(Parent,Child) :- father_of(Parent,Child).
parent_of(Parent,Child) :- mother_of(Parent,Child).

father_of(sam,henry).

mother_of(sarah,jim).

sister_of(jane,sam).
sister_of(sally,sarah).
sister_of(judy,sarah).

% Examples

aunt_of(jane,henry).
aunt_of(sally,jim).
aunt_of(judy,jim).
```

7.3.4 The complete example

The full listing of the example is given in Table 7.1. This is similar to the file `aunt.pl` found under the directory `examples4.4` in the CPROGOL4.4 distribution (see Section 7.2).

Note that a line beginning with the '%' character indicates a comment. Negative examples of *aunt_of* could have been included as follows.

```
:- aunt_of(henry,sally).
:- aunt_of(judy,sarah).
```

Note the use of ':-' to indicate negation.

7.3.5 The output

Assuming that the above listing is in a file called `aunt.pl` say, it would be given to PROGOL by typing the following on the command line.

```
progol aunt
```

Alternatively the session might be interactive in which case PROGOL would be executed with no arguments. The user could then type in modes, types and clauses or use Prolog input routines such as *consult* to read in files. See the manual page for further details. The output of executing the above command is listed in Table 7.2.

When PROGOL is given a file as input, it tries to develop a general rule for all predicates appearing in head mode declarations. In this case the only such predicate is *aunt_of*. It first ensures that the asserted clauses are consistent. In this case it reports that no contradictions were found so the input is consistent. It then constructs the most specific clause of the first positive example in the input. The first positive example is

```
aunt_of(jane,henry).
```

and its most specific clause is

```
aunt_of(A,B) :- parent_of(C,B), sister_of(A,C).
```

It then generates new clauses from this most specific clause and sees how many of the examples they prove. Here the clause

```
aunt_of(A,B).
```

explains 3 positive examples (400% inflation during positive-only learning and 3*4=12 is the second number in the list) and 11 random instances (the third element in the list). Similarly, the clause

```
aunt_of(A,B) :- parent_of(C,B).
```

Table 7.2. PROGOL output when learning the relation *aunt_of*.

```
CProgol Version 4.4

[Noise has been set to 100%]
[Example inflation has been set to 400%]
[The posonly flag has been turned ON]
[:- set(posonly)? - Time taken 0.00s]
[:- modeh(1,aunt_of(+person,+person))? - Time taken 0.00s]
[:- modeb(100,parent_of(-person,+person))? - Time taken 0.00s]
[:- modeb(100,parent_of(+person,-person))? - Time taken 0.00s]
[:- modeb(100,sister_of(+person,-person))? - Time taken 0.00s]
[Testing for contradictions]
[No contradictions found]
[Generalising aunt_of(jane,henry).]
[Most specific clause is]

aunt_of(A,B) :- parent_of(C,B), sister_of(A,C).

[Learning aunt_of/2 from positive examples]
[C:-0,12,11,0 aunt_of(A,B).]
[C:6,12,4,0 aunt_of(A,B) :- parent_of(C,B).]
[C:6,12,3,0 aunt_of(A,B) :- parent_of(C,B), sister_of(A,C).]
[C:6,12,3,0 aunt_of(A,B) :- parent_of(C,B), sister_of(A,D).]
[C:4,12,6,0 aunt_of(A,B) :- sister_of(A,C).]
[5 explored search nodes]
f=6,p=12,n=3,h=0
[Result of search is]

aunt_of(A,B) :- parent_of(C,B), sister_of(A,C).

[3 redundant clauses retracted]

aunt_of(A,B) :- parent_of(C,B), sister_of(A,C).

[Total number of clauses = 1]

[Time taken 0.02s]
```

explains the same number of positive examples but fewer (5) randomly constructed instances, and thus has a better measure of compression (6, the first number in the list). The fourth number is an optimistic estimate of the number of literals needed in the clause and the first is a measure based on the other three. See the next section and [7.8] for full details. PROGOL then adds the best generalisation to its clause base and retracts any clauses which have now been made redundant. Note that 3 clauses are retracted. Usually PROGOL would now move onto the next example and repeat this procedure. However, all examples are now covered, so PROGOL stops and shows its constructed definition.

7.4 The theory

The PROGOL system uses an approach to the general problem of ILP called mode directed inverse entailment (MDIE). In contrast to inverse resolution [7.10] and subsumption oriented approaches to induction [7.13], inverse entailment is based upon model-theory rather than resolution proof-theory. In this way a great deal of clarity and simplicity can be achieved. Furthermore by basing induction on a sounder theoretic footing, it is hoped that it is easier to develop completeness and consistency results. MDIE is a generalisation and enhancement of these previous approaches.

7.4.1 The general problem in ILP

The general problem in ILP can be summarized as follows. Given background knowledge B and examples E find the simplest consistent hypothesis H such that

$$B \wedge H \models E$$

If we rearrange the above using the law of contraposition we get the more suitable form

$$B \wedge \overline{E} \models \overline{H}$$

In general B, H and E can be arbitrary logic programs but if we restrict H and E to being single Horn clauses, $\overline{H}$ and $\overline{E}$ above will be ground skolemised unit clauses. If $\overline{\bot}$ is the conjunction of ground literals which are true in all models of $B \wedge \overline{E}$ we have

$$B \wedge \overline{E} \models \overline{\bot}$$

Since $\overline{H}$ must be true in every model of $B \wedge \overline{E}$ it must contain a subset of the ground literals in $\overline{\bot}$. Hence

$$B \wedge \overline{E} \models \overline{\bot} \models \overline{H}$$

and so

$$H \models \bot$$

A subset of the solutions for H can then be found by considering those clauses which θ-subsume $\bot$. The complete set of candidates for H could in theory be found from those clauses which imply $\bot$. As yet PROGOL does not attempt to find a fuller set of candidates (bypassing the undecidability of implication between clauses with bounds on the number of resolution steps in the Prolog interpreter). PROGOL searches the latter subset of solutions for H that *theta-subsume* $\bot$.

7.4.2 Mode declarations

In general $\bot$ can have infinite cardinality. PROGOL uses the head and body mode declarations together with other settings to build the most specific clause and hence to constrain the search for suitable hypotheses.

A mode declaration has either the form *modeh(n, atom)* or *modeb(n, atom).* where n, the recall, is an integer greater than zero or '*' and atom is a ground atom. Terms in the atom are either normal or place-marker. A normal term is either a constant or function symbol followed by a bracketed tuple of terms. A place-marker is either +type, -type or #type where type is a constant.

The recall is used to bound the number of alternative solutions for instantiating the atom. A recall of '*' indicates all solutions - in practice a large number. +type, -type, #type correspond to input variables, output variables and constants respectively.

Example mode declarations for a grammar learning problem are given in Table 7.3. The *modeh* declaration sets *s* as the target predicate, while the *modeb* declarations list the background knowledge predicates.

Table 7.3. PROGOL input file for learning grammars - settings and mode declarations.

```
% Grammar learning problem. Learns a simple English language
%       phrase grammar.

% Increase to 100 the resolution bound of the Prolog interpreter,
% i.e., the maximum number of unifications the interpreter will
% perform per call.
:- set(r,100)?

% Increase to 1000 the depth bound for the Prolog interpreter.
:- set(h,1000)?

% Set learning from positive examples only.
:- set(posonly)?

% Learn grammar rules with head s(In,Out) and body atoms
% representing determiners, prepositions, nouns, etc.
:- modeh(1,s(+wlist,-wlist))?
:- modeb(1,det(+wlist,-wlist))?
:- modeb(1,prep(+wlist,-wlist))?
:- modeb(1,noun(+wlist,-wlist))?
:- modeb(1,tverb(+wlist,-wlist))?
:- modeb(1,iverb(+wlist,-wlist))?
:- modeb(*,np(+wlist,-wlist))?
:- modeb(*,vp(+wlist,-wlist))?
```

Table 7.4. PROGOL input file for learning grammars - positive examples.

```
% Positive examples

s([the,man,walks,the,dog],[]).
s([the,dog,walks,to,the,man],[]).
s([a,dog,hits,a,ball],[]).
s([the,man,walks,in,the,house],[]).
s([the,man,hits,the,dog],[]).
s([a,ball,hits,the,dog],[]).
s([the,man,walks],[]).
s([a,ball,hits],[]).
s([every,ball,hits],[]).
s([every,dog,walks],[]).
s([every,man,walks],[]).
s([a,man,walks],[]).
s([a,small,man,walks],[]).
s([every,nice,dog,barks],[]).
```

PROGOL imposes a restriction upon the placement of input variables in hypothesised clauses. Suppose the clause is written as $h\ :-\ b_1,\ldots,b_n$ where h is the head atom and $b_i, 1 \leq i \leq n$ are the body atoms. Then every variable of +type in any atom b_i is either of +type in h or -type in some atom b_j where $1 \leq j < i$. This imposes a quasi-order on the body atoms and ensures that the clause is logically consistent in its use of input and output variables.

7.4.3 An example

In order to illustrate how PROGOL finds consistent hypotheses, the grammar learning example from Tables 7.3, 7.4, and 7.5 will be used (this is similar to the grammar example file in the CPROGOL4.4 release).

In Table 7.3, bounds on the dimensions of proofs carried out by the Prolog interpreter inside PROGOL, are set first. The resolution bound r (maximum number of unifications) is increased from its default[1] value of 400 to 1000 and the depth bound (maximum stack depth before forced backtracking) from its default of 30 to 100. Whenever the interpreter reaches a limit it gives a warning message and returns failure so the two sets of commands are just a precaution to avoid this happening.

The aim of the example is to find grammar rules which parse simple English sentences such as "The man walks the dog". The mode declarations express this and also define the input(+) and output(-) variables and the types of the parameters - in this case they all have the same type *wlist* (meaning word list). We define word lists recursively (see Table 7.5) in terms of a given set of words which appear in the examples and background knowledge.

[1] PROGOL's default settings can be found using the *settings* command when running PROGOL interactively.

Table 7.5. PROGOL input file for learning grammars - type declarations and background knowledge.

```
% Types

wlist([]).
wlist([W|Ws]) :- word(W), wlist(Ws).

word(a). word(at). word(ball). word(big). word(dog). word(every).
word(happy). word(hits). word(house). word(in). word(man).
word(nice). word(on). word(small). word(takes). word(the).
word(to). word(walks).

% Background knowledge

% The following represents the grammar rule NP -> DET NOUN
np(S1,S2) :- det(S1,S3), noun(S3,S2).
np(S1,S2) :- det(S1,S3), adj(S3,S4), noun(S4,S2).

% The following represents the grammar rule DET -> a
det([a|S],S).
det([the|S],S).
det([every|S],S).

vp(S1,S2) :- tverb(S1,S2).
vp(S1,S2) :- tverb(S1,S3), prep(S3,S2).

noun([man|S],S).
noun([dog|S],S).
noun([house|S],S).
noun([ball|S],S).

% Transitive and intransitive verbs.
tverb([hits|S],S).
tverb([takes|S],S).
tverb([walks|S],S).

iverb([barks|S],S).
iverb([hits|S],S).
iverb([takes|S],S).
iverb([walks|S],S).

prep([at|S],S).
prep([to|S],S).
prep([on|S],S).
prep([in|S],S).
prep([from|S],S).

adj([big|S],S).
adj([small|S],S).
adj([nice|S],S).
adj([happy|S],S).
```

In Table 7.4, 14 positive examples of simple English sentences are given. In practice this is a very small number but is sufficient for learning a few simple grammar rules. Table 7.5 lists the background knowledge. It contains lexical information (the predicates *adj*, *det*, *iverb*/*tverb*, and *prep* specify which words are adjectives, determiners, intransitive/transitive verbs and prepositions, respectively), as well as some grammatical knowledge defining noun phrases (*np*) and verb phrases (*vp*).

7.4.4 Construction of the most specific clause

The construction of the most specific clause $\perp$ proceeds as follows.

The first clause e to generalise is

```
s([the,man,walks,the,dog],[]).
```

From the head mode declaration

```
:- modeh(1,s(+wlist,-wlist))?
```

we have the trivial deduction

$$B \wedge \overline{e} \models \overline{s([the, man, walks, the, dog], [])}$$

From the body mode declaration

```
:- modeb(1,det(+wlist,-wlist))?
```

and replacing the input variable by [the,man,walks,the,dog] we have the deduction

$$B \wedge \overline{e} \models det([the, man, walks, the, dog], [man, walks, the, dog])$$

Note that this constructs the term [man,walks,the,dog] in place of the output variable. Using this new term with the body mode declaration

```
:- modeb(1,noun(+wlist,-wlist))?
```

and replacing the input variable by the new term we have the deduction

$$B \wedge \overline{e} \models noun([man, walks, the, dog], [walks, the, dog])$$

However, using other mode declarations in a similar way we can get the following deductions as well.

$$B \wedge \overline{e} \models np([man, walks, the, dog], [walks, the, dog])$$
$$B \wedge \overline{e} \models verb([walks, the, dog], [the, dog])$$
$$B \wedge \overline{e} \models vp([walks, the, dog], [the, dog])$$
$$B \wedge \overline{e} \models np([the, dog], [])$$

Putting these and all other similar deductions together we get

$$\begin{aligned} B \wedge \overline{e} \models\ & \overline{s([the, man, walks, the, dog], []) \wedge} \\ & det([the, man, walks, the, dog], [man, walks, the, dog]) \wedge \ldots \\ & \wedge np([the, dog], []) \end{aligned}$$

$\overline{\bot}$ is the right hand side of the above deduction. In effect a restricted minimal Herbrand model has been constructed for $B \wedge \overline{E}$, the mode declarations being used to guide the inclusion of predicates that might be of importance. To derive $\bot$, the above is first negated to give

$$\begin{aligned} & s([the, man, walks, the, dog], []) \vee \\ & \overline{det([the, man, walks, the, dog], [man, walks, the, dog])} \vee \ldots \vee \\ & \overline{np([the, dog], [])} \end{aligned}$$

and then the most specific clause can be constructed by replacing terms in the above by unique variables.

$$\begin{aligned} \bot = {} & s(A,B) \vee \overline{det(A,C)} \vee \overline{np(A,D)} \vee \overline{noun(C,D)} \vee \overline{tverb(D,E)} \vee \\ & \overline{iverb(D,E)} \vee \overline{vp(D,E)} \vee \overline{det(E,F)} \vee \overline{np(E,B)} \end{aligned}$$

ie $\bot$ is given by

```
s(A,B) :- det(A,C), np(A,D), noun(C,D), tverb(D,E),
          iverb(D,E), vp(D,E), det(E,F), np(E,B).
```

in Prolog notation (see Section 7.8 to see how to trace the construction of this clause).

7.4.5 Algorithm for constructing the most specific clause

The general algorithm for constructing the most specific clause is given in Table 7.6.

The final set $\bot$ is a disjunction of its members so the final clause will have the h constructed in step 2 as its head and the bs constructed in step 4 as atoms in its body.

Note that InTerms is not affected by variables corresponding to -type in step 3 because of the restriction mentioned earlier that an input variable in a body atom must either be an input variable in the head or an output variable in an earlier body atom.

The recall is used to determine how many times to call the Prolog interpreter for each instantiation of the clause in step 4. It may well be that the clause will succeed many times and produce many answer substitutions.

The maximum variable depth (default 3) determines how many times step 4 is executed. The Prolog interpreter inside PROGOL is also bounded in its number of resolution steps and in its depth.

Table 7.6. Algorithm for constructing $\bot$.

Let e be the clause $a\ :-\ b_1, \ldots, b_n$.
Then $\overline{e}$ is $\overline{a} \wedge b_1 \wedge \ldots \wedge b_n$.
$hash : Terms \rightarrow N$ is a function uniquely mapping terms to natural numbers.

1. Add $\overline{e}$ to the background knowledge
2. $InTerms = \emptyset$, $\bot = \emptyset$
3. Find the first head mode declaration h such that h subsumes a with substitution θ
   For each v/t in θ,
      if v corresponds to a #type, replace v in h by t
      if v corresponds to a +type or -type, replace v in h by v_k
         where v_k is the variable such that $k = hash(t)$
      If v corresponds to a +type, add t to the set InTerms.
   Add h to $\bot$.
4. For each body mode declaration b
      For every possible substitution θ of variables corresponding to +type
            by terms from the set InTerms
         Repeat recall times
            If Prolog succeeds on goal b with answer substitution θ'
               For each v/t in θ and θ'
                  If v corresponds to #type, replace v in b by t
                  otherwise replace v in b by v_k where $k = hash(t)$
                  If v corresponds to a -type, add t to the set InTerms
               Add $\overline{b}$ to $\bot$
5. Increment the variable depth
6. Goto step 4 if the maximum variable depth has been achieved.

7.4.6 Searching the subsumption lattice

Thus, for each example, the search for suitable hypotheses is limited to the bounded sub-lattice

$$\Box \preceq H \preceq \bot$$

where $\preceq$ denotes θ-subsumption and $\Box$ is the empty (false) clause. The sub-lattice has a most general element top which is $\Box$ and a most specific or least general element bottom which is $\bot$. The construction of the most specific clause $\bot$ reduces the search space significantly.

When generalising an example E relative to background knowledge B, PROGOL constructs $\bot$ and searches from general to specific through the sub-lattice of single clause hypotheses H such that $\Box \preceq H \preceq \bot$. The search space is thus bounded both above and below. The search is therefore better constrained than in other approaches to general to specific searching. For the purpose of searching this lattice of clauses ordered by θ-subsumption, PROGOL employs a refinement operator. Since $H \preceq \bot$, it must be the case that there exists a substitution θ such that $H\theta \subseteq \bot$. Thus for each literal l in H, there exists a literal l' in $\bot$ such that $l\theta = l'$. The PROGOL refinement operator has simply to keep track of θ and a list of those literals l' in $\bot$ which have a corresponding literal l in H. Any clause H which subsumes $\bot$ corresponds to a subset of the literals in $\bot$ with substitutions applied. In Progol we have the added provision that the head atom must always be included in order to obtain a sensible generalisation.

For the most specific clause in the example above, Progol starts from the empty clause

```
false :- true.
```

and first refines it to

```
s(A,B).
```

This clause is then tested to see how many of the positive and negative examples it predicts. The refinements of this clause are then considered. Progol now refines this clause by adding the first body atom det(A,C) to the clause.

```
s(A,B) :- det(A,C).
```

The new clause is then tested against the examples and possible refinements are again considered. Progol is searching a tree whose root node is □. Children of a node are the possible refinements that can be made to the clause at that node. Certain nodes will be pruned by the use of the user-defined prune statements which delimit the hypothesis space. In this example none of the subsumed clauses are accepted so Progol then backtracks to try

```
s(A,B) :- np(A,C).
```

In this example Progol has to search 21 nodes before finding the most compressive hypothesis

```
s(A,B) :- np(A,C), vp(C,D), np(D,B).
```

Note also that the refinement operator considers only variable/variable substitutions which map hypothesised clauses to subsets of $\bot$. General subsumption can be achieved when necessary by the use of the equality operator in modeb declarations. See [7.8] for further details.

7.4.7 The search algorithm

To search the subsumption lattice Progol applies an A*-like algorithm to find the clause with maximal compression. A simple outline of this algorithm is given in Table 7.7.

Table 7.7. Algorithm for searching the lattice.

Algorithm for searching $\square \preceq C \preceq \bot$

e is the example being generalised

1. $Open = \{\square\}$, $Closed = \emptyset$
2. $s = best(Open)$, $Open = Open - \{s\}$, $Closed = Closed \cup \{s\}$
3. if $prune(s)$ goto 5
4. $Open = (Open \cup refinements(s)) - Closed$
5. if $terminated(Closed, Open)$ return $best(Closed)$
6. if $Open = \emptyset$ return e (no generalisation)
7. goto 2

The algorithm calculates the following for each candidate clause s

$$p_s = number\ of\ positive\ examples\ correctly\ deducible\ from\ s$$
$$n_s = number\ of\ negative\ examples\ incorrectly\ deducible\ from\ s$$
$$c_s = length\ of\ clause\ s - 1$$
$$h_s = number\ of\ further\ atoms\ to\ complete\ the\ clause$$
$$f_s = p_s - (n_s + c_s + h_s)$$

h_s is calculated by inspecting the output variables in the clause and determining whether they have been defined. For example, the clause $s(A, B).$, would have $h_s = 3$ because it requires at least three literals from $\bot$ to construct a chain of atoms connecting A to B. This is found from a static analysis of $\bot$.

f_s is a measure of how well a clause s explains all the examples with preference given to shorter clauses.

The function $best(S)$ returns a clause $s \in S$ with the highest f value in S.

$prune(s)$ is true if $n_s = 0$ and $f_s > 0$. In this case it is not worth considering refinements of s as they cannot possibly do better since any refinement will add another atom to the body of the clause and so cannot have a higher value of p than s does. It also cannot improve upon n_s as the latter is zero.

$terminated(S, T)$ is true if $s = best(S)$, $n_s = 0$, $f_s > 0$ and for each t in T it is the case that $f_s \geq f_t$. In other words none of the remaining clauses nor any potential refinements of them can possibly produce a better outcome than the current one.

This algorithm is guaranteed to terminate and to return the clause (if it exists) which has both maximum explanatory power and high compression as measured by its length.

In the worst case the algorithm will consider all clauses in the subsumption ordering. For further details on bounds for the algorithm and a more theoretical discussion of the PROGOL system see [7.8].

Table 7.8. Covering algorithm.

B is the background knowledge, E the set of examples.

1. If $E = \emptyset$ return B
2. Let e be the first example in E
3. Construct clause $\perp$ for e
4. Construct clause H from $\perp$
5. Let $B = B \cup H$
6. Let $E' = \{e : e \in E \textit{ and } B \models e\}$
7. Let $E = E - E'$
8. Goto 1

7.4.8 The covering algorithm

PROGOL uses a simple set cover algorithm to deal with multiple examples. It repeatedly generalises examples in the order found in the PROGOL source file and adds the generalisation to the background knowledge. Examples which are now redundant relative to the new background knowledge are then removed. This is shown in Table 7.8.

For the grammar learning problem, only one pass through this algorithm is needed to cover all the examples. This results in the single hypothesised general clause solution.

7.4.9 The output

The output of PROGOL on the grammar learning example is given in Table 7.9. The names of the variables may not always match those in the description since PROGOL generates the name of a variable from its number. The clauses will be equivalent up to alphabetic renaming though.

Once PROGOL has determined the most specific clause $\perp$, it lists those clauses which subsume it and which might be part of any final hypothesis it reaches. For each clause the 4 integers output correspond to f, p, n and h. These are a measure of predictive power/compression, the number of positive examples explained, the number of negative examples incorrectly explained and the number of further atoms to complete the clause, respectively.

Note that f is defined as

$$f = P(p - (n + c + h))/p$$

where c is now the length of the clause and P is the total number of positive examples. The multiplication by P and division by p are just a way of normalising the measure while still leaving f a monotonically increasing function of p. The value of f output is rounded to the nearest integer. Higher values of f indicate the most promising candidate clauses for the final hypothesis.

Table 7.9. PROGOL output on the grammar learning example.

```
CProgol Version 4.4

[:- set(r,100)? - Time taken 0.00s]
[:- set(h,1000)? - Time taken 0.00s]
[Noise has been set to 100%]
[Example inflation has been set to 400%]
[The posonly flag has been turned ON]
[:- set(posonly)? - Time taken 0.00s]
[:- modeh(1,s(+wlist,-wlist))? - Time taken 0.00s]
[:- modeb(1,det(+wlist,-wlist))? - Time taken 0.00s]
[:- modeb(1,prep(+wlist,-wlist))? - Time taken 0.00s]
[:- modeb(1,noun(+wlist,-wlist))? - Time taken 0.00s]
[:- modeb(1,tverb(+wlist,-wlist))? - Time taken 0.00s]
[:- modeb(1,iverb(+wlist,-wlist))? - Time taken 0.00s]
[:- modeb(100,np(+wlist,-wlist))? - Time taken 0.00s]
[:- modeb(100,vp(+wlist,-wlist))? - Time taken 0.00s]
[Testing for contradictions]
[No contradictions found]
[Generalising s([the,man,walks,the,dog],[]).]
[Most specific clause is]

s(A,B) :- det(A,C), np(A,D), noun(C,D), tverb(D,E), iverb(D,E),
          vp(D,E), det(E,F), np(E,B).

[Learning s/2 from positive examples]
[C:-3,56,55,3 s(A,B).]
[C:42,56,9,3 s(A,B) :- det(A,C).]
[C:50,56,2,2 s(A,B) :- np(A,C).]
[C:51,52,1,1 s(A,B) :- np(A,C), tverb(C,D).]
[C:50,52,1,1 s(A,B) :- np(A,C), tverb(C,D), iverb(C,D).]
[C:50,52,1,1 s(A,B) :- np(A,C), tverb(C,D), iverb(C,E).]
[C:50,52,1,1 s(A,B) :- np(A,C), tverb(C,D), vp(C,D).]
[C:50,52,1,1 s(A,B) :- np(A,C), tverb(C,D), vp(C,E).]
[C:35,16,1,1 s(A,B) :- np(A,C), tverb(C,D), det(D,E).]
[C:39,16,1,0 s(A,B) :- np(A,C), tverb(C,D), np(D,B).]
[C:35,16,1,1 s(A,B) :- np(A,C), tverb(C,D), np(D,E).]
[C:51,56,1,1 s(A,B) :- np(A,C), iverb(C,D).]
[C:50,52,1,1 s(A,B) :- np(A,C), iverb(C,D), vp(C,D).]
[C:50,52,1,1 s(A,B) :- np(A,C), iverb(C,D), vp(C,E).]
[C:35,16,1,1 s(A,B) :- np(A,C), iverb(C,D), det(D,E).]
[C:39,16,1,0 s(A,B) :- np(A,C), iverb(C,D), np(D,B).]
[C:35,16,1,1 s(A,B) :- np(A,C), iverb(C,D), np(D,E).]
[C:51,52,1,1 s(A,B) :- np(A,C), vp(C,D).]
[C:42,24,1,1 s(A,B) :- np(A,C), vp(C,D), det(D,E).]
[C:44,24,1,0 s(A,B) :- np(A,C), vp(C,D), np(D,B).]
[C:42,24,1,1 s(A,B) :- np(A,C), vp(C,D), np(D,E).]
[21 explored search nodes]
f=44,p=24,n=1,h=0
[Result of search is]

s(A,B) :- np(A,C), vp(C,D), np(D,B).

[6 redundant clauses retracted]
```

Also note that the example illustrated is perhaps of one of the most basic kinds that PROGOL can cope with. It has no constant declarations, functions, lists, integrity constraints, recursion and has few examples and background clauses. It has been used for illustrative purposes only.

Table 7.9. Continued.

```
[Generalising s([the,man,walks],[]).]
[Most specific clause is]
s(A,B) :- det(A,C), np(A,D), noun(C,D), tverb(D,B), iverb(D,B), vp(D,B).
[Learning s/2 from positive examples]
[C:-12,48,55,2 s(A,B).]
[C:34,32,9,2 s(A,B) :- det(A,C).]
[C:46,32,2,1 s(A,B) :- det(A,C), np(A,D).]
[C:40,24,2,1 s(A,B) :- det(A,C), np(A,D), noun(C,D).]
[C:40,24,2,1 s(A,B) :- det(A,C), np(A,D), noun(C,E).]
[C:46,28,1,0 s(A,B) :- det(A,C), np(A,D), tverb(D,B).]
[C:44,28,1,1 s(A,B) :- det(A,C), np(A,D), tverb(D,E).]
[C:44,28,1,0 s(A,B) :- det(A,C), np(A,D), tverb(D,B), iverb(D,B).]
[C:44,28,1,0 s(A,B) :- det(A,C), np(A,D), tverb(D,B), iverb(D,E).]
[C:44,28,1,0 s(A,B) :- det(A,C), np(A,D), tverb(D,B), vp(D,B).]
[C:44,28,1,0 s(A,B) :- det(A,C), np(A,D), tverb(D,B), vp(D,E).]
[C:47,32,1,0 s(A,B) :- det(A,C), np(A,D), iverb(D,B).]
[C:46,32,1,1 s(A,B) :- det(A,C), np(A,D), iverb(D,E).]
[C:44,28,1,0 s(A,B) :- det(A,C), np(A,D), iverb(D,B), vp(D,B).]
[C:44,28,1,0 s(A,B) :- det(A,C), np(A,D), iverb(D,B), vp(D,E).]
[C:46,28,1,0 s(A,B) :- det(A,C), np(A,D), vp(D,B).]
[C:44,28,1,1 s(A,B) :- det(A,C), np(A,D), vp(D,E).]
[C:42,24,2,1 s(A,B) :- det(A,C), noun(C,D).]
[C:44,24,1,0 s(A,B) :- det(A,C), noun(C,D), tverb(D,B).]
[C:42,24,1,1 s(A,B) :- det(A,C), noun(C,D), tverb(D,E).]
[C:42,24,1,0 s(A,B) :- det(A,C), noun(C,D), tverb(D,B), iverb(D,B).]
[C:42,24,1,0 s(A,B) :- det(A,C), noun(C,D), tverb(D,B), iverb(D,E).]
[C:42,24,1,0 s(A,B) :- det(A,C), noun(C,D), tverb(D,B), vp(D,B).]
[C:42,24,1,0 s(A,B) :- det(A,C), noun(C,D), tverb(D,B), vp(D,E).]
[C:44,24,1,0 s(A,B) :- det(A,C), noun(C,D), iverb(D,B).]
[C:42,24,1,1 s(A,B) :- det(A,C), noun(C,D), iverb(D,E).]
[C:42,24,1,0 s(A,B) :- det(A,C), noun(C,D), iverb(D,B), vp(D,B).]
[C:42,24,1,0 s(A,B) :- det(A,C), noun(C,D), iverb(D,B), vp(D,E).]
[C:44,24,1,0 s(A,B) :- det(A,C), noun(C,D), vp(D,B).]
[C:42,24,1,1 s(A,B) :- det(A,C), noun(C,D), vp(D,E).]
[C:50,48,2,1 s(A,B) :- np(A,C).]
[C:48,28,1,0 s(A,B) :- np(A,C), tverb(C,B).]
[C:46,28,1,1 s(A,B) :- np(A,C), tverb(C,D).]
[C:46,28,1,0 s(A,B) :- np(A,C), tverb(C,B), iverb(C,B).]
[C:46,28,1,0 s(A,B) :- np(A,C), tverb(C,B), iverb(C,D).]
[C:46,28,1,0 s(A,B) :- np(A,C), tverb(C,B), vp(C,B).]
[C:46,28,1,0 s(A,B) :- np(A,C), tverb(C,B), vp(C,D).]
[C:44,28,1,0 s(A,B) :- np(A,C), tverb(C,B), iverb(C,B), vp(C,B).]
[C:44,28,1,0 s(A,B) :- np(A,C), tverb(C,B), iverb(C,B), vp(C,D).]
[C:44,28,1,0 s(A,B) :- np(A,C), tverb(C,B), iverb(C,D), vp(C,B).]
[C:44,28,1,0 s(A,B) :- np(A,C), tverb(C,B), iverb(C,D), vp(C,D).]
[C:44,28,1,0 s(A,B) :- np(A,C), tverb(C,B), iverb(C,D), vp(C,E).]
[C:49,32,1,0 s(A,B) :- np(A,C), iverb(C,B).]
[C:47,32,1,1 s(A,B) :- np(A,C), iverb(C,D).]
[C:46,28,1,0 s(A,B) :- np(A,C), iverb(C,B), vp(C,B).]
[C:46,28,1,0 s(A,B) :- np(A,C), iverb(C,B), vp(C,D).]
[C:48,28,1,0 s(A,B) :- np(A,C), vp(C,B).]
[C:50,44,1,1 s(A,B) :- np(A,C), vp(C,D).]
[48 explored search nodes]
f=49,p=32,n=1,h=0
[Result of search is]

s(A,B) :- np(A,C), iverb(C,B).

[8 redundant clauses retracted]

s(A,B) :- np(A,C), vp(C,D), np(D,B).
s(A,B) :- np(A,C), iverb(C,B).

[Total number of clauses = 2]
[Time taken 0.32s]
```

7.5 Estimating accuracy and significance

Once PROGOL has learnt a set of rules from examples, the system permits a quantifiable analysis of predictive accuracy. The methods used to achieve this vary according to the number of test examples. Suppose we are given the training file gram_trn.pl in Table 7.10: this is almost the example of the last section with a reduction in the positive examples. The types and background knowledge are the same as in the previous example and have thus been omitted. Given this reduction PROGOL fails to find the same hypothesis as before and produces some incorrect classifications.

Table 7.10. The PROGOL input file gram_trn.pl.

```
:- set(r,100)?
:- set(h,1000)?
:- set(posonly)?

% Learn grammar rules with head s(In,Out) and body atoms
% representing determiners, prepositions, nouns, etc.
:- modeh(1,s(+wlist,-wlist))?
:- modeb(1,det(+wlist,-wlist))?
:- modeb(1,prep(+wlist,-wlist))?
:- modeb(1,noun(+wlist,-wlist))?
:- modeb(1,tverb(+wlist,-wlist))?
:- modeb(1,iverb(+wlist,-wlist))?
:- modeb(*,np(+wlist,-wlist))?
:- modeb(*,vp(+wlist,-wlist))?

%%%%%%%%%%%%%%%%%%%
% Types

% ...

%%%%%%%%%%%%%%%%%%%%%%
% Background knowledge

% ...

%%%%%%%%%%%%%%%%%%%
% Positive examples

s([the,man,walks,the,dog],[]).
s([the,dog,walks,to,the,man],[]).
s([a,dog,hits,a,ball],[]).
s([the,man,walks,in,the,house],[]).

s([the,man,walks],[]).
```

Table 7.11. The PROGOL test file gram_tst.pl.

```
% Positive test examples

s([the,man,hits,the,dog],[]).
s([a,ball,hits,the,dog],[]).
s([a,ball,hits],[]).
s([every,ball,hits],[]).
s([every,dog,walks],[]).
s([every,man,walks],[]).
s([a,man,walks],[]).
s([a,small,man,walks],[]).
s([every,nice,dog,barks],[]).

% Negative test examples

:- s([every,man],[]).
:- s([a,man],[]).
:- s([a,small,man],[]).
:- s([every,nice,dog],[]).
```

Table 7.12. The output of PROGOL when testing on gram_trn.pl.

```
|- consult(gram_trn)?

|- generalise(s/2)?

s(A,B) :- np(A,C), vp(C,D), np(D,B).
s(A,B) :- np(A,C), tverb(C,B).

|- test(gram_tst)?

[False negative:]s([every,nice,dog,barks],[]).

[PREDICATE s/2]

Contingency table=     ________A________~A
                     P|        8|        0|        8
                      |(     5.5)|(    2.5)|
                    ~P|        1|        4|        5
                      |(     3.5)|(    1.5)|
                       ~~~~~~~~~~~~~~~~~~~~
                               9         4        13
[Overall accuracy= 92.31% +/- 7.39%]
[Chi-square = 5.87]
 [Without Yates correction = 9.24]
[Chi-square probability = 0.0154]
```

Once PROGOL has found a rule for *s* it can be given a set of test examples to see how accurate its predictions are. Suppose the test file `gram_tst.pl` is as given in Table 7.11. We can call PROGOL interactively by just typing `progol` on the command line and then consult the file `gram_trn.pl` to read in the definitions and examples, generalise the predicate in question and then test PROGOL's hypothesis against the test file. Note that in interactive mode we have to explicitly tell PROGOL to find a general rule. This interaction with PROGOL is given in Table 7.12.

The *consult* command is used to add the definitions and clauses to PROGOL's knowledge base. The *generalise* command then tells PROGOL to find a general rule for *s*. The argument to this command has the form *predicate/arity*. The second rule that PROGOL finds incorrectly has *tverb* instead of *iverb*. This rule is then tested using the test command which expects a filename as its sole argument. The result of testing are as follows: out of the 9 positive examples in the test all but one were correctly predicted by the rules. The 4 negative examples in the test are all correctly predicted by the rules. The contingency table above shows this - P stands for predicted by the rule, $\sim P$ not predicted, A stands for actual positive examples, $\sim A$ for negative ones. A statistical chi-square test has been applied to the data and an overall accuracy and chi-square probability calculated.

To improve accuracy, one could then try adding the test examples to the learning examples and see if the system can come up with an improved rule. In this case the test rules are a little more general and PROGOL does come up with a new rule that explains all the examples.

A further testing technique provided by the system is the leave-one-out method. When there are not many examples to learn from and test, an overall accuracy measure can be calculated by putting one example aside for testing and learning a rule from the remaining examples. This is then repeated for all the examples. The procedure to do this is to consult the relevant files and then to call the built-in procedure *leave* with an argument of the form *predicate/arity*. Note that generalise must not be used; otherwise the rule is learnt from all the examples.

Another technique that can be used is that of layered learning. In this case PROGOL is first allowed to form a general rule and then it is given a file of examples from which it tests and tries to improve its rule. A number of the examples are tested against the general rule and when the first false positive or negative is encountered PROGOL will improve its rule to take account of this exception. The process is then repeated but now the exceptions are accumulated until at least 5 clauses have been tested before PROGOL improves its rule from them. The algorithm then repeats with 25, 125, ... clauses to be tested before trying to improve the rule. This command is issued as follows.

```
|- layer(filename)?
```

7.6 Declarative bias

In the following two sub-sections we describe the main mechanisms in PROGOL that can be used to control the search[2]. These are a) integrity constraints and b) prune statements. Their effects are illustrated in Figure 7.1. Note that with integrity constraints if the hypothesis H is disallowed then so are all more general clauses. Conversely, with prune statements, if H is pruned, then so are all more specific clauses in the search. Good use of these declarative bias mechanisms (especially prune statements) can be extremely effective for reducing the amount of search performed by PROGOL.

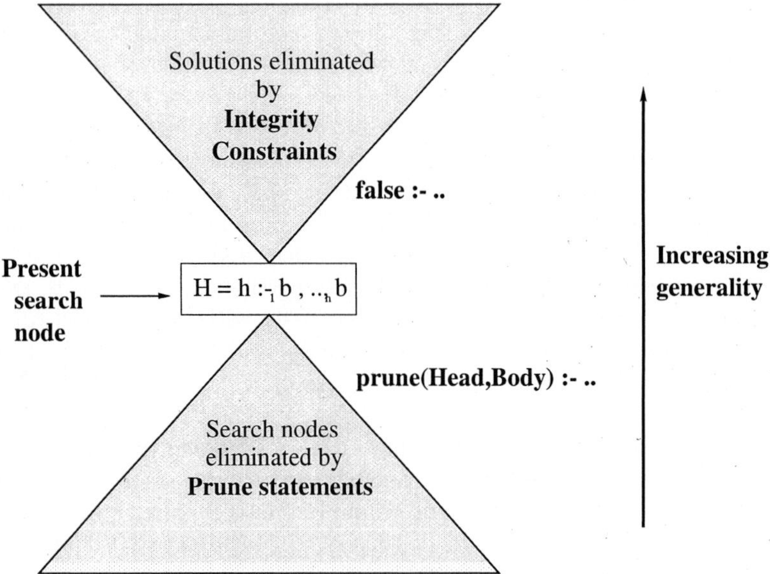

Fig. 7.1. Controlling search using integrity constraints and prune statements.

7.6.1 Integrity constraints

PROGOL allows the user to add integrity constraints into the clause base. These are represented as headless clauses, though are stored internally as clauses with head *false*. For example, if we were concerned with the predicates *person*, *male* and *female*, we might include the following clauses.

```
:- person(X), male(X), female(X).
```

This headless clause is stored internally as

[2] Section 7.7 describes how to set resource bounds which can also be used to control the search.

```
false :- person(X), male(X), female(X).
```

and can be read as saying "false is provable if there is a person who is both male and female", i.e., no person can be both male and female. If the Prolog query *false?* succeeds it means that the integrity constraints have been violated. The integrity constraints can be listed in interactive mode using the query *listing(false)?*. Note that negative examples are just a special case of integrity constraints. If PROGOL detects that a hypothesis under consideration is inconsistent with this integrity constraint then it will disallow its acceptance.

Closed world integrity constraint. A more advanced example of an integrity constraint is as follows.

```
:- hypothesis(male(X),Body,_), person(X), Body,
   not(clause(male(X),true)).
```

The PROGOL built-in predicate *hypothesis* returns the present hypothesis under consideration if one exists and fails otherwise. The three arguments of hypothesis are 1) hypothesis head, 2) hypothesis body and 3) the unique clause number of the hypothesised clause. The integrity constraint can be read as saying that "false is provable (i.e., a contradiction has been found) if a) the present hypothesis is male(X) :- Body and b) X is a person and c) Body is provable and d) male(X) is not a unit clause (positive example) in the clause base". Such an integrity constraint is often called a "Closed World Assumption", meaning that the hypotheses must not make predictions about examples which have not been seen. The use of such a closed world assumption ensures that PROGOL will not overgeneralise, which is useful for instance in the case in which no negative examples are available. However, the integrity constraint eliminates overgeneralisation by ensuring that PROGOL will not generalise at all with respect to the positive examples!

Output completeness integrity constraint. The following is a less severe integrity constraint, which has been referred to as "input completeness" [7.7]. The constraint requires that the hypothesised predicate must act as functions (i.e., $f : X \to Y$ where there is at most one $f(x) \in Y$ for every $x \in X$) with respect to the training examples. In [7.7] this type of constraint is used for learning rules for the construction of the past tense of English. Thus for every verb it is assumed that there is a unique past tense. The following PROGOL integrity constraint encodes this constraint.

```
:- hypothesis(past(X,Y),Body,_),
        clause(past(X,Z),true), Body, not(Y==Z).
```

The constraint can be read as saying "false if a) the hypothesis is past(X,Y) :- Body, and b) there is a unit clause (i.e., example) past(X,Z) and c) when the Body is proved the substitution for Y differs from Z". In other words the hypothesised clause must not make a prediction for the past tense of a verb in the example set which differs from the actual past tense of that verb.

Generative integrity constraint. The examples above illustrate the use of integrity constraints for testing semantic properties of hypotheses. It is also quite usual to use them to test syntactic properties of clauses. One such widely used syntactic property of clauses is whether they are *generative* [7.11], otherwise called *range-restricted.* A generative clause is one in which each variable appears in at least two different atoms. It can be shown that if all clauses in a logic program are range restricted then all the atoms which can be derived from it are ground. The following integrity constraint and associated definitions encode the requirement of generativeness for hypothesised clauses.

```
:- hypothesis(Head,Body,_), nongenerative((Head,Body)).

nongenerative(C) :-    % C is a comma-separated list of atoms
        in(A1,C), var(V1,A1),
        not((in(A2,C), A1\==A2, var(V2,A2), U1==V2)).

var(V,V) :- var(V).
var(V,T) :- arg(N,T,T1), N>0, var(V,T1).
```

The built-in predicate *in* tests the membership of its first argument within the comma separated list represented by its second argument. The predicate *nongenerative* succeeds if there is a variable *V1* in atom *A1* of clause *C* where *V1* does not appear in any other atom *A2* in *C*.

7.6.2 Prune statements

Integrity constraints reject certain hypotheses, but only after the PROGOL interpreter has constructed the hypothesis to be tested. If pruning is used instead, clauses can be taken out of consideration with little execution time overhead. Consider for example the following prune statement.

```
prune(Head,Body) :- Head, not(Body).
```

This statement will reject an hypothesis generated by PROGOL if it is found that with a certain substitution the Head can be proved but the Body cannot. Put another way, the prune rule says that for every instance of the head the body must be provable. This means that the target rule set consists of a single clause which covers all of the examples. Thus any hypothesised clause which does not do so can be ignored, as can all more specific clauses. In this way the entire search tree below such an hypothesis can be pruned.

Prune statements are extremely useful for stating which kinds of clause should not be considered in the search. For instance, suppose you wanted to disallow self-recursive clauses. This can be achieved using the following simple prune rule.

```
prune(Head,Body) :- in(Head,Body).
```

The prune statement eliminates any clause whose Head unifies with an atom in its Body. Similarly a single level of mutual recursion can be eliminated using the following prune statement.

```
prune(Head,Body) :-
        in(Atom,Body), clause(Atom,Body1),
        in(Head,Body1).
```

7.7 Setting resource bounds

It is necessary to put finite limits on a variety of time and space resources used within PROGOL's search in order to ensure that it terminates efficiently. This is done largely using internal settings of various named bounds. These bounds all have default values which can be viewed using the built-in command *settings?* and can be altered using the built-in commands *set* and *unset.*

Below is a list of the named resource bounds followed by their default values in brackets and a short description of each.

h (Default=30). This is the maximum depth of any proof carried out by the interpreter, i.e., the maximum stack depth before backtracking occurs. When the limit is exceeded a warning of the form *[WARNING: depth-bound failure - use set(h,..)]* is issued. This is not an error. It is necessary to have such failures when learning recursive rules.

r (Default=400). This is the maximum depth of resolutions (unifications) allowed in any proof carried out by the interpreter. Once this limit is exceeded the entire proof is failed (unlike *h* in which backtracking occurs when the bound is reached). and the warning *[WARNING: depth-bound failure - use set(r,..)]* is issued. Again this is not an error. It is necessary to have such failures when learning recursive rules.

nodes (Default=200). This is a bound on the number of search nodes expanded in the search of the lattice of clauses. According to sample complexity results in the paper [7.9] this value should be set to around 1.6 times the number of examples in the training set in order to minimise expected error.

c (Default=4). This is the maximum number of atoms in the body of any hypothesised clause. Increasing this value potentially increases the maximum clause search size exponentially.

i (Default=3). This is a bound on the number of iterations carried out in the construction of the bottom clause. Each iteration introduces new variables based on the instantiation of output terms associated with modeb declarations.

7.8 Debugging Progol input files

Note that writing background knowledge, integrity constraints and prune statements involves programming. It should by now be clear that each of these components of a Progol input file is itself a piece of Prolog code. Thus it is important to have debugging tools which allow the user to test and correct this code. The available debugging tools are essentially the same as those found in a standard Prolog interpreter (i.e., `trace` and `spy`) [7.1], though there is an additional mechanism in CProgol4.4 to test how a particular bottom clause is being constructed. This mechanism can be illustrated with reference to the worked example from Section 7.4.4. Suppose that we have loaded the file `grammar.pl` and wish to trace the construction of the most specific clause associated with the example sentence "The man walks the dog". We proceed as follows.

```
|- trace?
|- s([the,man,walks,the,dog],[])!
[Testing for contradictions]
(0) Call: false
(0) Fail: false
[No contradictions found]
```

Note the use of '!' after the example rather than '?'. This tells Progol to construct the most specific clause for this example. The first thing which Progol does is to test if the given example contradicts any integrity constraint. This is done by testing the goal `false?`, which fails, leading to the diagnosis `[No contradictions found]`. Next the *modeh* declaration is matched as follows.

```
(0) Call: s(_0,_1)=s([the,man,walks,the,dog],[])
(0) Done: s([the,man,walks,the,dog],[])=s([the,man,walks,the,dog],[])
```

The first and second arguments are found to be the terms `[the, man, walks, the, dog]` and `[]`. Next these terms are type checked as follows.

```
(0) Call: wlist([the,man,walks,the,dog]) s
(0) Done: wlist([the,man,walks,the,dog])
```

The user's response of 's' causes the tracing of the sub-proof to be skipped, leading to the result that `[the, man, walks, the, dog]` succeeds as a `wlist`. Having determined that the new term is of type `wlist`, Progol then uses the modeb declarations to test this term with respect to the background knowledge as follows.

```
(0) Call: det([the,man,walks,the,dog],_1) s
(0) Done: det([the,man,walks,the,dog],[man,walks,the,dog])
```

Here it is shown that `the` is a determiner followed by the phrase `[man, walks, the, dog]`, which in term becomes a new term which in turn will be tested using the modeb declarations and associated background predicate definitions.

As is usual with Prolog interpreters, tracing can be turned off by responding with `n`, and it is possible to leap to the next spy-point by responding with `l`. Spy-points are set on individual predicate definitions using the `spy` command. Thus turning off tracing using `notrace?` and calling `spy(det/2)?` above would ensure that the trace only started when the arity 2 predicate *det* was called. All spy points can be turned off using `nospy?`.

Notice that a command such as `s([the, man, walks, the, dog], [])!` above only leads to the construction of the bottom clause. If you want to trace the entire search through the lattice, it is necessary to first issue the command `set(searching)?`.

7.9 Summary

The design methodology for PROGOL was to present the user with a standard Prolog interpreter augmented with inductive capabilities. The syntax for examples, background knowledge and hypotheses is the Dec-10 Prolog syntax with the usual augmentable set of prefix, postfix and infix operators. Headless Horn clauses are used to represent negative examples and integrity constraints. Indeed it is possible for PROGOL to learn headless constraints from headless ground unit clauses by the use of a modeh declaration for the predicate *false*. The standard library of primitive predicates described in Clocksin and Mellish [7.1] is built into PROGOL and available as background knowledge. PROGOL constructs new clauses by generalising from the examples in the Prolog database using an inverse entailment algorithm. Results from the theory of ILP guarantee that PROGOL conducts an admissible search through the space of generalisations, finding the maximally compressive set of clauses from which all the examples can be inferred. The choice of engineering a complete Prolog interpreter was taken in order to make induction a first-class and efficient operation on the same footing as deductive theorem proving. PROGOL has been used successfully in experiments in learning to predict mutagenic molecules [7.5, 7.15], in drug design [7.4, 7.3], and in protein shape prediction [7.12, 7.16].

Acknowledgments

The first author would like to thank his wife, Thirza Castello-Cortes, and daughter Clare, who have provided unfailing support during the writing of this chapter. This work was supported partly by the Esprit Long Term Research Action ILP II (project 20237), EPSRC grant GR/K57985 on Experiments with Distribution-based Machine Learning and an EPSRC Advanced Research Fellowship held by the author.

The second author would like to thank his wife, Valmai Alysia, for much valued help in proof-reading various versions of this chapter.

References

7.1 W. F. Clocksin and C. S. Mellish. *Programming in Prolog.* Springer, Berlin, 1981.

7.2 J. Cussens, D. Page, S. Muggleton, and A. Srinivasan. Using Inductive Logic Programming for Natural Logic Processing. In *ECML'97 - Workshop Notes on Empirical Learning of Natural Language Processing Tasks*, pages 25–34. Laboratory of Intelligent Systems, Faculty of Informatics and Statistics, University of Economics, Prague, Czech Republic, 1997.

7.3 P. Finn, S. Muggleton, D. Page, and A. Srinivasan. Pharmacophore discovery using the inductive logic programming system Progol. *Machine Learning*, 30: 241–271, 1998.

7.4 R. D. King, S. Muggleton, R. Lewis, and M. J. E. Sternberg. Drug design by machine learning: The use of inductive logic programming to model the structure-activity relationships of trimethoprim analogues binding to dihydrofolate reductase. *Proceedings of the National Academy of Sciences*, 89(23): 11322–11326, 1992.

7.5 R. D. King, S. Muggleton, A. Srinivasan, and M. J. E. Sternberg. Structure-activity relationships derived by machine learning: the use of atoms and their bond connectives to predict mutagenicity by inductive logic programming. *Proceedings of the National Academy of Sciences*, 93: 438–442, 1996.

7.6 R. D. King and A. Srinivasan. Prediction of rodent carcinogenicity bioassays from molecular structure using inductive logic programming. *Environmental Health Perspectives*, 104(5): 1031–1040, 1996.

7.7 R. J. Mooney and M. E. Califf. Induction of first-order decision lists: Results on learning the past tense of English verbs. *Journal of Artificial Intelligence Research*, 3: 1–24, 1995.

7.8 S. Muggleton. Inverse entailment and Progol. *New Generation Computing*, 13: 245–286, 1995.

7.9 S. Muggleton. Learning from positive data. *Machine Learning*, 2001. Forthcoming.

7.10 S. Muggleton and W. Buntine. Machine invention of first-order predicates by inverting resolution. In *Proceedings of the Fifth International Conference on Machine Learning*, pages 339–352. Morgan Kaufmann, San Mateo, CA, 1988.

7.11 S. Muggleton and C. Feng. Efficient induction of logic programs. In S. Muggleton, editor, *Inductive Logic Programming*, pages 281–298. Academic Press, London, 1992.

7.12 S. Muggleton, R. D. King, and M. J. E. Sternberg. Protein secondary structure prediction using logic-based machine learning. *Protein Engineering*, 5(7): 647–657, 1992.

7.13 S-H. Nienhuys-Cheng and R. de Wolf. *Foundations of Inductive Logic Programming.* Springer, Berlin, 1997.

7.14 A. Srinivasan. The Aleph Manual. Technical Report, Computing Laboratory, Oxford University, 2000. Available at `http://web.comlab.ox.ac.uk/oucl/research/areas/machlearn/Aleph/`

7.15 A. Srinivasan, S. Muggleton, R. D. King, and M. J. E. Sternberg. Theories for mutagenicity: A study of first-order and feature based induction. *Artificial Intelligence*, 85(1,2): 277–299, 1996.

7.16 M. Turcotte, S. Muggleton, and M. J. E. Sternberg. Application of inductive logic programming to discover rules governing the three-dimensional topology of protein structure. In *Proceedings of the Eighth International Workshop on Inductive Logic Programming*, pages 53–64. Springer, Berlin, 1998.

8. Discovery of Relational Association Rules

Luc Dehaspe[1] and Hannu Toivonen[2]

[1] Department of Computer Science, Katholieke Universiteit Leuven
Celestijnenlaan 200A, B-3001 Leuven, Belgium

[2] Nokia Research Center
P.O. Box 407, FIN-00045 Nokia Group, Finland

Abstract

Within KDD, the discovery of frequent patterns has been studied in a variety of settings. In its simplest form, known from association rule mining, the task is to discover all frequent item sets, i.e., all combinations of items that are found in a sufficient number of examples. We present algorithms for relational association rule discovery that are well-suited for exploratory data mining. They offer the flexibility required to experiment with examples more complex than feature vectors and patterns more complex than item sets.

8.1 Introduction

Discovery of recurrent patterns in large data collections has become one of the central topics in data mining. In tasks where the goal is to uncover structure in the data and where there is no preset target concept, the discovery of relatively simple but frequently occurring patterns has shown good promise.

Association rules [8.2] are a basic example of this kind of setting. A prototypical application example is in market basket analysis: first find out which items tend to be sold together, the frequent item set discovery phase, and next postprocess these frequent patterns into rules about the conditional probability that one set of items will be in the basket given another set already there, the association rule discovery phase. The motivation for such an application is the potentially high business value of the discovered patterns. At the heart of the task is the problem of determining all combinations of items that occur frequently together, where "frequent" is defined as "exceeding a user-specified frequency threshold". The use of a frequency threshold for filtering out non-interesting patterns is natural for a large number of data mining problems. Patterns that are rare, e.g., that concern only a couple of customers, are probably not reliable nor useful for the user.

Problem settings that are close to the problem of discovering association rules include the use of item type hierarchies [8.14, 8.15, 8.31], the discovery of episodes in event sequences [8.21, 8.23], and the search of sequential patterns from series of transactions [8.4, 8.32]. For all these cases the pattern language is more complex than in the market basket application, and specialized algorithms exist for the tasks.

We present a powerful inductive logic programming algorithm, WARMR [8.11], for a large subfamily of this type of tasks. WARMR discovers frequent Prolog queries that succeed with respect to a "sufficient" number of examples. In other words, the pattern language consists of Prolog queries, and WARMR outputs those that are "frequent" in a given Prolog (or relational) database. The Prolog formulation is very general, as it allows the use of variables and multiple relations in patterns, and it thus significantly extends the expressive power of patterns that can be found.

The flexibility of WARMR is a strong advantage over previous algorithms for the discovery of frequent patterns. Each discovery task is specified to WARMR in terms of a declarative language bias definition. The language bias declarations determine which Prolog queries are admissible. With different languages (and databases) WARMR can be adapted to diverse tasks, including the settings mentioned above and also more complex novel problems, without requiring changes to the implementation. WARMR thus supports truly explorative data mining: pattern types can be modified and experimented with very flexibly with a single tool.

This chapter is organized as follows. We start in Section 2 by introducing the relational patterns and explain how they relate to the original item sets and association rules. In Section 3, we discuss two evaluation measures that are useful to filter and rank output rules. In Section 4, we introduce a formalism for specifying the class of patterns to be discovered. In Section 5, we describe WARMR, an algorithm that discovers frequent Prolog queries and relational association rules. A sample run with WARMR is shown in Section 6. Finally, in Section 7, we conclude with a brief discussion.

8.2 From association rules to query extensions

Essentially, the type of database we consider is a relational database (see, e.g., [8.13]), and the type of patterns we consider reduce to SQL queries. We say that a query / pattern matches the database if the set of tuples returned by the query is not empty.

Consider the relational database *RD* in Table 8.1 that consists of three relations.

Table 8.1. Relational database *RD* with customer information.

CUSTOMER
ID
allen
bill
carol
diana

PARENT	
IDSR	IDJR
allen	bill
allen	carol
bill	zoe
carol	diana

BUYS	
ID	ITEM
allen	wine
bill	cola
bill	pizza
diana	pizza

As an example of a pattern, let us consider SQL query SQ_1:

```
SELECT  CUSTOMER.Id, PARENT.IdJr
FROM    CUSTOMER, PARENT, BUYS
WHERE   CUSTOMER.Id = PARENT.IdSr
AND     PARENT.IdJr = BUYS.Id
AND     BUYS.Item='cola'
```

i.e., "find couples (customer,child) where the child buys cola".

We say SQL query / pattern SQ_1 matches relational database RD because there exists a tuple, that is <*allen,bill*>, in the resulting relation.

In practice we will use Prolog [8.8, 8.34] for the representation of data and patterns. More specifically, databases correspond to Prolog programs and patterns reduce to queries evaluated w.r.t. those programs. We say the pattern matches the database if the query succeeds with the program loaded.

Table 8.2. Prolog database D with customer information.

customer(allen).	*parent(allen,bill).*	*buys(allen,wine).*
customer(bill).	*parent(allen,carol).*	*buys(bill,cola).*
customer(carol).	*parent(bill,zoe).*	*buys(bill,pizza).*
customer(diana).	*parent(carol,diana).*	*buys(diana,pizza).*

Assume we are given query Q_1:

$$?\text{-}\ customer(X),\ parent(X,Y),\ buys(Y,cola),$$

and the Prolog database D shown in Table 8.2. We say query / pattern Q_1 matches database D because there exists an answer $\{X = allen,\ Y = bill\}$ for which the pattern succeeds. Observe that Prolog database D and Prolog query Q_1 are equivalent to respectively relational database RD and SQL query RQ_1.

Queries can be viewed as "relational item sets". Compare example query Q_1 to itemset:

$$\{customer,\ parent,\ child,\ cola\_buyer\}$$

The "market-basket interpretation" of this pattern would be that in the example a customer, a parent, a child, and a cola buyer occur. This is in fact also partly the meaning of query pattern Q_1. The variables X and Y in Q_1 however add extra information: customer and parent are the same (variable X), child and cola buyer are the same (variable Y), and parent and child belong to the same family. We could drop those variables and construct a query equivalent to the itemset above. This illustrates the fact that queries are indeed a more expressive variant of item sets.

Now let us consider the relational version of association rules. We introduce new terminology to describe these rules. This can be motivated by two observations. First, at least some people in the data mining community are not happy with the name *association rules* [8.20]. Second, especially in first-order logic, association rules in their traditional notation can easily be confused with Prolog clauses. Therefore we call "relational association rules" query extensions, and use a new notation.

Consider again query Q_1, and assume we have a method to compute that this query holds for 25% of our customers (we explain this method in the next section):

?- *customer*(*X*), *parent*(*X*, *Y*), *buys*(*Y*,*cola*).
FREQ: 0.25

i.e., "25% of our customers are parents of children who buy cola" (Allen).

We may also have statistics on another query Q_2:

?- *customer*(*X*), *parent*(*X*, *Y*).
FREQ: 0.75

i.e., "75% of our customers are parents" (Allen, Bill, and Carol).

In the original itemset framework we would construct from the combination of two itemsets

{*customer*,*parent*,*child*,*cola_buyer*}
FREQ: 0.25

{*customer*,*parent*,*child*}
FREQ: 0.75

an association rule

{*customer*,*parent*,*child*} → {*cola_buyer*}
FREQ: 0.25
CONF: $\frac{0.25}{0.75} = 0.33$

In a similar fashion, we would like to obtain from the combination of Prolog queries Q_1 and Q_2 a relational association rule with 33% confidence.

Let us first consider two solutions that are obvious but not correct. The first obvious choice would be to construct a Prolog clause:

buys(*Y*,*cola*) :- *customer*(*X*), *parent*(*X*, *Y*).

However, as a property of customers, this clause should be read as "if a customer has a child, then this child will buy cola". Notice Allen does not meet this rule: his child Carol does not buy cola. In fact, Diana is the only customer who does meet this rule, and then only trivially: she is not a parent. Depending on whether we want to count these trivial cases, the confidence of the clause would be 0% or 25%, but never 33% as required. Clearly, Prolog clauses do not correspond to the relational association rules we are looking for.

In a second attempt we use a format close to the original one:

$$?\text{-}\ customer(X),\ parent(X,Y). \rightarrow ?\text{-}\ buys(Y,cola).$$

The question then is how to interpret this formula. The right hand side could be interpreted as "someone buys cola", which holds for all customers, much like "the moon circles the earth".

The following is a first correct, albeit lengthy, version:

$$?\text{-}\ customer(X), parent(X,Y). \rightarrow ?\text{-}\ customer(X), parent(X,Y), buys(Y, cola).$$

i.e., "if a customer has a child, then that customer also has a child that buys cola". Notice that in this case the rule does hold for Allen: he has a child, for instance Carol, and he also has a child that buys cola, that is Bill. If we ignore the customers without children, for whom the rule holds trivially, then we find indeed that its confidence equals 33%.

Finally, we introduce a new notation for the above rule that avoids repetition of conditions, and thus preserves the brevity of the original association rules. Like the above rule, *query extension* QE_1

$$?\text{-}\ customer(X),\ parent(X,Y) \rightsquigarrow buys(Y,cola).$$
FREQ: 0.25
CONF: 0.33

should be read as "if a customer has a child, then that customer also has a child that buys cola".

Formally, a *query extension* is an existentially quantified implication

$$?\text{-}\ l_1, \ldots, l_m. \rightarrow ?\text{-}\ l_1, \ldots, l_m, l_{m+1}, \ldots, l_n.$$

with $1 \leq m < n$. By their definition, query extensions are closely connected to queries, actually they are two queries, where one is longer than *–extends–* the other. To avoid confusion with clauses (which are also implications) and shorten notation we will write this query extension as

$$?\text{-}\ \mathrm{l}_1, \ldots, \mathrm{l}_m \rightsquigarrow \mathrm{l}_{m+1}, \ldots, \mathrm{l}_n.$$

We call query

$$?\text{-}\ l_1, \ldots, l_m.$$

the *body* of the query extension, and the subquery

$$l_{m+1}, \ldots, l_n.$$

the *head* of the query extension.

It should be stressed that in the case of query extensions, the head does not correspond to the conclusion (as with clauses). Following standard terminology, we rather look at the unshortened notation, and call query

$$?\text{-}\ l_1, \ldots, l_m, l_{m+1}, \ldots, l_n.$$

the *conclusion* of the query extension.

To simplify the discussion, and circumvent the pitfalls of negation, we will restrict ourselves to so-called *range-restricted* queries, and query extensions. A *range-restricted query* is a query in which all variables that occur in negative literals also occur in at least one positive literal. A *range-restricted query extension* is a query extension such that both the body and the conclusion are range-restricted queries. The patterns below marked with * are not range-restricted (the responsible **Y** variable is shown in bold), the unmarked patterns are range-restricted.

Queries:

?- *buys(X,pizza)*, ¬*friend(X,Y)*, *buys(Y,cola)*.
*?- *buys(X,pizza)*, ¬*friend(X,***Y***)*.

Query extensions:

?- buys(X,pizza), buys(Y,cola) ⇝ ¬friend(X,Y).
*?- buys(X,pizza) ⇝ ¬friend(X,**Y**).
*?- buys(X,pizza), ¬friend(X,**Y**) ⇝ buys(X,cola).

8.3 Evaluation measures

8.3.1 Frequency and confidence

The frequency of a query extension QE is defined as the frequency of the query that makes up the conclusion of QE. The confidence of a query extension QE is defined as the frequency of the conclusion of QE divided by the frequency of the body of QE. Both evaluation metrics are copied from the association rule literature. In this section we concentrate on how to compute the frequency of queries, which is at the heart of both statistics.

Let us reconsider the following query:

Q_1: ?- *customer(X)*, *parent(X,Y)*, *buys(Y,cola)*.

Query Q_1 can be interpreted as a boolean attribute Q_1^A of customers: each customer either is or is not "parent of a cola buyer". To find out for a particular customer, we substitute the variable X in Q_1 with the customer's identifier, and evaluate against the database. For instance, query

?- *customer(allen)*, *parent(allen,Y)*, *buys(Y,cola)*.

i.e., "customer *allen* is parent of a cola buyer Y", succeeds with $Y = bill$. Hence, attribute Q_1^A has value 1 for customer *allen*. Observe substitutions of X with the three remaining customer names *bill, carol, diana* all result in failing queries, such that for these customers attribute Q_1^A has value 0.

In general, to cast a Prolog query Q as an attribute Q^A we need two things:

1. a separate relation in the database with example identifiers; in our example relation CUSTOMER with identifiers *allen,bill,carol,diana*; and
2. the atom corresponding to this relation (here *customer*(X)) has to occur in query Q.

We then define the *frequency* of a query Q as the sum of value Q^A for all examples, i.e., the number of example identifiers for which query Q succeeds. Since we usually report the relative frequencies, we divide this number by the total number of examples.

Given the above interpretation of conjunctive queries as boolean attributes, and given four conjunctive queries with the obligatory *customer*(X) atom:

Q_1: ?- *customer*(X), *parent*(X,Y), *buys*(Y,*cola*).
Q_2: ?- *customer*(X), *parent*(X,Y).
Q_3: ?- *customer*(X), *parent*(X,Y), *buys*(Y,*wine*).
Q_4: ?- *customer*(X), *buys*(X,Y).

Table 8.3 contains the attribute-value description with frequencies, of our Prolog database D.

Table 8.3. Attribute-value description, with frequencies, of Prolog database D.

CUSTOMER_AVL				
ID	Q_1^A	Q_2^A	Q_3^A	Q_4^A
allen	1	1	0	1
bill	0	1	0	1
carol	0	1	0	0
diana	0	0	0	1
FREQ	0.25	0.75	0	0.75

In our framework the obligatory *customer*(X) atom is essential, as it determines *what* is counted. Each binding of the variables in this atom uniquely identifies an entity. Therefore, we call *customer*(X) the *key* parameter of the relational pattern discovery task. This parameter has been absent from previous formulations of the frequent pattern discovery task because it has always followed unambiguously from the application context.

The original absence of the *key* parameter can be traced back to the limited knowledge representation framework in which frequent pattern discovery has traditionally been casted. Reconsider the canonical market basket problem. There is no confusion possible about the object of counting: market baskets (also called *transactions*). Compare this to our customers database. We can do something similar and count baskets. But we might as well focus on customers – as we have done above – or even children of customers. With the *key* parameter we can change this focus while leaving the other inputs, in particular the database, untouched.

Before we move on to the a third evaluation measure, let us once more establish the link with relational database terminology. In the SQL syntax the absolute frequency of Q with respect to a relational database can be obtained with the following query, inspired by [8.18, 8.5]:

```
SELECT  count(distinct *)
FROM    SELECT  fields that correspond to the variables in key
        FROM    relations in Q
        WHERE   conditions expressed in Q
```

For instance, for query Q_1, the following SQL query would return absolute frequency 1.

```
SELECT  count(distinct *)
FROM    SELECT  CUSTOMER.Id
        FROM    CUSTOMER, PARENT, BUYS
        WHERE   CUSTOMER.Id = PARENT.IdSr
        AND     PARENT.IdJr = BUYS.Id
        AND     BUYS.Item='cola'
```

8.3.2 Deviation

We sometimes use a third quality criterion for query extensions which is less common in the association rule literature: the *deviation* label on a query extension indicates the unusualness of the dependency between the body and the conclusion. Essentially, we measure the statistical significance of the confidence with one-tailed test using the overall frequency of the head as the null hypothesis. In the rest of this section we explain this evaluation measure in more detail.

Consider the following query extension QE_2:

?- *customer*(X), *parent*(X, Y) $\rightsquigarrow$ *buys*(X,*pizza*).
FREQ: 0.25
CONF: 0.75

It is not clear from these statistics whether there is anything special about customers that are parents. It might well be that, overall, about three quarters of the customers buy pizza. In that case, the rule QE_2 may not be worth noticing[1]. If on the other hand on average as much as 99% or as few as 1% of the customers turn out to buy pizza, customers-parents to deviate from the expected as they buy unusually less or more pizza[2].

Various deviation measures have been proposed in a data mining context, e.g., in EXPLORA [8.16, 8.17] and MIDOSindexmidos@MIDOS [8.38]. We use the binomial distribution to measure the deviation of the confidence of a rule from the mean.

Assume an experiment that consists of N trials, where in each trial some event either happens or fails to happen. Let p be the probability that the event will happen in a single trial, and $q = 1 - p$ be the probability that the event will fail to happen in a single trial. We can then compute the probability that the event will happen in exactly X trials (and fail to happen in $N - X$ trials) with the following probability function:

$$bino\,(p, N, X) = \binom{N}{X} p^X q^{N-X}$$

Since $\sum_{X=0}^{N} bino\,(p, N, X) = 1$, function *bino* defines a discrete probability distribution, called the *binomial distribution* (or *Bernoulli distribution*) for X.

Given N and p, the probability that an event will happen in Y or *more* trials is $\sum_{X=Y}^{N} bino\,(p, N, X)$. If Y is greater then mean pN, this probability is less than 0.5, and we say result Y *deviates positively* from mean pN. The closer probability $\sum_{X=Y}^{N} bino\,(p, N, X)$ approaches 0, the higher the positive deviation.

Given N and p, the probability that an event will happen in Y or *less* trials is $\sum_{X=0}^{Y} bino\,(p, N, X)$. If Y is less then mean pN, this probability is less than 0.5, and we say result Y *deviates negatively* from mean pN. The closer probability $\sum_{X=0}^{Y} bino\,(p, N, X)$ approaches 0, the higher the negative deviation.

We now translate these definitions to the present context. The experiment consists in repeatedly drawing an object matched by the body of the query extension where N is the absolute frequency of the body. Following [8.16, 8.17], we call this set the *subgroup*. In QE_2, the subgroup would be the customers with kids, and since there are 3 of them, $N = 3$.

Again following [8.16, 8.17], we call the set of objects matched by the head of the query extension the *target group*. Each time we draw an element from the subgroup, this element can either be in or out the target group. The probability p that event "in target group" will happen in a single trial "draw from subgroup" is assumed to be identical to the probability that event "in target group" will happen in a single trial "draw from total population".

[1] A difference of say 71%-75% may still be "important" if the number of observed orders is very large. Our deviation measure should (and will) account for this.

[2] The inverse of the previous comment holds here. If only say four customers are observed, even a difference 33%-75% may be nonessential.

That is, p is set to the frequency of the head of the query extension. In QE_2 this is the fraction of customers that buy pizza, which happens to be 0.5.

We then observe the result of the experiment. The number of successful trials corresponds to the number of elements in the intersection of subgroup and target group, i.e., to the absolute frequency of the query extension. In QE_2, there is one customer that has kids and buys pizza.

Finally, we can compare the outcome of the experiment with the assumed mean and measure to what extent it deviates (positively or negatively). In QE_2, we expect $0.5 * 3 = 1.5$ elements in the intersection. Obviously, our observed result 1 deviates negatively. We quantify this deviation as follows: if we draw 3 elements from the set of all customers, then the probability that this selection will contain 1 or less customers who buy pizza is $\sum_{X=0}^{1} bino\,(0.5, 3, X) = 0.5$. This means that, if we repeat the experiment 100 times, a result of 1 or less is bound to come up in about 50 cases. In other words, rule QE_2 is not at all remarkable.

It should be stressed at this point that it is necessary to be very careful with any statistical interpretation of the deviation test. In statistical decision theory, the 0.5 deviation would be interpreted as the significance level of the hypothesis that the proportion of target group elements differs in the subgroup and the total population. The problem is that in frequent query discovery we look at many target groups. Due to what statisticians call the *multiplicity effect* [8.29], some of them are bound to come up seemingly "significant". Therefore we use the deviation measure rather as a ranking criterion. Its purpose is to filter uninteresting patterns, rather than to corroborate interesting ones. The closer the deviation is to 0.5 the less interesting it is.

8.4 Declarative language bias

The representation of patterns as Prolog queries requires a formalism to constrain the query language $\mathcal{L}$ to a set of meaningful and useful patterns. With association rules the definition of $\mathcal{L}$ is straightforward: $\mathcal{L}$ is simply 2^I, the collection of all subsets of the set I of items. Srikant, Vu, and Agrawal [8.33] describe a technique to impose and exploit user-defined constraints on combinations of items, but otherwise the definition of $\mathcal{L}$ has received little attention in the frequent pattern discovery literature. In inductive logic programming, on the other hand, this issue has been studied extensively in the subfield of declarative language bias. This is motivated by huge, often infinite, search spaces, that require a tight specification of patterns worth considering. Several formalisms have been proposed for adding language bias information in a declarative manner to the search process (for an overview, see [8.1, 8.27]). Our formalism, WARMODE, is an adaptation to WARMR of the RMODE format developed for TILDE [8.6] which, in turn, is based on the formalism originally developed for PROGOL [8.25].

We will demonstrate later how the WARMODE notation can be used to constrain $\mathcal{L}$ to some interesting classes of patterns. The required subset of WARMODE is described below.

8.4.1 The WARMODE basics

Let us first look at the simple case where $\mathcal{L}$ contains no variables, i.e., only ground queries are allowed. Under these circumstances, the WARMODE notation extends the straightforward $\mathcal{L} = 2^I$ bias to Prolog queries: given a set *Atoms* of ground atoms, the language $\mathcal{L}$ consists of 2^{Atoms}, i.e., of all possible combinations of the atoms. For example,

Atoms = {*parent(a,b), buys(a,cola)*}

defines $\mathcal{L} = 2^{Atoms} =$

{ *?- true.*
?- parent(a,b).
?- buys(a,cola).
?- parent(a,b),buys(a,cola).}

When variables are allowed in $\mathcal{L}$, the power set idea can be extended to a set of literals, as done, for instance, in [8.36] and [8.10]. However, this solution is inconvenient for two reasons. First, we might want to define infinite languages. For instance, in our running examples, we might want to go up the family tree and allow queries

?- parent(X_1,X_2)*, parent*(X_2,X_3)*, parent*(X_3,X_4)*, ...*

of arbitrary length. Thereto, an atom with predicate symbol in *Atoms* (e.g., *parent*) should be allowed several times in the query and not just once as in 2^{Atoms}. Second, we do not want to control the exact names of the variables in the query, as we do for constants, but rather the sharing of names between variables. For example, query

?- rectangle(Width,Height), Width<Height.

is equivalent to query

?- rectangle(X,Y), X<Y.

but not to query

?- rectangle(Width,Height),X<Y.

In the WARMODE framework, non-ground atoms in *Atoms* are allowed to occur multiple times in the query, as long as their variables obey the so-called *mode constraints.* These are declared for each variable argument of each atom by means of three *mode-labels* +, −, and ±:

+ the variable is strictly *input*, i.e., bound before the atom is called
− the variable is strictly *output*, i.e., bound by the atom
± the variable can be both input and/or output

In our approach the atoms of the query are evaluated one by one, following an ordering that is consistent with the mode declarations. The intuition is that the evaluation of some atoms, such as $X < Y$ in the example above, presupposes the binding of certain *input* variables to a constant. On the other hand some atoms, e.g., *rectangle(Width,Height)*, are allowed or required to introduce new *output* variables bound during evaluation of the atom. A query is then mode conform if an ordering of atoms exists such that every input variable occurs in one of the previous atoms, and no output variable does. Some examples of queries that are consistent and inconsistent with mode declarations are listed in Table 8.4. Throughout this chapter we use the notational convention that the atoms of a query are evaluated from left to right.

Table 8.4. Examples of WARMODE definitions and queries (dis)allowed in the corresponding languages.

WARMODE definition *Atoms*	example pattern $\in \mathcal{L}$	example pattern $\notin \mathcal{L}$
$\{p(-,-),q(-)\}$	?- $p(X,Y),q(Z)$	?- $p(X,Y),q(Y)$
$\{p(+,-),q(-)\}$	?- $q(X),p(X,Y),p(Y,Z)$	?- $p(X,Y)$
$\{p(+,\pm),q(-)\}$	?- $q(X),p(X,X),p(X,Y)$	?- $q(X),p(Y,X)$
$\{p(\pm,a),q(\pm)\}$	?- $p(X,a),q(X),q(Z),p(Z,a)$	?- $p(X,Y)$

8.4.2 Typing in WARMODE

Additional constraints on the sharing of variable names can be imposed via type declarations. The WARMODE convention is to append these to the mode declarations in *Atoms*. A query is then type conform if and only if arguments that share a variable name either have identical types or at least one of the arguments is untyped. For example, with mode and type declarations

$$Atoms = \{parent(-p,-c),buys(+c,cola)\},$$

query

?- *parent(X,Y),buys(Y,cola).*

is in $\mathcal{L}$, but not

?- *parent(X,Y),buys(X,cola).*

since the first arguments of predicates *parent* and *buys* have incompatible types p, c. Notice the difference between constants (*cola*) and types (p,c) in declarations *Atoms*: types are preceded by a mode label and constants are not.

8.4.3 The WARMODE key

As we have seen, the frequent query discovery task requires the specification of a *key* atom which is obligatory in all queries. Within WARMODE notation this is done with *key* = *KeyAtom*, where *KeyAtom* is a mode and type declaration as defined above. Obviously, the key atom declaration should not contain any + mode labels.

For an example of a non-ground language, consider

key = *parent(–,–)*

and

Atoms = {*buys(+,cola)*}.

These declarations together define $\mathcal{L}$ =

{ *?- parent(X,Y).*
?- parent(X,Y), buys(X,cola).
?- parent(X,Y), buys(Y,cola).
?- parent(X,Y), buys(X,cola), buys(X,cola).
?- parent(X,Y), buys(X,cola), buys(Y,cola).
...}.

Notice the fourth query in the above list is logically redundant.

8.4.4 Logical redundancy and WARMODE

The WARMODE notation is only meant to capture application-specific constraints on $\mathcal{L}$. These are usually supplemented by a number of general constraints hardwired in the data mining algorithm in which WARMODE is embedded. Logical non-redundancy is such an application-neutral and algorithm-specific constraining principle. For instance, the last WARMODE definition in Table 8.4 allows

?- p(X,a), q(X), p(Y,a), q(Y), q(Y).

which is logically equivalent to the shorter

?- p(X,a),q(X).

Within WARMR the first pattern would be filtered away in the candidate generation phase, as explained below in Section 8.5.2.

8.5 Query (extension) discovery with WARMR

Design of algorithms for frequent pattern discovery has turned out to be a popular topic in data mining (for a sample of algorithms, see [8.2, 8.3, 8.19,

8.30, 8.35]). Almost all algorithms are on some level based on the same idea of levelwise search, known in data mining from the APRIORI algorithm [8.3]. We first review the generic levelwise search method and its central properties and then introduce the algorithm WARMR [8.10] for finding frequent queries. To conclude this section, we recall how this method fits in the two-phased discovery of frequent and confident rules.

8.5.1 The levelwise algorithm

The levelwise algorithm [8.22] is based on a breadth-first search in the lattice spanned by a specialization relation $\preceq$ between patterns, cf. [8.24], where $p1 \preceq p2$ denotes "pattern $p1$ is more general than pattern $p2$", or "$p2$ is more specific than pattern $p1$".

The method looks at a level of the lattice at a time, starting from the most general patterns. The method iterates between candidate generation and candidate evaluation phases: in *candidate generation*, the lattice structure is used for pruning non-frequent patterns from the next level; in the *candidate evaluation* phase, frequencies of candidates are computed with respect to the database. Pruning is based on monotonicity of $\preceq$ with respect to frequency: if a pattern is not frequent then none of its specializations is frequent. So while generating candidates for the next level, all the patterns that are specializations of infrequent patterns can be pruned. For instance, in the APRIORI algorithm for frequent itemsets, candidates are generated such that all their subsets (i.e., generalizations) are frequent.

The levelwise approach has two crucial useful properties:

- Assuming all candidates of a level are tested in single database pass, the database is scanned at most $d + 1$ times, where d is the maximum level (size) of a frequent pattern. This is an important factor when mining large databases.
- The time complexity is in practice linear in the product of the size of the result times the number of examples, assuming matching patterns against the data is fast.

8.5.2 The WARMR algorithm

The algorithm in Table 8.5, steps (5–10), shows WARMR's main loop as an iteration of candidate evaluation in step (6) and candidate generation in step (9). The manipulation of set $\mathcal{I}$ of infrequent queries in steps (3) and (7) is necessary for the generation phase. This is discussed below, together with some other features that distinguish WARMR from APRIORI.

Specialization relation. The subset specialization relation used in most frequent pattern discovery settings can in some restricted cases also be used for structuring a space of Prolog queries, as done in [8.36, 8.37] and [8.10]. In general however, the subset condition is too strong. For instance, we would like to consider query

Table 8.5. Algorithm WARMR .

`Inputs:`	Database **r**; WARMODE language $\mathcal{L}$ and *key* ; threshold *minfreq*
`Outputs:`	All queries $Q \in \mathcal{L}$ with frequency $\geq$ *minfreq*

1. Initialize level $d := 1$
2. Initialize the set of candidate queries $\mathcal{Q}_1 := \{$*?- key*$\}$
3. Initialize the set of infrequent queries $\mathcal{I} := \emptyset$
4. Initialize the set of frequent queries $\mathcal{F} := \emptyset$
5. While $\mathcal{Q}_d$ not empty
6.   Find frequency of all queries $Q \in \mathcal{Q}_d$ using WARMR -EVAL
7.   Move the queries $\in \mathcal{Q}_d$ with frequency below *minfreq* to $\mathcal{I}$
8.   Update $\mathcal{F} := \mathcal{F} \cup \mathcal{Q}_d$
9.   Compute new candidates $\mathcal{Q}_{d+1}$ from $\mathcal{Q}_d$, $\mathcal{F}$ and $\mathcal{I}$ using WARMR -GEN
10.   Increment d
11. Return $\mathcal{F}$

?- likes(Z,Z),buys(Z,cola).

as a specialization of query

?- likes(X,Y),likes(Y,X).

although the first query is not a subset of the second.

The most obvious general-purpose definition of the subsumption relation is based on logical implication: $Query1 \preceq Query2$ if and only if $Query2 \models Query1$. Logical implication could detect for instance that *?- likes(X,Y), likes(Y,X)* is a generalization of *?- likes(Z,Z),buys(Z,cola)*. However, due to the high computational cost of the logical implication check, inductive logic programming algorithms often rely on a stronger variant coined θ-subsumption by Plotkin [8.28][3]. $Query1$ θ-subsumes $Query2$ (abbreviated to: $Query1$ ts $Query2$) if and only if there exists a (possibly empty) substitution of the variables of $Query2$, such that every atom of the resulting query occurs in $Query1$, i.e., $Query1 \supseteq Query2\theta$. For instance,

?- likes(Z,Z),buys(Z,cola). ts *?- likes(X,Y),likes(Y,X).*

with $\theta = \{X/Z, Y/Z\}$.

For all queries Q_g and Q_s:

$$Q_s \supseteq Q_g \quad \begin{array}{|c|}\hline \not\Leftarrow \\ \Rightarrow \\ \hline\end{array} \quad Q_s \text{ ts } Q_g$$

$$Q_s \text{ ts } Q_g \quad \begin{array}{|c|}\hline \not\Leftarrow \\ \Rightarrow \\ \hline\end{array} \quad Q_s \models Q_g$$

Consider for instance queries:

---

[3] Plotkin's definition as used everywhere else in the ILP literature, including other chapters in this book, applies to clauses, i.e., disjunctions. We have adapted it to queries, which are conjunctions. More details on this adaptation can be found in [8.9], pp. 80–83.

?- likes(Z,Z),buys(Z,cola).	$\supseteq$ ts $\models$	*?- likes(Z,Z).*
?- likes(Z,Z),buys(Z,cola).	$\not\supseteq$ ts $\models$	*?- likes(X,Y),likes(Y,X).*
?- rich(X), not rich(f(f(X))).	$\not\supseteq$ t̸s $\models$	*?- rich(X), not rich(f(X)).*

The examples above illustrate θ-subsumption is weaker than the subset relation, but stronger than logical implication. To understand the third example it helps to interpret $f(X)$ as father of X. Then the formula says that if there is rich person whose grandfather is not rich, it follows that there is a rich person (the same as before or his/her father) whose father is not rich.

Candidate evaluation. Algorithm WARMR -EVAL presented in Table 8.6 adapts our earlier notion of frequency of a single query Q to the levelwise approach, which matches a set of patterns against one example at a time. The example is here represented by θ_k, the substitution for the key variables obtained in step (2) by running *?- key* against the database. The algorithm, in step (2.b), applies a fixed substitution θ_k to the subsequent queries Q_j drawn from $\mathcal{Q}$, and increments an associated counter q_j in case $Q_j\theta_k$ succeeds with respect to the database.

If we execute the latter evaluation with respect to $\mathbf{r}$, we still need one pass through the database per query, instead of one pass per level. The solution adopted in WARMR -EVAL, step (2.a), is based on the assumption that there exists a relatively small subset $\mathbf{r}_k$ of $\mathbf{r}$, such that the evaluation of any $Q\theta_k$ only involves tuples from $\mathbf{r}_k$. Readers familiar with relational database technology might notice a similar assumption underlies the definition of a cluster index. In many cases $\{\mathbf{r}_k\}$ is a partition on $\mathbf{r}$. The algorithm then makes a single pass through the data in the sense that the key values θ_k are retrieved one by one, the subsequent subdatabases $\mathbf{r}_k$ are activated once in (2.a), and all queries are evaluated locally with respect to $\mathbf{r}_k$ in (2.b). An experimental evaluation of this localization of information in a related data mining task can be found in [8.7].

Consider as an example our database D with customer information. Each subset $\mathbf{r}_k$ of this database would contain a fact *customer(cid$_k$)* $\leftarrow$ and zero or more facts *parent(wid$_k$,atype)* $\leftarrow$. This subset of the database indeed suffices for solving queries $Q\theta_k$ built with predicates *customer* and *parent*. But what about the facts of the form *buys(person, item)* $\leftarrow$? Since *person* might refer both to parent or child, these facts are relevant for many keys wid_k and involved in solving queries in many examples θ_k. As a consequence, they cannot be assigned to one $\mathbf{r}_k$ exclusively. The solution is to either duplicate

Table 8.6. Algorithm WARMR -EVAL .

`Inputs:`	Database **r**; set of queries $\mathcal{Q}$; WARMODE *key*
`Outputs:`	The frequencies of queries $\mathcal{Q}$

1. For each query $Q_j \in \mathcal{Q}$, initialize frequency counter $q_j := 0$
2. For each answer θ_k resulting from the execution of query *?- key* w.r.t. database **r**:
   (a) Isolate the relevant fraction of the database $\mathbf{r}_k \subseteq \mathbf{r}$
   (b) For each query $Q_j \in \mathcal{Q}$, do the following:
   If query $Q_j\theta_k$ succeeds w.r.t. $\mathbf{r}_k$, increment counter q_j

the information, or put these facts in a separate section of the database which is always activated.

Candidate generation. To generate candidates, WARMR -GEN presented in Table 8.7 employs at step (2) a classical specialization operator under θ-subsumption [8.28, 8.26]. A specialization operator ρ maps queries $\in \mathcal{L}$ onto sets of queries $\in 2^{\mathcal{L}}$, such that for any $Query1$ and $\forall\, Query2 \in \rho(Query1)$, $Query1$ θ-subsumes $Query2$. The operator used in WARMR -GEN essentially adds one atom to the query at a time, as allowed by WARMODE declarations.

Mode and type declarations on variables may cause an atom to be added for the first time only deep down the lattice. For instance, in our running example, atom *buys(Y,cola)* should only be added once *parent(X,Y)* is in Q. This complicates pruning significantly. We can no longer require that all generalizations of a candidate are frequent as some of the generalizations, such as

?- customer(X),buys(Y,cola).

might simply not be in the language of admissible patterns. Instead, WARMR -GEN at step (2.i) requires candidates not to be θ-subsumed by any infrequent query. In step (2.ii), we also require that candidates and frequent queries are mutually inequivalent under θ-subsumption. This way a potentially huge set of redundant solutions is eliminated.

Table 8.7. Algorithm WARMR -GEN .

`Inputs:`	WARMODE language $\mathcal{L}$; infrequent queries $\mathcal{I}$; frequent queries $\mathcal{F}$; frequent queries $\mathcal{Q}_d$ for level d
`Outputs:`	Candidate queries $\mathcal{Q}_{d+1}$ for level $d+1$

1. Initialize $\mathcal{Q}_{d+1} := \emptyset$
2. For each query $Q_j \in \mathcal{Q}_d$, and for each immediate specialization $Q'_j \in \mathcal{L}$ of Q_j:
   Add Q'_j to $\mathcal{Q}_{d+1}$, unless:
   (i) Q'_j is more specific than some query $\in \mathcal{I}$, or
   (ii) Q'_j is equivalent to some query $\in \mathcal{Q}_{d+1} \cup \mathcal{F}$

8.5.3 Two-phased discovery of frequent and confident rules

Frequent patterns are commonly not considered useful for presentation to the user as such. Their popularity is mainly based on the fact that they can be efficiently post-processed into rules that exceed given confidence and frequency threshold values. The best known example of this two-phased strategy is the discovery of association rules [8.2], and closely related patterns include episodes [8.23] and sequential patterns [8.4]. For all these patterns, the threshold values offer a natural way of pruning weak and rare rules.

As observed in [8.2] for association rules, confident and frequent query extensions can be found effectively in two steps. In the first step one determines the set of all frequent queries, and in the second produces query extensions whose confidence exceeds the given threshold. As explained above, query extensions are in fact couples of queries where one extends the other. The construction of query extensions therefore simply consists of finding all such couples in the list of frequent queries. Formally, we look for couples of queries (B, C) such that conclusion C θ-subsumes body B. As an example, reconsider the combination of two queries (Q_2, Q_1):

Q_2: ?- *customer*(X), *parent*(X,Y).
FREQ: 0.75
Q_1: ?- *customer*(X), *parent*(X,Y), *buys*(Y,*cola*).
FREQ: 0.25

where Q_1 extends/θ-subsumes Q_2, to query extension QE_1:

?- *customer*(X), *parent*(X,Y) $\leadsto$ *buys*(Y,*cola*).
FREQ: 0.25
CONF: 0.33

If all frequent queries and their frequencies are known as a result of the first step, then this easy second step is guaranteed to output all frequent and confident query extensions. This is illustrated with a sample run of WARMR on the customer database in the next section.

8.6 A sample run

In this section we show WARMR at work on the *customer* example used throughout this chapter.

This application is defined by two input files: the data and the pattern language. The Prolog database for this application is presented in Table 8.2, the language bias declarations in Table 8.8. Frequency and confidence thresholds are set to 10%, the default values in WARMR.

Given these inputs, WARMR outputs 25 frequent queries. These are all listed, with their frequency in Table 8.9. Notice we have omitted the *customer(A)* prefix to save space.

Table 8.8. Input file with declarative language bias settings in WARMODE format.

```
warmode_key(customer(-)).
warmode(parent(+,-)).
warmode(buys(+,cola)).
warmode(buys(+,pizza)).
warmode(buys(+,wine)).
```

Table 8.9. All queries in the *customer* domain with frequency above 0.10 (prefix *?- customer(A)* omitted).

frequent query: ?- customer(A),...	freq
parent(A,B).	0.75
buys(A,cola).	0.25
buys(A,pizza).	0.50
buys(A,wine).	0.25
parent(A,B),parent(B,C).	0.25
parent(A,B),buys(B,cola).	0.25
parent(A,B),buys(A,cola).	0.25
parent(A,B),buys(B,pizza).	0.50
parent(A,B),buys(A,pizza).	0.25
parent(A,B),buys(A,wine).	0.25
buys(A,cola),buys(A,pizza).	0.25
parent(A,B),parent(B,C),buys(B,cola).	0.25
parent(A,B),parent(B,C),buys(C,pizza).	0.25
parent(A,B),parent(B,C),buys(B,pizza).	0.25
parent(A,B),parent(B,C),buys(A,wine).	0.25
parent(A,B),buys(B,cola),buys(B,pizza).	0.25
parent(A,B),buys(B,cola),buys(A,wine).	0.25
parent(A,B),buys(A,cola),buys(A,pizza).	0.25
parent(A,B),buys(B,pizza),buys(A,wine).	0.25
parent(A,B),parent(B,C),buys(B,cola),buys(B,pizza).	0.25
parent(A,B),parent(B,C),buys(B,cola),buys(A,wine).	0.25
parent(A,B),parent(B,C),buys(C,pizza),buys(A,wine).	0.25
parent(A,B),parent(B,C),buys(B,pizza),buys(A,wine).	0.25
parent(A,B),buys(B,cola),buys(B,pizza),buys(A,wine).	0.25
parent(A,B),parent(B,C),buys(B,cola),buys(B,pizza),buys(A,wine).	0.25

In the second phase, these 25 queries are used to generate relational association rules, or rather, query extensions. In total, WARMR outputs 93 rules with frequency and confidence above the 10% threshold. Table 8.10 lists the 35 query extensions with two literals in the body: the omitted *customer(A)* and one of *parent(A,B)*, *buys(A,cola)*, *buys(A,pizza)* or *buys(A,wine)*.

Table 8.10 only shows frequency and confidence, not the deviation measure introduced above. Since the rules are discovered on a very small data sample, none of them are interesting according to this measure. This example does illustrate the importance of an additional ranking criterion: even with such a small database and language, about 100 rules are generated. In cases where this list gets much longer, association rules (relational or not) can no

Table 8.10. All query extension in the *customer* domain with 2-literal bodies and with frequency and confidence above 0.10 (prefix *?- customer(A)* omitted).

frequent and confident query extension: ?- customer(A),...	**freq**	**conf**
parent(A,B) ⤳ parent(B,C).	0.25	0.33
parent(A,B) ⤳ buys(B,cola).	0.25	0.33
parent(A,B) ⤳ buys(A,cola).	0.25	0.33
parent(A,B) ⤳ buys(B,pizza).	0.5	0.67
parent(A,B) ⤳ buys(A,pizza).	0.25	0.33
parent(A,B) ⤳ buys(A,wine).	0.25	0.33
parent(A,B) ⤳ parent(B,C),buys(B,cola).	0.25	0.33
parent(A,B) ⤳ parent(B,C),buys(C,pizza).	0.25	0.33
parent(A,B) ⤳ parent(B,C),buys(B,pizza).	0.25	0.33
parent(A,B) ⤳ parent(B,C),buys(A,wine).	0.25	0.33
parent(A,B) ⤳ buys(B,cola),buys(B,pizza).	0.25	0.33
parent(A,B) ⤳ buys(B,cola),buys(A,wine).	0.25	0.33
parent(A,B) ⤳ buys(A,cola),buys(A,pizza).	0.25	0.33
parent(A,B) ⤳ buys(B,pizza),buys(A,wine).	0.25	0.33
parent(A,B) ⤳ parent(B,C),buys(B,cola),buys(B,pizza).	0.25	0.33
parent(A,B) ⤳ parent(B,C),buys(B,cola),buys(A,wine).	0.25	0.33
parent(A,B) ⤳ parent(B,C),buys(C,pizza),buys(A,wine).	0.25	0.33
parent(A,B) ⤳ parent(B,C),buys(B,pizza),buys(A,wine).	0.25	0.33
parent(A,B) ⤳ buys(B,cola),buys(B,pizza),buys(A,wine).	0.25	0.33
parent(A,B) ⤳ parent(B,C),buys(B,cola),buys(B,pizza),buys(A,wine).	0.25	0.33
buys(A,cola) ⤳ parent(A,B).	0.25	1.00
buys(A,cola) ⤳ buys(A,pizza).	0.25	1.00
buys(A,cola) ⤳ parent(A,B),buys(A,pizza).	0.25	1.00
buys(A,pizza) ⤳ parent(A,B).	0.25	0.50
buys(A,pizza) ⤳ buys(A,cola).	0.25	0.50
buys(A,pizza) ⤳ parent(A,B),buys(A,cola).	0.25	0.50
buys(A,wine) ⤳ parent(A,B).	0.25	1.00
buys(A,wine) ⤳ parent(A,B),parent(B,C).	0.25	1.00
buys(A,wine) ⤳ parent(A,B),buys(B,cola).	0.25	1.00
buys(A,wine) ⤳ parent(A,B),buys(B,pizza).	0.25	1.00
buys(A,wine) ⤳ parent(A,B),parent(B,C),buys(B,cola).	0.25	1.00
buys(A,wine) ⤳ parent(A,B),parent(B,C),buys(C,pizza).	0.25	1.00
buys(A,wine) ⤳ parent(A,B),parent(B,C),buys(B,pizza).	0.25	1.00
buys(A,wine) ⤳ parent(A,B),buys(B,cola),buys(B,pizza).	0.25	1.00
buys(A,wine) ⤳ parent(A,B),parent(B,C),buys(B,cola),buys(B,pizza).	0.25	1.00

longer be considered summaries of the data, unless they are presented to the user in some reasonable order.

8.7 Discussion

We have presented a general Prolog formulation of the frequent pattern discovery problem: find out which Prolog queries in a user-defined language succeed frequently in a given database. These queries can then be combined into relational association rules, or rather query extensions, with frequency and confidence above the user-specified threshold. We outlined WARMODE, a declarative formalism for specifying the language bias, i.e., the search space $\mathcal{L}$ of admissible or potentially interesting Prolog queries. We also gave an algorithm, WARMR, for solving such tasks, and presented a sample run of WARMR on a small application.

In this chapter we have used a toy application which has the advantage that all examples can be verified by the reader. The scientific and commercial potential of WARMR and frequent query discovery is demonstrated in [8.12, 8.11, 8.9] where applications are discussed in telecommunications and biochemistry. In both cases the network structure of the data calls for the additional expressive power offered by WARMR.

WARMR, which is available for academic purposes upon request, is a flexible tool that can be used by both users and developers as an explorative data mining tool: pattern types can be modified in a flexible way, and thus a number of frequent pattern discovery settings can be easily experimented with without changes in the implementation. It is not inconceivable that for any setting addressed with WARMR, specialized algorithms can be developed that will outperform WARMR by several orders of magnitude. This is for instance the case with the market basked application and the APRIORI algorithm designed specifically for this task. However, a generic tool such as WARMR is complementary to these specialized algorithms and offers several advantages both to users and developers.

To the users WARMR offers mainly two types of flexibility. First, the user can jump from one frequent pattern discovery setting to another with just minor changes to the language bias and background knowledge. Individual pattern types that turn out to be of particular interest can then be mined in a second stage with specialized algorithms. An additional danger with using a specialized algorithm as a first approach is that any information which cannot be used within this method is bound to be ignored or even cut away in a preprocessing step.

A second type of flexibility comes with the possibility to add background knowledge. Background knowledge has at least two functions in the process of knowledge discovery in databases: it can be used to (1) add information in the form of general rules, but also (2) to change with minor effort the view on the data, without going through the typically laborious preprocessing of the raw data themselves. Again, once the experiments converge on some specific setting, efficiency can be cranked up by reorganization of the data into some very specific format.

On the other hand, for the developers of specialized algorithms WARMR can function as a benchmark, and as a verification/validation method: the special algorithm should run significantly faster, and produce the same output.

Acknowledgments

At the time of writing this chapter, Luc Dehaspe was Fellow of the K.U.Leuven Research Council. Hannu T.T. Toivonen was at Rolf Nevanlinna Institute, University of Helsinki, while this work was done. He was supported by the Academy of Finland.

The authors are grateful to Luc De Raedt and Heikki Mannila for comments on the paper and for many fundamental ideas and discussions, to Hendrik Blockeel, Bart Demoen, and Wim Van Laer for their share in the implementation of WARMR.

References

8.1 H. Adé, L. De Raedt, and M. Bruynooghe. Declarative bias for specific-to-general ILP systems. *Machine Learning*, 20(1/2): 119–154, 1995.

8.2 R. Agrawal, T. Imielinski, and A. Swami. Mining association rules between sets of items in large databases. In *Proceedings of the ACM SIGMOD International Conference on Management of Data*, pages 207–216. ACM Press, New York, 1993.

8.3 R. Agrawal, H. Mannila, R. Srikant, H. Toivonen, and A. I. Verkamo. Fast discovery of association rules. In U. Fayyad, G. Piatetsky-Shapiro, P. Smyth, and R. Uthurusamy, editors, *Advances in Knowledge Discovery and Data Mining*, pages 307–328. AAAI Press, Menlo Park, CA, 1996.

8.4 R. Agrawal and R. Srikant. Mining sequential patterns. In *Proceedings of the Eleventh International Conference on Data Engineering*, pages 3–14. IEEE Computer Society Press, Los Alamitos, CA, 1995.

8.5 H. Blockeel and L. De Raedt. Relational knowledge discovery in databases. In *Proceedings of the Sixth International Workshop on Inductive Logic Programming*, pages 199–212. Springer, Berlin, 1996.

8.6 H. Blockeel and L. De Raedt. Top-down induction of first order logical decision trees. *Artificial Intelligence*, 101(1-2): 285–297, 1998.

8.7 H. Blockeel, L. De Raedt, N. Jacobs, and B. Demoen. Scaling up inductive logic programming by learning from interpretations. *Data Mining and Knowledge Discovery*, 3(1): 59–93, 1999.

8.8 I. Bratko. *Prolog Programming for Artificial Intelligence*, 2nd edition. Addison-Wesley, Wokingham, England, 1990.

8.9 L. Dehaspe. *Frequent Pattern Discovery in First-Order Logic.* PhD thesis. Department of Computer Science, Katholieke Universiteit Leuven, Belgium, 1998. Available at `http://www.cs.kuleuven.ac.be/~ldh/`.

8.10 L. Dehaspe and L. De Raedt. Mining association rules in multiple relations. In *Proceedings of the Seventh International Workshop on Inductive Logic Programming*, pages 125–132. Springer, Berlin, 1997.

8.11 L. Dehaspe and H. Toivonen. Discovery of frequent datalog patterns. *Data Mining and Knowledge Discovery*, 3(1): 7–36, 1999.

8.12 L. Dehaspe, H. Toivonen, and R. D. King. Finding frequent substructures in chemical compounds. In *Proceedings of the Fourth International Conference on Knowledge Discovery and Data Mining*, pages 30–36. AAAI Press, Menlo Park, CA, 1998.

8.13 R. Elmasri and S. B. Navathe. *Fundamentals of Database Systems*, 2nd edition. Benjamin/Cummings, Redwood City, CA, 1989.

8.14 J. Han and Y. Fu. Discovery of multiple-level association rules from large databases. In *Proceedings of the Twenty-first International Conference on Very Large Data Bases*, pages 420–431. Morgan Kaufmann, San Mateo, CA, 1995.

8.15 M. Holsheimer, M. Kersten, H. Mannila, and H. Toivonen. A perspective on databases and data mining. In *Proceedings of the First International*

Conference on Knowledge Discovery and Data Mining, pages 150–155. AAAI Press, Menlo Park, CA, 1995.

8.16 W. Kloesgen. Problems for knowledge discovery in databases and their treatment in the statistics interpreter EXPLORA. *International Journal of Intelligent Systems*, 7(7): 649–673, 1992.

8.17 W. Kloesgen. EXPLORA: A multipattern and multistrategy discovery assistant. In U. Fayyad, G. Piatetsky-Shapiro, P. Smyth, and R. Uthurusamy, editors, *Advances in Knowledge Discovery and Data Mining*, pages 249–271. AAAI Press, Menlo Park, CA, 1996.

8.18 G. Lindner and K. Morik. Coupling a relational learning algorithm with a database system. In *Proceedings of the MLnet Familiarization Workshop on Statistics, Machine Learning and Knowledge Discovery in Databases*, FORTH, Heraklion, Greece, 1995.

8.19 H. Lu, R. Setiono, and H. Liu. Neurorule: A connectionist approach to data mining. In *Proceedings of the Twenty-first International Conference on Very Large Data Bases*, pages 478–489. Morgan Kaufmann, San Mateo, CA, 1995.

8.20 H. Mannila. Database methods for data mining. Tutorial notes, Fourth International Conference on Knowledge Discovery and Data Mining. Technical report, AAAI Press, Menlo Park, CA, 1998.

8.21 H. Mannila and H. Toivonen. Discovering generalized episodes using minimal occurrences. In *Proceedings of the Second International Conference on Knowledge Discovery and Data Mining*, pages 146–151. AAAI Press, Menlo Park, CA, 1996.

8.22 H. Mannila and H. Toivonen. Levelwise search and borders of theories in knowledge discovery. *Data Mining and Knowledge Discovery*, 1(3): 241–258, 1997.

8.23 H. Mannila, H. Toivonen, and A. I. Verkamo. Discovery of frequent episodes in event sequences. *Data Mining and Knowledge Discovery*, 1(3): 259–289, 1997.

8.24 T. Mitchell. Generalization as search. *Artificial Intelligence*, 18: 203–226, 1982.

8.25 S. Muggleton. Inverse entailment and Progol. *New Generation Computing*, 13, 1995.

8.26 S. Muggleton and L. De Raedt. Inductive logic programming : Theory and methods. *Journal of Logic Programming*, 19, 20: 629–679, 1994.

8.27 C. Nédellec, H. Adé, F. Bergadano, and B. Tausend. Declarative bias in ILP. In L. De Raedt, editor, *Advances in Inductive Logic Programming*, pages 82–103. IOS Press, Amsterdam, 1996.

8.28 G. Plotkin. A note on inductive generalization. In *Machine Intelligence*, pages 153–163. Edinburgh University Press, Edinburgh, 1970.

8.29 S. L. Salzberg. On comparing classifiers: pitfalls to avoid and a recommended approach. *Data Mining and Knowledge Discovery*, 1(3): 317–328, 1997.

8.30 A. Savasere, E. Omiecinski, and S. Navathe. An efficient algorithm for mining association rules in large databases. In *Proceedings of the Twenty-first International Conference on Very Large Data Bases*, pages 432–444. Morgan Kaufmann, San Mateo, CA, 1995.

8.31 R. Srikant and R. Agrawal. Mining generalized association rules. In *Proceedings of the Twenty-first International Conference on Very Large Data Bases*, pages 407–419. Morgan Kaufmann, San Mateo, CA, 1995.

8.32 R. Srikant and R. Agrawal. Mining sequential patterns: Generalizations and performance improvements. In *Proceedings of the Fifth International Conference on Extending Database Technology*, pages 3–17. Springer, Berlin, 1996.

8.33 R. Srikant, Q. Vu, and R. Agrawal. Mining association rules with item constraints. In *Proceedings of the Third International Conference on Knowledge Discovery and Data Mining*, pages 67–73. AAAI Press, Menlo Park, CA, 1997.

8.34 L. Sterling and E. Shapiro. *The art of Prolog.* MIT Press, Cambridge, MA, 1986.

8.35 H. Toivonen. Sampling large databases for association rules. In *Proceedings of the Twenty-second International Conference on Very Large Data Bases*, pages 134–145. Morgan Kaufmann, San Mateo, CA, 1996.

8.36 I. Weber. Discovery of first-order regularities in a relational database using offline candidate determination. In *Proceedings of the Seventh International Workshop on Inductive Logic Programming*, pages 288–295. Springer, Berlin, 1997.

8.37 I. Weber. A declarative language bias for levelwise search of first-order regularities. In *Proceedings of Fachgruppentreffen Maschinelles Lernen.* Technischer Bericht 98/11, Technische Universität, Berlin, 1998. `http://www.informatik.uni-stuttgart.de/ifi/is/Personen/Irene/fgml98.ps.gz`.

8.38 S. Wrobel. An algorithm for multi-relational discovery of subgroups. In *Proceedings of the First European Symposium on Principles of Data Mining and Knowledge Discovery*, pages 78–87. Springer, Berlin, 1997.

9. Distance Based Approaches to Relational Learning and Clustering

Mathias Kirsten[1], Stefan Wrobel[2], and Tamás Horváth[1]

[1] German National Research Center for Information Technology, GMD - AiS.KD Schloß Birlinghoven, D-53754 Sankt Augustin, Germany

[2] School of Computer Science, IWS, University of Magdeburg Universitätsplatz 2, D-39016 Magdeburg, Germany

Abstract

Within data analysis, *distance-based* methods have always been very popular. Such methods assume that it is possible to compute for each pair of objects in a domain their mutual distance (or similarity). In a distance-based setting, many of the tasks usually considered in data mining can be carried out in a surprisingly simple yet powerful way. In this chapter, we give a tutorial introduction to the use of distance-based methods for relational representations, concentrating in particular on predictive learning and clustering. We describe in detail one relational distance measure that has proven very successful in applications, and introduce three systems that actually carry out relational distance-based learning and clustering: RIBL2, RDBC and FORC. We also present a detailed case study of how these three systems were applied to a domain from molecular biology.

9.1 Introduction

Within data analysis, *distance-based* methods have always been very popular [9.23]. The central assumption of distance-based methods is that it is possible, for a particular domain under consideration, to specify for each pair of objects their mutual distance (or similarity). In certain settings, these distance values can be assumed to be given as part of the application, e.g., when humans are asked to rate distances between pairs of choices as done in certain psychological experiments. More often, however, in data analysis we are starting with objects described in a certain representation language (our *instance space* X) and use appropriate algorithms to compute distances for these objects.

Once we have such distances between object pairs, many of the tasks usually considered in data mining can be carried out in a disarmingly simple yet powerful way.

9.1.1 Predictive learning

Let us first consider how it is possible to perform predictive learning from examples given only distances between pairs of objects. Instead of forming an explicit hypothesis in the form of first-order clauses, we simply store all available examples. Later, when we are asked to classify an unseen object, we compare this new object to all previously seen objects, and predict the function value associated with the *nearest neighbor* or our query object, i.e., the example object that has the smallest distance to our query object.

Considering only one object makes this algorithm very sensitive to noise, so generally, the method of choice is k-nearest neighbor, where the k nearest objects get to vote by majority on the prediction. Since this method of learning is strongly based on the instances available as examples, it is often referred to as *instance-based learning.* Many authors also call it *lazy learning*, emphasizing the fact that the learner does not bother to explicitly form a hypothesis, but lazily waits until an object to be classified comes along, and only then forms a (local, query-specific) hypothesis.

9.1.2 Clustering

The task of cluster analysis is, roughly speaking, the task of grouping a set of instances $\mathcal{I}$ from an instance space X into different subsets (clusters) such that objects belonging to the same cluster are maximally similar to each other while simultaneously being maximally dissimilar to objects from other clusters. Finding a clustering that is guaranteed to be optimal in terms of a chosen quality measure is an infeasible task, as it would require an exhaustive search through the space of all possible groupings of objects.

Hence, the available distance-based clustering algorithms use heuristic strategies, two of the more prominent of which are bottom-up *agglomerative* clustering strategies, and iterative *k-means* clustering . Hierarchical agglomerative clustering (c.f. [9.3]) constructs a hierarchy of nested clusterings by repeatedly merging the two existing clusters that are closest to each other, starting with all the individual instances as one-element clusters. The *k-means* method [9.15], on the other hand, first randomly selects k cluster centers, then repeatedly (a) assigns each instance to the cluster of the nearest cluster center, and (b) computes the new cluster center for each of the k current clusters, until a termination criterion is satisfied.

In this chapter, we give a tutorial introduction into the use of the above-described distance-based methods for learning and clustering. As a key component, in Section 9.2, we first explain what are the special challenges in defining a distance for first-order representations, and show in detail one first-order distance measure that has proven very successful in applications. We then turn to three systems that actually carry out first-order distance-based learning and clustering: RIBL2 (Section 9.3) is an instance-based learner, while RDBC and FORC (Sections 9.4 and 9.5) are agglomerative and k-means clustering systems, respectively. In each of the system's descriptions, we also provide more detail as to the workings of the chosen basic method, and discuss the particular adaptations to first-order problems. Finally, in Section 9.6, we present a detailed case study of how these three systems were applied to a domain from molecular biology. We conclude with a few general remarks on the use of distance-based methods in first-order data mining.

9.2 A first-order distance measure

In this section we give a description of the first-order distance measure used in our distance-based multi-relational algorithms RIBL2, RDBC, and FORC. This measure was developed and first used for the instance-based learning system RIBL2 [9.2, 9.8] and has since also become the basis of RDBC [9.12] and FORC [9.13]. This measure, an extended variant of the one originally introduced in [9.5], based on [9.1] has proven very successful in several applications, see, e.g., [9.4, 9.7]. For alternative first-order distance measures the reader is referred to, e.g., [9.9, 9.16, 9.17, 9.18, 9.19]. The description below is informal, for an algorithmical definition of our distance measure and for its detailed discussion the reader is referred to [9.8].

9.2.1 First-order instances and their representations

In our algorithms, first-order *instances* are ground atoms of a distinguished predicate symbol and are represented as sets of ground atoms about the instances, called first-order *cases*. To illustrate these notions, we give a running example that will be used throughout this section.

Example 9.2.1. Suppose we want to describe prepackaged culinary present sets from different vendors in a relational language. Assume that about each set, we know its price and delivery mode (personal, mail, or pick-up). In addition, we know the contents of each package which for simplicity we assume consist of different red wines and cheeses. We also know something about different vineyards.

The *instances* in our example correspond to the culinary present sets, and are given in our relational language by ground atoms of a ternary predicate P. For a particular culinary set, the three arguments of its corresponding P-atom define the set's identifier, price, and delivery mode, respectively. Each culinary set (or instance) is then further described by specifying the cheeses and wines it contains. In our example, cheeses and wines are described by ground atoms of the quaternary predicates 'cheese' and 'wine', respectively. The four arguments of a ground 'cheese'-atom describing a cheese of a culinary present set give the set's identifier, and the type, weight, and origin of the cheese, respectively. Similarly, a wine of a given set is described by a ground atom of the quaternary predicate 'wine', where the arguments give the wine's culinary set, vineyard, vintage year, and bottle size, respectively. In our example, each vineyard is described by the quaternary predicate 'vineyard', where the arguments define the vineyard's identifier, popularity, size, and country, respectively.

Let us assume that $I = P(set1, 125, personal)$ is an instance in our example and is further described by the ground atoms

cheese($set1, camembert, 150, france$)
wine($set1, mouton, 1988, 0.75$)
wine($set1, gallo, 1995, 0.5$)
vineyard($gallo, famous, large, usa$)
vineyard($mouton, famous, small, france$) .

The set consisting of I and the above five ground atoms is the *case* of I.

Our example indicates that a *type* defining the set of possible values can be assigned to each argument of the predicate symbols. Furthermore, the case of an instance is usually not explicitly given. Instead, a set $\mathcal{B}$ of ground atoms, also referred to as *background knowledge*, is given, and for a given instance I, a subset of $\mathcal{B}$ is computed that defines (together with I) the case of I. This approach allows for a more succinct representation, as ground atoms may occur in more than one case.

More precisely, when applying one of our algorithms, the user has first to define the vocabulary of the predicate symbols to be used in the relational representation. Among these symbols there is a distinguished predicate symbol P that will be used to represent the elements of the set X of all possible *instances*, also called *instance space.* The other predicate symbols, called *background predicates*, are used to describe the user's background knowledge about the task. Each predicate symbol is associated (also by the user) with a declaration defining the types and modes of its arguments. The type of an argument is one of the basic types from {number, discrete, list, term, name}, while its mode defines whether or not the argument must be compared with corresponding arguments during the distance computation. The type 'name' is used for arguments that contain identifiers representing objects.

Example 9.2.2. The vocabulary of Example 9.2.1 consists of P and the background predicates 'cheese', 'wine', and 'vineyard' that are associated with the following declaration:

$P(a_1 : \text{name},\ a_2 : \text{number},\ a_3 : \text{discrete})$
cheese(a_1 : name, a_2 : discrete, a_3 : number, a_4 : discrete)
wine(a_1 : name, a_2 : discrete, a_3 : number, a_4 : number)
vineyard(a_1 : name, a_2 : discrete, a_3 : discrete, a_4 : discrete) .

In order to compute a case of an instance $I \in X$, we rely on the instance I, the background knowledge $\mathcal{B}$, and a non-negative integer $d \geq 0$ called *depth bound.* A case of I with respect to $\mathcal{B}$ and d is defined as the largest set $\{I\} \cup \mathcal{B}'$ such that $\mathcal{B}' \subseteq \mathcal{B}$ and for every background atom $B \in \mathcal{B}'$ it holds that there are $d' < d$ and $B_0, B_1, \ldots, B_{d'-1}, B_{d'}$ satisfying

- $B_0 = I$, $B_{d'} = B$, and $B_i \in \mathcal{B}'$ for $i = 1, \ldots, d' - 1$,

– B_i and B_{i+1} contain at least one common identifier for every $i = 0, \dots, d' - 1$.

Example 9.2.3. To illustrate the above definition let us assume that the background knowledge $\mathcal{B}$ in Example 9.2.1 consists of the following facts:

% set ID:name, cheese type:discrete, weight:number, origin:discrete
cheese($set1, camembert, 150, france$)
cheese($set25, roquefort, 200, france$)
cheese($set25, ricotta, 100, italy$)

% set ID:name, vineyard ID:name, year:number, bottle size:number
wine($set1, mouton, 1988, 0.75$)
wine($set1, gallo, 1995, 0.5$)
wine($set25, mouton, 1995, 0.75$)

% vineyard ID:name, popularity:discrete, size:number, country:discrete
vineyard($gallo, famous, large, usa$)
vineyard($mouton, famous, small, france$)

Then the case defined by the instance $P(set1, 125, personal)$ with respect to $\mathcal{B}$ and $d = 2$ is given in Figure 9.1.

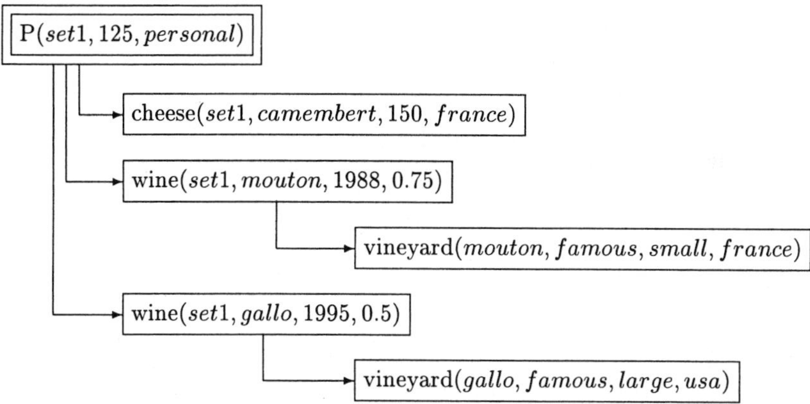

Fig. 9.1. The case defined by I_1 with respect to $\mathcal{B}$ and $d = 2$ in Example 9.2.1.

9.2.2 The distance measure

Before giving an informal description of the algorithm for computing the distance between two instances, we first give an example to illustrate how this algorithm works.

Example 9.2.4. Suppose we are given

- the instances $I_1 = P(set1, 125, personal)$, $I_2 = P(set25, 195, mail)$,
- the background knowledge $\mathcal{B}$ of Example 9.2.3,
- the depth bound $d = 2$,

and want to compute the distance between cases of I_1 and I_2 with respect to $\mathcal{B}$ and d.

This distance is computed as the normalized sum of the distances between the instances' corresponding arguments, i.e.,

$$\text{dist}(I_1, I_2, \mathcal{B}, d) = \frac{1}{3}\Big(\text{dist}(set1, set25) + \text{dist}(125, 195) + \\ + \text{dist}(personal, mail)\Big). \quad (9.1)$$

For computing the last two terms, standard propositional distance measures can be applied, i.e.,

$$\text{dist}(125, 195) = |195 - 125|/500 = 0.14 \quad (9.2)$$

(here the denominator denotes the highest possible price) and

$$\text{dist}(personal, mail) = 1 \quad (9.3)$$

(as they are different). In order to compute the distance between the object identifiers $set1$ and $set25$, from the cases of $set1$ and $set25$ we first collect the sets of literals containing $set1$ and $set25$, respectively, and partition them by the predicate symbols. That is, we obtain the sets

$$\begin{aligned} L_{set1,cheese} &= \{cheese(set1, camembert, 150, france)\} \\ L_{set1,wine} &= \{\text{wine}(set1, mouton, 1988, 0.75), \\ &\qquad \text{wine}(set1, gallo, 1995, 0.5)\} \\ L_{set25,cheese} &= \{\text{cheese}(set25, roquefort, 200, france), \\ &\qquad \text{cheese}(set25, ricotta, 100, italy)\} \\ L_{set25,wine} &= \{\text{wine}(set25, mouton, 1995, 0.75)\} \end{aligned}$$

Now the distance D_{cheese} between the sets $L_{set1,cheese}$ and $L_{set25,cheese}$ is defined by

$$D_{\text{cheese}} = \frac{\min_{l \in L_{set25,cheese}} \text{dist}\Big(\text{cheese}(set1, camembert, 150, france), l\Big)}{|L_{set25,cheese}|} \quad (9.4)$$

$$= \frac{\min\Big((1 + \frac{|150-200|}{300} + 0)/3, (1 + \frac{|150-100|}{300} + 1)/3\Big)}{2} \quad (9.5)$$

$$= 0.1944 \ .$$

We note that in (9.4) we normalize by the cardinality of the *larger* set $L_{set25,cheese}$, and by 3 the two terms in (9.5), as we do not compare again the first arguments containing the set identifiers. Applying that

$$\begin{aligned}
\text{dist}(mouton, gallo) &= \\
&= \frac{\min\left(\frac{\text{dist}(famous,famous)+\text{dist}(small,large)+\text{dist}(france,usa)}{3}\right)}{1} \\
&= 0.6666 \ ,
\end{aligned}$$

the distance D_{wine} between $L_{set1,wine}$ and $L_{set25,wine}$ is computed in a similar manner, i.e.,

$$\begin{aligned}
D_{\text{wine}} &= \frac{\min\limits_{l \in L_{set1,wine}} \text{dist}\Big(\text{wine}(set25, mouton, 1995, 0.75), l\Big)}{|L_{set1,wine}|} \\
&= \frac{\min\left((0 + \frac{|1995-1988|}{100} + 0)/3, (0.6666 + 0 + \frac{|0.75-0.5|}{1})/3\right)}{2} \\
&= 0.0117 \ .
\end{aligned}$$

Using the above values, the distance between object identifiers $set1$ and $set25$ is then computed by

$$\begin{aligned}
\text{dist}(set1, set25) &= \frac{D_{\text{cheese}} + D_{\text{wine}}}{2} \\
&= 0.1031 \ .
\end{aligned}$$

Combining (9.2), (9.3), and the above equation with (9.1), we finally obtain

$$\text{dist}(I_1, I_2, \mathcal{B}, d) = (0.1031 + 0.14 + 0)/3 = 0.081 \ .$$

As shown by the above example, the algorithm for computing the distance between two instances is given

- a vocabulary of the above form associated with an argument type and mode declaration for each of its predicate symbols,
- the two instances I_1 and I_2 from the instance space X (i.e., two P-atoms),
- background knowledge $\mathcal{B}$ consisting of background atoms,
- a depth bound $d \geq 0$ (used to control the recursion) .

I_1, I_2, and the atoms in $\mathcal{B}$ are all ground (i.e., they do not contain variables). For the two instances I_1 and I_2, the algorithm first computes their corresponding *cases* with respect to the given background knowledge $\mathcal{B}$ and the depth bound d as described in the previous subsection. To compare I_1 and I_2, the distance function recursively descends into their corresponding cases, beginning with the ground P-atoms I_1 and I_2. In general, when comparing two ground atoms, each with the same predicate symbol, one first computes the distance between their corresponding arguments that have been marked for comparison in the mode declaration, then sums up these distances, and normalizes finally this sum with the number of marked arguments. Depending on their common type, two arguments are compared as follows.

number and discrete: For arguments of these basic types, some corresponding elementary distance function is called.

list and term: For arguments of these types, our measure employs edit distances. Intuitively, with an edit distance, we are given a set of edit operations such as insert, delete, and change with an associated cost function that tells us how expensive it is to delete or insert an element of the underlying alphabet, or how expensive it is to change one into another. The edit distance is then defined as the smallest cost of a sequence of such operations that turns the first object into the second. For both of lists and terms, there are efficient algorithms for computing the edit distance (see, e.g., [9.22, 9.24]).

name: When comparing two arguments of this type, the algorithm first checks whether the depth of the recursion has already reached the input parameter d (i.e., the depth bound d is used to ensure limited effort for the recursive computation). If yes, the arguments are compared as that of type discrete. Otherwise, all further facts from the cases about the respective named objects are first collected. Each of these two sets is then partitioned by the predicate symbols, and for every predicate symbol we compare its corresponding subsets from the two sets as follows. For each fact from the smallest set we compute recursively its distance from the larger set (i.e., the distance between this fact and its nearest fact from the larger set), then sum up these distances and normalize with the cardinality of the larger set.

9.3 Instance-based learning with RIBL2

RIBL2 is a first-order instance-based learner applying the classical k-nearest neighbor (kNN) classifier in a relational representation, using the distance measure described in the preceding section. In more detail, a k-nearest neighbor algorithm determines its predictions as follows. To simplify the discussion, note that our distance measure takes on values in the range $[0,1]$, so we can equivalently work with the function

$$\mathrm{sim}(x,y) := 1 - \mathrm{dist}(x,y)$$

which is 1 for identical objects and falls towards 0 with increasing distance between objects. Given a set of examples E and a new instance x_{new} from X to be classified, let us call $N \subseteq E$ the set of k nearest neighbors of x_{new} according to *sim*. Further details depend on whether we are predicting a discrete or continuous variable. For a *discrete* variable, denote by

$$N_{y_j} := \{e = (x, y_j) \in N\}.$$

those neighbors that are voting for value y_j. The algorithm will then predict

$$h(x_{new}) := argmax_{y_j} \sum_{e=(x,y_j)\in N_{y_j}} sim(x, x_{new})^a,$$

where a is a parameter that determines how much an influence is allocated to farther-away neighbors. If we are predicting a continuous variable, we use a weighted average instead:

$$h(x_{new}) := \frac{\sum_{e=(x,y)\in N} sim(x, x_{new})^a \cdot y}{\sum_{e=(x,y)\in N} sim(x, x_{new})^a}.$$

In order to determine k, most k-nearest neighbor algorithm internally estimate the accuracy that can be achieved (e.g., by cross-validation), and then choose the value of k that has ranked highest for all future classifications. Despite this very simple principle, k-nearest neighbor methods are known to be competitive with and often superior to other approaches in terms of predictive accuracy. Thus, they are attractive choices whenever an explicit hypothesis is not needed.

9.4 Hierarchical agglomerative clustering with RDBC

Finding a clustering that is guaranteed to be optimal in terms of a chosen quality measure is an infeasible task, as it would require an exhaustive search through the space of all possible clusterings. Hence, the available distance-based clustering algorithms use heuristic strategies. The RDBC algorithm employs one of the most basic distance-based clustering methods, namely hierarchical agglomerative clustering (c.f. [9.3]).

9.4.1 Forming a cluster hierarchy

Together with the distance measure described in Section 9.2, hierarchical agglomerative clustering does not build flat partitions of the data but iteratively constructs a list of nested clusterings $\mathcal{C}_1, \ldots, \mathcal{C}_n$. Figure 9.2 shows an example together with the typical visualization of such a clustering in a *dendrogram.*

The construction of the hierarchical clustering is done in a bottom-up manner by starting with the clustering $\mathcal{C}_1$ where each case forms its own one-elemental cluster. The next clustering is built by combining the two closest respectively most similar clusters to a new cluster. In this case, the distance between two clusters is simply the distance between the two cases, as each cluster consists of a single element. For all subsequent clusterings the step of combining the two most similar clusters is repeated until finally the clustering ($\mathcal{C}_n$) consists of a single cluster only. When joining two clusters to a new one, the distances between the new cluster and all other clusters can be computed from the already known distances of the current clusters. Depending on the chosen computation rule we get different variants of the hierarchical agglomerative clustering:

$$dist(C_1 \cup C_2, C_3) := f(dist(C_1, C_3), dist(C_2, C_3)).$$

RDBC includes the four most popular variants: $f = min$, the so called *single-link* clustering, $f = max$ the *complete-link* clustering, and for $f = average$ the *average-link* clustering. A somewhat more complicated choice of f results in *Ward's method*, which guarantees that the sum of (squared) distances is minimized for each aggregation step (this does not necessarily hold for the final result though).

So, suppose we have a set of instances $I_1, \ldots, I_5$ together with their pair-wise distance information computed by RIBL's distance measure:

	I_1	I_2	I_3	I_4	I_5
I_1	0	0.12	0.60	0.64	0.77
I_2		0	0.70	0.80	0.84
I_3			0	0.35	0.45
I_4				0	0.40
I_5					0

Employing the *complete-link* variant of RDBC on this data yields the cluster hierarchy shown in Figures 9.2 and 9.3.

9.4.2 Turning the cluster hierarchy into flat partitions

The chosen agglomerative approach renders a distance-based hierarchy of cases where the distance between subtrees decreases as we move from root towards the leaves. For many applications, such a hierarchical tree is all that is required. If however, the clustering task requires the selection of a single set of clusters, we need a technique for turning the tree into such a set. Since we are trying to minimize intra-cluster distance (respectively maximize intra-cluster similarity), it is natural to base cluster selection from the tree on this criterion. RDBC therefore uses a search procedure that "cuts up" the tree according to growing intra-cluster similarity thresholds, and selects the threshold and its corresponding group of clusters that minimizes average intra-cluster distance. Thus the exact number of clusters does not have to be predefined by the user but is automatically computed according to the quality function. More precisely, in RDBC the quality of a set of clusters $\mathcal{C}$ is defined as:

$$\mathcal{C}_1 = \{(I_1), (I_2), (I_3), (I_4), (I_5)\}$$
$$\mathcal{C}_2 = \{(I_1, I_2), (I_3), (I_4), (I_5)\}$$
$$\mathcal{C}_3 = \{(I_1, I_2), (I_3, I_4), (I_5)\}$$
$$\mathcal{C}_4 = \{(I_1, I_2), (I_3, I_4, I_5)\}$$
$$\mathcal{C}_5 = \{(I_1, I_2, I_3, I_4, I_5)\}$$

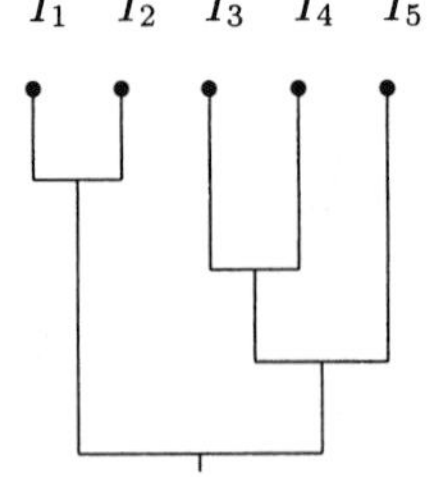

Fig. 9.2. Hierarchical clustering and its visualization in a dendrogram.

$$qual(\mathcal{C}) = \frac{\sum_{C \in \mathcal{C}} dist_{intra}(C)}{|\, \mathcal{C} \,|},$$

where $dist_{intra}(C)$ denotes the intra-cluster distance of a cluster C, which is the averaged sum of all distances between element pairs of the cluster, and is hence a measure of the cluster's compactness.

It should be noted, that the above quality measure cannot be applied directly as it reaches its maximum for a partitioning where each cluster consists of a single case. RDBC solves the problem by moving all single-element clusters into an *outlier cluster*, and the optimization of $qual(\mathcal{C})$ takes place on the resulting clustering.

So, coming back to our example involving $I_1, \ldots, I_5$, the transformation of the hierarchy into a set of partitions firstly requires to find a threshold for which the above quality function $qual(\mathcal{C})$ is maximal.

All potential thresholds are evaluated by removing those connections from the hierarchical tree structure which lay below the given threshold. Thus, the result is a clustering forest rather than a tree. A cluster (respectively a partition) is then the set of instances belonging to a single tree[1].

In the example, cutting up the hierarchy at $d = 0.5$ results in two partitions $C_1 = \{I_1, I_2\}$ and $C_2 = \{I_3, I_4, I_5\}$. The quality for this clustering is calculated as

$$qual(\{C_1, C_2\}) = \frac{dist(I_1,I_2)+(dist(I_3,I_5)+dist(I_4,I_5)+dist(I_4,I_5))/3}{2}$$
$$= 0.26$$

Finding the best threshold in the example is left as an exercise to the reader.

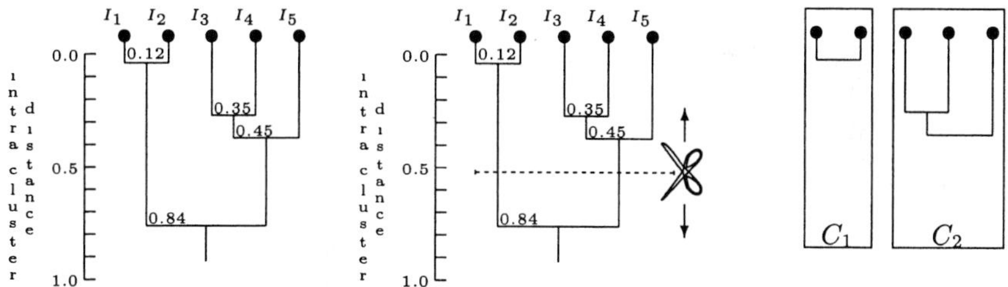

Fig. 9.3. Transforming a hierarchical clustering into flat partitions.

9.5 FORC - *k*-means for multirelational data

A very popular clustering technique for finding clusterings of a predefined size k is the *k-means* method [9.15]. FORC is the first clustering system to

[1] Isolated instances without any connections are not handled as one-elemental clusters but are gathered into the *outlier* cluster.

carry this algorithm over to first-order, relying on the distance measure of Section 9.2.

The basic idea of k-means is as follows. Firstly, an initial state has to be constructed by selecting or creating k cluster centers from X. In practice this is mostly done by randomly drawing k elements from the set of cases. With this initial state a two-step iteration is started:

1. Assign each case to the cluster of the nearest cluster center.
2. Compute the new cluster center for each of the k current cluster.
3. Go to step 1 unless clustering quality does not increase any more.

Thus, in each iteration the centers of the current clusters are determined in order to subsequently rebuild the k clusters on basis of the new centers. Figure 9.4 shows an example of this process for $X = \mathbb{R}^2$ (the corresponding cluster centers are pictured by the large circle respectively box).

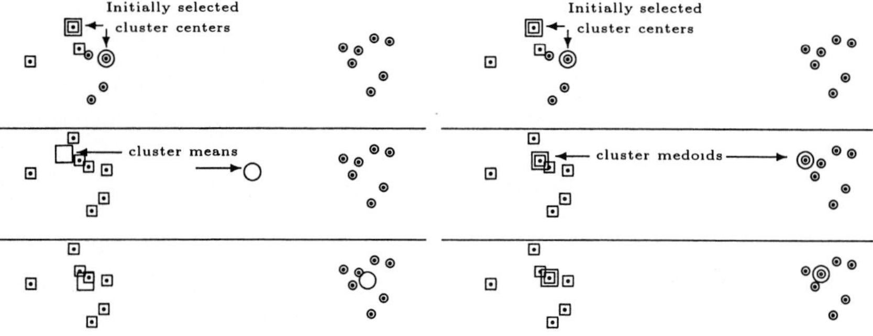

Fig. 9.4. Example for k-means and k-medoids in the $\mathbb{R}^2$ space.

Obviously, k-means requires a distance measure on the given instance space X, for which as pointed out above, the distance measure described in Section 9.2 is used. However, k-means also requires that in X the center of a set of cases can be computed, i.e., the center of a set of points is the point whose summed distances to the cases of the set is minimal. This is most easy if $X = \mathbb{R}^n$, i.e., for strictly numerical instance spaces. There the center of a set $C \subseteq X$ is simply given by its mean:

$$center(C) := \frac{1}{|C|} \sum_{c \in C} c$$

The corresponding (and usual) distance measure is the n-dimensional Euclidean metric but the Mahalanobis distance is [9.6, 9.14] also used.

For first-order data the problem is more challenging, since similarly obvious solutions for finding a center are not at hand. In fact, on multirelational data there is no longer any geometric or intuitive notion of the phrase "cluster center". In addition, an analytic solution for the construction of the cluster center does not exist, therefore we have to make do with approximation.

FORC employs a solution that alleviates the problem of how to construct a previously not existing cluster center, which is to take an existing case as an approximation for the cluster center. This should be the most central instance of the set, i.e., the instance whose sum of squared distances to all other instances in the set is minimal and which is thus closest to the (unknown) cluster center. This approach is already known as the k-medoids method [9.10, 9.11], and first results of FORC obtained on the mutagenesis data set [9.21] using RIBL's distance measure show k-medoids' promising potential on first-order data [9.13].

Obviously, the whole process relies on distance information only, and is hence applicable to any representation for which an appropriate distance measure is available (e.g., a first-order representation + RIBL's distance measure). On the downside, when making no prior assumptions about the representation used, the computational complexity suffers from the expensive search for the medoid which requires time proportional to the squared number of instances in a cluster.

9.6 A case study in mRNA signal structures

In this section we present a case study designed to illustrate how relational distance-based learning and clustering algorithms can be applied in a real-world domain: the mRNA signal structure prediction and the automatic discovery of uncharacterized mRNA signal structure classes, respectively. More detail on this domain of application, and on the results of using RIBL for predictive learning in this domain, is given in [9.8]. For further applications, see [9.4, 9.7]. In the following, we first define the prediction and clustering problems arising in the mRNA signal structure application domain, describe the relational representation used for both problems, and briefly discuss the kinds of results that can be obtained.

9.6.1 The mRNA prediction and clustering problems

During the process of protein synthesis, messenger RNA (mRNA) molecules, a special kind of RNA molecules, mediate between the sites of information storage (i.e., DNA, which is functionally inactive) and protein synthesis. In general, an RNA is a linear sequence of the four bases (adenine – A, cytosine – C, guanine – G, and uracil – U) folded and coiled into specific conformations. The three-dimensional conformation of an RNA contains local regions of short complementary base-pairing, i.e., the two stable Watson-Crick base pairs A–U, G–C, and the weak G–U. The structure of RNA molecules is usually described at the following levels of abstraction:

primary structure: It is the linear sequence of the four bases that make up an RNA.

secondary structure: It is the collection of the base pairs occurring in the RNA's three-dimensional structure.

tertiary structure: It is the three-dimensional conformation of the RNA.

In addition, for the primary structures of messenger RNA molecules it also holds that they can be divided into three parts; they consist of a coding region (in their middle) defining the amino acid sequence of the target protein to be synthesized, and two non-coding regions (in their two ends) that contain regulatory motifs. Such regulatory motifs, called *signal structures*, control gene expression at the posttranscriptional level of the protein synthesis. Depending on their regulatory roles during the process of protein synthesis, signal structures are grouped into different signal structure *classes*. In this field of research, there are (at least) the following two challenging computational tasks:

- Detect (possibly new) signal structures belonging to known signal structure classes on a given uncharacterized mRNA.
- Find uncharacterized signal structure classes.

As a first step towards the direction of finding good algorithmical solutions to the above two problems, we consider the following two related problems, respectively. (In the description below, X denotes the set of all signal structures.)

signal structure prediction task: Given a set C of signal structure classes, a set $i \subseteq X$ of characterized signal structures (i.e., each with known class) belonging to one of the classes from C, and a signal structure $I \in X$ of unknown class from C, the aim is to give a good prediction of the class of I.

automatic discovery of uncharacterized signal structure classes: In this task, we are given a set $i \subseteq X$ of uncharacterized signal structures (i.e., with unknown class information), the aim is to find a clustering $(I_1, \ldots, I_k)$ of i with k as small as possible such that each cluster is a subset of some (possibly uncharacterized) signal structure class.

Now we turn to the problem of how to represent the components of the above computational tasks. When viewing the secondary structures of mRNA molecules, one can observe a set of different patterns [9.25], like hairpin loops, internal loops, etc, that make them up (see also Figure 9.5). These patterns are also referred to as *structure elements*. Furthermore, it also holds that mRNA's secondary structures can be viewed as trees (for more details see e.g., [9.20]), where the nodes represent the structure elements. As signal structures are (continuous) subsequences of the primary structures, this observation holds for them as well.

As an example, we consider a signal structure belonging to the class Selenocysteine Insertion Sequence (SECIS). Its primary structure is

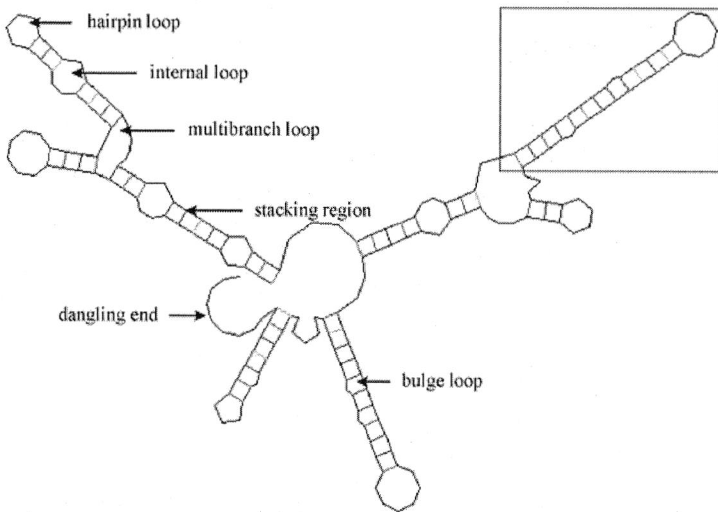

Fig. 9.5. mRNA structure with marked structure elements.

AUGACGAGCUAGGCAAUAGGCGAGACCUAGACUUAGCUCUGAG ,

and its secondary structure is given by the following complementary base pairs

```
AUGACGAGCUAGGCAAUAGGCG
 | | ||||||||   |||| A
GAGU CUCGAUUCAG AUCCAG
```

In Figure 9.6, we give the structure elements that make up this secondary structure, and also the tree structure of these structure elements. As shown in the figure, the inner nodes on the longest path from the root all correspond to the pattern 'stem'. The root itself on this path is a dummy node, i.e., it does not represent any part of the signal structure. The other endpoint of the path corresponds to the hairpin loop (i.e., CGAGA) in the signal structure. This example indicates the following description of signal structures in our relational representation language.

The instances of this application are the signal structures. Using their tree structures, a signal structure, an instance from X, is described by the binary predicate

$$\text{structure}(struct\_id, tree\_structure)$$

where the first argument of type 'name' defines its identifier, and the second argument of type 'term' is used to describe its tree structure. For instance, the signal structure given in Figure 9.6 can be described by the ground atom

$$\text{structure}(s_1, dummy(se_{1,1}, se_{1,2}(se_{1,3}, se_{1,4}(\ldots), se_{1,11}), se_{1,12}))$$

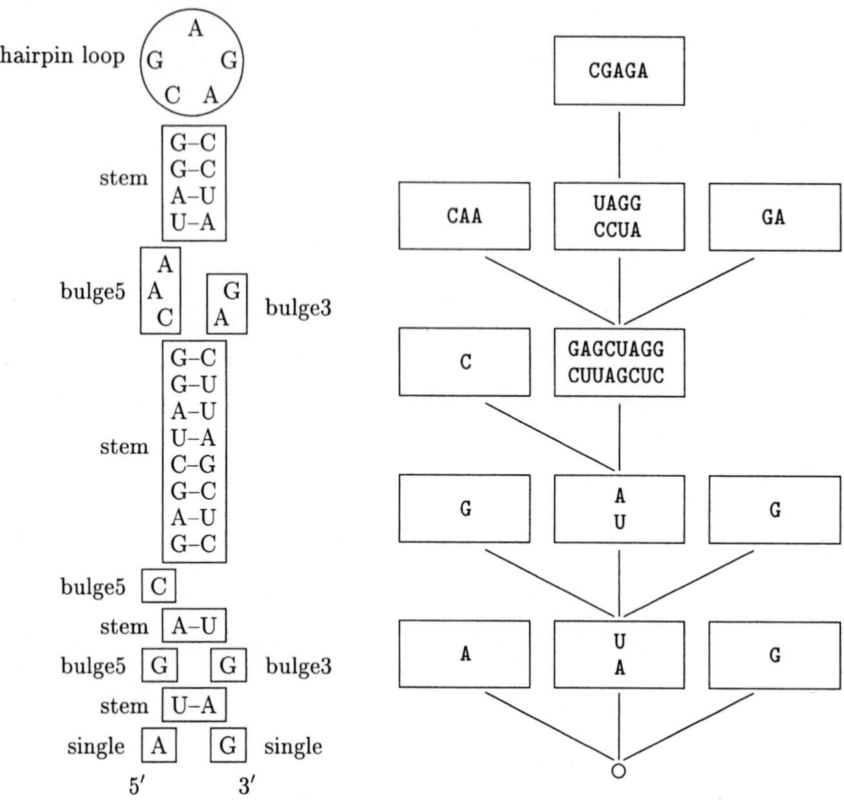

Fig. 9.6. The representation and the tree structure of a SECIS signal structure.

where 'dummy' is a constant denoting the root for each signal structure, and $se_{1,1}, \ldots, se_{1,12}$ are further identifiers denoting the structure elements of s_1. (We note that the identifiers $se_{1,1}, \ldots, se_{1,12}$ are assigned to the descendants of the root of the tree given in Figure 9.6 according to preorder tree traversal.) The structure elements of the given signal structures are further described in the background knowledge. Depending on their types, structure elements are either helical (e.g., stems) or single (e.g., hairpin loop, bulge5, bulge3). A pattern of type single is described by the background predicate

$$\text{single}(se\_id, type, sequence)$$

where the argument *se_id* of type 'name' is an identifier denoting the structure element, *type* of type 'discrete' gives the structure element's pattern, and *sequence* of type 'list' defines the primary structure of the structure element. Patterns of type helical are described by the background predicate

$$\text{helical}(se\_id, type, sequence_1, sequence_2)$$

where the arguments have similar semantics as in the case of predicate single. For instance, the structure elements of the signal structure in our example could be described by the background atoms:

$$\begin{aligned}
&\text{single}(se_{1,1}, \text{single}, [a]),\\
&\text{helical}(se_{1,2}, \text{stem}, [u], [a]),\\
&\text{single}(se_{1,3}, \text{bulge5}, [g]),\\
&\text{helical}(se_{1,4}, \text{stem}, [a], [u]),\\
&\text{single}(se_{1,5}, \text{bulge5}, [c]),\\
&\text{helical}(se_{1,6}, \text{stem}, [g, a, g, c, u, a, g, g], [c, u, u, a, g, c, u, c]),\\
&\text{single}(se_{1,7}, \text{bulge5}, [c, a, a]),\\
&\text{helical}(se_{1,8}, \text{stem}, [u, a, g, g], [c, c, u, a]),\\
&\text{single}(se_{1,9}, \text{hairpin}, [c, g, a, g, a]),\\
&\text{single}(se_{1,10}, \text{bulge3}, [g, a]),\\
&\text{single}(se_{1,11}, \text{bulge3}, [g]),\\
&\text{single}(se_{1,12}, \text{single}, [g]).
\end{aligned}$$

9.6.2 Results

For our experiments we have chosen a small domain consisting of 66 signal structures over 5 signal structure classes. For predictive learning, the optimal number of neighbors (k) has been estimated by *leave-one-out* cross-validation. The results for the different values of k are given in the following table (for more details see also [9.8]).

Table 9.1. Results for estimating the optimal number of neighbors (k).

#instances		#correctly classified for										best	accuracy
	$k=$	1	2	3	4	5	6	7	8	9	10	k	(%)
66		**63**	63	59	60	62	60	60	61	60	61	1	95.4

Clustering experiments were conducted with RDBC. Using all 66 instances for clustering, RDBC returned ten clusters (see Figure 9.7). While some of the actual classes were rediscovered by RDBC (s_tar, s_ire, s_histone, and s_secise2), the clustering of class s_secise1 appears to be hard. Whereas the mixed cluster C_8 made up of s_secise1, and s_secise2 may be caused by a folding which those instances have in common, cluster C_6 is hard to explain. In addition the misplaced instances s_secise1(5) and s_secise1(10) in cluster C_2 would better fit into cluster C_5.

To further evaluate the quality of RDBC's results, we used the computed clusters for prediction. Two experiments were conducted and the prediction accuracy was determined via a leave-one-out-approach. In both experiments class information was not used during the clustering stage. In the first experiment all instances except the test instance were available for building the clusters. The test instance's class was then determined by a majority vote over instances of the most similar of the obtained clusters (according to the distance measure). In contrast to that, for the second setup we included the test instance during clustering. As in the first experiment, we employed majority voting for determining the test instance's class[2]. Table 9.2 gives the accuracies obtained by our two experiments.

Table 9.2. RDBC's accuracies on the mRNA domain.

RDBC with test instance	RDBC without test instance
87.9%	83.3%

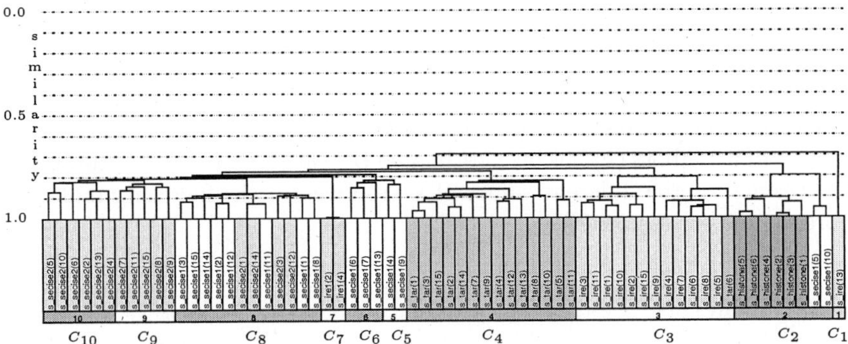

Fig. 9.7. Sample cluster hierarchy on the mRNA data. Shaded bars at the bottom mark the clusters found by RDBC.

9.7 Conclusion

First-order distance-based learning and clustering methods offer a lot of potential for applications involving complex and structured objects. For learning, distance-based "lazy" methods are known to often outperform their non-lazy counterparts; for clustering, a distance-based formulation is naturally at the heart of cluster quality. As shown in this chapter and the work on which

[2] In nine cases the majority vote resulted in a draw between two equally frequent classes. We resolved this by assigning the class of the instance most similar to our test instance. For this nearest-neighbor voting only instances of the cluster belonging to one of the two equally frequent classes were taken into account.

it is based, it is possible to define useful distance measures for first-order objects, making the strengths of distance-based methods available to first-order problems. Whenever an explicit generalization is not needed, the methods presented here are very strong contenders in terms of solution quality.

However, a few words of caution are also in order with respect to scalability. Firstly, instance-based learning methods must store all (or, when optimized variants are used, at least some) of their instances until classification time. Thus, the compression effect that can be obtained with explicit hypotheses is not achieved here. Secondly, computing distances between first-order objects is not as cheap an operation as computing a Euclidean distance (see [9.8] for details). While still faster than most generalization algorithms, this distance computation must be carried out at classification time, so such methods are not suitable if very rapid classification is a primary concern. As to the clustering methods, once the distance matrix is computed, they largely inherit the known computational properties of standard distance-based clustering algorithms; however, convergence of k-means may be somewhat slower since no exact cluster centers are computed.

In sum, all three of the methods described here are most suitable for domains with no more than several thousands of objects; more research is needed to scale them to the very large databases sometimes found in KDD.

References

9.1 G. Bisson. Conceptual Clustering in a First-Order Logic Representation. In *Proceedings of the Tenth European Conference on Artificial Intelligence*, pages 458–462. John Wiley and Sons, Chichester, 1992.

9.2 U. Bohnebeck, T. Horváth, and S. Wrobel. Term comparisons in first-order similarity measures. In *Proceedings of the Eighth International Conference on Inductive Logic Programming*, pages 65–79. Springer, Berlin, 1998.

9.3 W. Dillon and M. Goldstein. *Multivariate analysis*, pages 157–208. John Wiley and Sons, Chichester, 1984.

9.4 S. Džeroski, S. Schulze-Kremer, K. Heidtke, K. Siems, and D. Wettschereck. Diterpene structure elucidation from ^{13}C NMR spectra with machine learning. In N. Lavrač, E. Keravnou, and B. Zupan, editors, *Intelligent Data Analysis in Medicine and Pharmacology*, pages 207–225. Kluwer, Boston, 1997.

9.5 W. Emde and D. Wettschereck. Relational Instance-Based Learning. In *Proceedings of the Thirteen International Conference on Machine Learning*, pages 122–130. Morgan Kaufmann, San Francisco, CA, 1996.

9.6 J. Hartigan. *Clustering Algorithms*, pages 58–73. John Wiley and Sons, Chichester, 1975.

9.7 T. Horváth, Z. Alexin, T. Gyimóthy, and S. Wrobel. Application of different learning methods to Hungarian part-of-speech tagging. In *Proceedings of the Ninth International Workshop on Inductive Logic Programming*, pages 128–139. Springer, Berlin, 1999.

9.8 T. Horváth, S. Wrobel, and U. Bohnebeck. Relational instance-based learning with lists and terms. *Machine Learning*, 43(1/2): 53–80, 2001.

9.9 A. Hutchinson. Metrics on Terms and Clauses. In *Proceedings of the Ninth European Conference on Machine Learning*, pages 138–145. Springer, Berlin, 1997.

9.10 L. Kaufmann and P. J. Rousseeuw. Clustering by means of medoids. In Y. Dodge, editor, *Statistical Data Analysis based on the L_1 Norm*, pages 405–416. Elsevier, Amsterdam, 1987.

9.11 L. Kaufmann and P. J. Rousseeuw. *Finding Groups in Data: an Introduction to Cluster Analysis.* John Wiley and Sons, Chichester, 1990.

9.12 M. Kirsten and S. Wrobel. Relational Distance-Based Clustering. In *Proceedings of the Eighth International Conference on Inductive Logic Programming*, pages 261–270. Springer, Berlin, 1998.

9.13 M. Kirsten and S. Wrobel. Extending k-means clustering to first-order representations. In *Proceedings of the Tenth International Conference on Inductive Logic Programming*, pages 112–129. Springer, Berlin, 2000.

9.14 P. Mahalanobis. On the generalized distance in statistics. *Proceedings of the Indian National Institute Science*, 2: 49–55, Calcutta, 1936.

9.15 J. McQueen. Some methods of classification and analysis of multivariate observations. In *Proceedings of Fifth Berkeley Symposium on Mathematical Statistics and Probability*, pages 281–293, 1967.

9.16 S.-H. Nienhuys-Cheng. Distance Between Herbrand Interpretations: A Measure for Approximations to a Target Concept. In *Proceedings of the Seventh International Workshop on Inductive Logic Programming*, pages 213–226. Springer, Berlin, 1997.

9.17 S.-H. Nienhuys-Cheng. Distances and limits on Herbrand interpretations. In *Proceedings of the Eighth International Conference on Inductive Logic Programming*, pages 250–260. Springer, Berlin, 1998.

9.18 J. Ramon and M. Bruynooghe. A framework for defining distances between first-order logic objects. *Proceedings of the Eighth International Conference on Inductive Logic Programming*, pages 271–280. Springer, Berlin, 1998.

9.19 M. Sebag. Distance Induction in First Order Logic. In *Proceedings of the Seventh International Workshop on Inductive Logic Programming*, pages 264–272. Springer, Berlin, 1997.

9.20 B. A. Shapiro and K. Zhang. Comparing Multiple RNA Secondary Structures Using Tree Comparisons. *Computer Applications in Biosciences*, 6(4): 309–318, 1990.

9.21 A. Srinivasan, S. Muggleton, and R. D. King. Comparing the use of background knowledge by inductive logic programming systems. In *Proceedings of the Fifth International Workshop on Inductive Logic Programming*, pages 199–230. Department of Computer Science, Katholieke Universiteit Leuven, Belgium, 1995.

9.22 E. Ukkonen. Algorithms for Approximate String Matching. *Information and Control*, 64: 100–118, 1985.

9.23 D. Wettschereck and D. Aha. Weighting Features. In *Proceedings of the First International Conference on Case-Based Reasoning*, pages 347–358. Springer, Berlin, 1995.

9.24 K. Zhang and D. Shasha. Simple Fast Algorithms for the Editing Distance Between Trees and Related Problems. *SIAM Journal on Computing*, 18(6): 1245–1262, 1989.

9.25 M. Zuker and P. Stiegler. Optimal Computer Folding of Large RNA Sequences Using Thermodynamics and Auxiliary Information. *Nucleic Acids Research*, 9(1): 133-148, 1981.

Part III

From Propositional to Relational Data Mining

10. How to Upgrade Propositional Learners to First Order Logic: A Case Study

Wim Van Laer[1] and Luc De Raedt[2]

[1] Department of Computer Science, Katholieke Universiteit Leuven
Celestijnenlaan 200A, B-3001 Leuven, Belgium

[2] Institut für Informatik, Albert-Ludwigs-Universität Freiburg
Am Flughafen 17, D-79110 Freiburg, Germany

Abstract

We describe a methodology for upgrading existing attribute-value learners towards first-order logic. This method has several advantages: one can profit from existing research on propositional learners (and inherit its efficiency and effectiveness), relational learners (and inherit its expressiveness) and PAC-learning (and inherit its theoretical basis). Moreover there is a clear relationship between the new relational system and its propositional counterpart. This makes the ILP system easy to use and understand by users familiar with the propositional counterpart. We demonstrate the methodology on the ICL system which is an upgrade of the propositional learner CN2.

10.1 Introduction

Current machine learning systems are often distinguished on the basis of their representation, which can either be propositional or first-order logic. Systems belonging to the first category are often called attribute-value learners, systems of the second category are called relational learners or inductive logic programming systems.

In this paper, we shall argue that it is effective to develop first-order learners that have existing propositional machine learning systems as a special case. Advantages include: ease of understanding and use of the first-order system by users familiar with the propositional case; potential exploitation of all expertise (and heuristics) available for the propositional learner; a clear relation between the propositional learner and its first-order variant, resulting in e.g., identical results on identical (propositional) data.

Given this viewpoint we develop a methodology for upgrading propositional learners towards first-order logic and demonstrate it at work. This methodology is perhaps the most important lesson learned during the development of several inductive logic programming systems and results (including [10.21], Tilde [10.9, 10.7], ICL [10.23], Claudien [10.20], Warmr [10.26]) of the machine learning group in Leuven. The methodology starts from an existing propositional learner and provides a recipe for upgrading it towards the use of first-order logic. The recipe involves the use of examples which correspond to sets of ground facts (interpretations), the adaptation of the representation of hypotheses towards Prolog, the employment of θ-subsumption to structure and search the space of hypotheses, the introduction of a declarative bias, and otherwise recycles as much as possible from the original system. Following the methodology, it should be easy to turn virtually any propositional symbolic learner into an inductive logic programming system.

To show how the methodology works, we demonstrate it on upgrading the well-known CN2 [10.14, 10.13] learning algorithm towards ICL [10.23]. In Section 10.6 we give an overview of other systems that follow the same methodology.

The paper is structured as follows: we first elaborate on the characteristics of the propositional and the first-order knowledge representation and we show how the relational representation can overcome limitations of the propositional representation. After describing the propositional learner CN2, we present our methodology for upgrading a propositional learner and illustrate each step w.r.t. CN2 resulting in the relational learner ICL. We also present some experimental results with ICL that show that the methodology is worthwhile. In the last section we discuss some related results and conclude.

10.2 Knowledge representation

10.2.1 Attribute-value learning

Consider Figure 10.1. Each example or scene can be described by a fixed number of attributes: `shape-left`, `size-left`, `color-left`, `shape-right`, `size-right`, `color-right` and `class`. The data-set can be summarized in one table as in Figure 10.1, where each row (or in relational database terms, each tuple) represents one example. Many well-known systems like C4.5 [10.54, 10.55] and CN2 [10.14, 10.13] are based on this *attribute-value representation* (also called propositional representation), and are as such called attribute-value learners. Also, data mining mainly focusses on learning from single tables.

The examples in Figure 10.2 however, cannot easily be described by a fixed number of features. A scene or example consists of a variable number of geometrical objects (such as lines, points, squares, triangles, circles,...), each having a number of different properties (e.g., white, black, small, large, horizontal,...), and a variable number of relations between objects. Representing these scenes with a fixed set of attribute-value pairs results in a number of problems (cf. [10.19]). *First*, one should fix the maximum number of objects in a scene. Given a bound b on the number of objects, one could then list attributes $A_{i,1}, ..., A_{i,j}$ characterizing each object i. Some of these attributes will yield nil values since not all scenes may possess the same

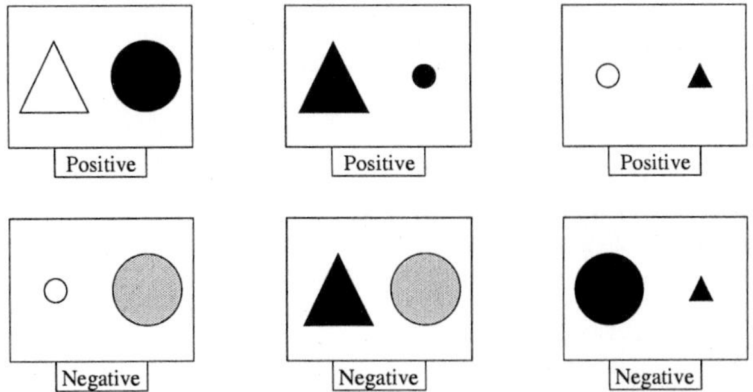

Fig. 10.1. A simple classification problem. One scene (example) consists of a left-side and a right-side object. Each scene is tagged with a class (positive or negative).

number of objects and not all attributes/properties are meaningful for each object. *Secondly*, one should also order the objects in a scene, which is more problematic. Indeed, reconsider scene 1 in Figure 10.1. Its representation in Table 10.1 assumes that the order is from left to right. In general, the objects will be essentially unordered (as in Figure 10.2). Without determining the order of objects within a scene there is an exponential number of equivalent representations of a scene (in the number of objects). Scene 1 of Figure 10.1 corresponds to one such representation, another representation (based on a different order) would swap the left and right object. For similar reasons, the representation of rules will also require one such ordering. These two problems prohibit an efficient encoding of first-order problems into the attribute-value representations employed by typical machine learning systems. *Thirdly*, one should provide an attribute for each possible relation between each pair of objects (in a specific order). E.g., the first object is left of the second object, the first object is left of the third object,... Again, the number of such attributes (relations) grows exponentially in the number of objects available.

Though the above problem is a toy-problem, it is very similar to real-life problems in e.g., the field of molecular biology (see [10.41, 10.11]) where essentially the same representational problems arise. Data consists of a set of

Table 10.1. An attribute-value representation for Figure 10.1 (with `id` a unique identifier for each example).

`id`	`shape-left`	`size-left`	`color-left`	`shape-right`	`size-right`	`color-right`	`class`
1	triangle	large	white	circle	large	black	positive
2	triangle	large	black	circle	small	black	positive
3	circle	small	white	triangle	small	black	positive
4	circle	small	white	circle	large	grey	negative
5	triangle	large	black	circle	large	grey	negative
6	circle	large	black	triangle	small	black	negative

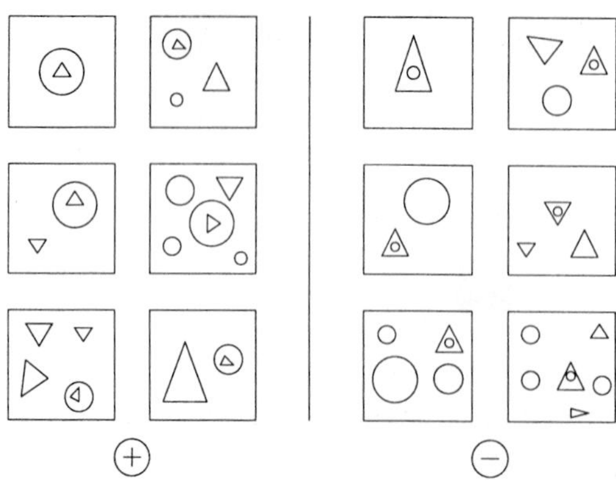

Fig. 10.2. A more complex classification problem: Bongard problem 47, developed by the Russian scientist M. Bongard in [10.10]. It consists of 12 scenes (or examples), six of class $\oplus$ and six of class $\ominus$. The goal is to discriminate between the two classes.

molecules, each of which is composed of several atoms with specific properties (like charge). Similar to a scene, there exists a number of relations between atoms (like bonds, structures,...).

10.2.2 First-order representations

The above sketched problems can be overcome using a *relational/first-order representation.* We propose the following representation for examples:

an example is a set of ground facts

Ground facts are tuples in a relational database.

From a logical point of view this is called a (Herbrand) *interpretation* because the facts represent all atoms which are true for the example, thus all facts not in the example are assumed to be false. From a computational view this can be seen as a a small relational database or a Prolog knowledge base, so we can make use of a Prolog interpreter for querying an example.

To illustrate this representation, let's reexamine the Bongard problem in Figure 10.2. The upper left scene consists of a small triangle, pointing up, and which is in a large circle. This scene can be specified as follows:

{`class`(*positive*), `object`(*o1*), `object`(*o2*), `shape`(*o1*, *circle*), `size`(*o1*, *large*), `shape`(*o2*, *triangle*), `size`(*o2*, *small*), `pointing`(*o2*, *up*), `in`(*o2*, *o1*)}.

The other scenes can be encoded in the same way. The number of objects in one scene is not limited, and objects are not ordered (they could be called *a* and *b* instead of *o*1 and *o*2). Different objects can have different properties, e.g., a triangle can be pointing up or down, but this property makes no sense for a circle. And finally, the number of relations between objects is unlimited.

This first-order representation is more general and more expressive than the attribute-value representation which is a special case of it. Indeed, an attribute-value table with k attributes can be mapped to a set of interpretations/examples as follows:

> For each example (a tuple/row in the attribute-value table), construct the fact {example(val_1, ..., val_k).}, where val_i is the value of the ith attribute of the example in the table. Then each of these facts is the interpretation of the corresponding example.
> Instead of mapping to one fact, an alternative is to map each row or example to a set of k facts {$\text{att}_1(val_1)$, ..., $\text{att}_k(val_k)$} where val_i is the value for the ith attribute.

For instance, the first example in Figure 10.1 can be represented by

{`example`(*triangle*, *large*, *white*, *circle*, *large*, *black*, *positive*)}
or
{`shape-left`(*triangle*), `size-left`(*large*), `color-left`(*white*),
`shape-right`(*circle*), `size-right`(*large*), `color-right`(*black*),
`class`(*positive*)}

At this point, it is worth noting that in the attribute-value representation each example must have a single value for a given attribute. Therefore, if we know that the value of e.g., `size-left`=*large*, we also know that `size-left`≠*small*. This corresponds to making a kind of closed world assumption at the level of each example (cf. [10.37]). Due to the use of Prolog (and the implicit negation as failure), the meaning of each example in the above representation is correctly captured! (i.e., if `color-left`(*black*) then ?-`color-left`(X) will only succeed for X=*black*.)

This framework and use of Prolog is quite similar to what happens in the older work on structural matching (e.g., [10.38, 10.39, 10.69, 10.70, 10.71]).

10.2.3 Background knowledge

It is useful to use not only factual knowledge in the examples, but also Prolog rules (or definite clauses). If these rules are common to all the examples, they are referred to as background knowledge. Such knowledge can take various forms: e.g., abstraction of specific values into a taxonomy or interval, deriving new properties from a combination of existing ones, summarizing or aggregating values of several facts/tuples into a single value, etc.

For our Bongard problem in Figure 10.2, we could add the following definitions:

```
polygon(X) :- triangle(X).
polygon(X) :- rectangle(X).
number_objects(NO) :- setof(O, object(O), LO), length(LO, NO).
```

The first two clauses state that a polygon can be either a triangle or a rectangle. The last clause calculates the number of objects in an example by creating a set of all objects and counting the number of elements in this set.

As background knowledge is visible for each example, all the facts that can be derived from the background knowledge and an example are part of the extended example[1]. When querying an example, it suffices to assert the background knowledge and the example; the Prolog interpreter will do the necessary derivations.

10.2.4 Note

The above representation of examples is known in the literature as *learning from interpretations* [10.21, 10.18]. It is only one of the possible representations used within inductive logic programming. More details on the relation among various inductive logic programming settings can be found in [10.18].

10.3 The propositional learner CN2

CN2 is a well-known attribute-value learning system which is described in [10.14, 10.13]. Originally, it induced an ordered list of rules using entropy as its search heuristic [10.14]. Two improvements to the algorithm are described in [10.13]: the use of the Laplace error estimate as evaluation function and the generation of unordered rules instead of ordered rules. In the rest of the paper we will only consider the algorithm for learning unordered rules.

Informally, CN2's problem specification is: given a set of (AV) examples E (represented as described in Section 10.2.1) and a set of classes C (each example belongs to one class), find an unordered set of rules of the form `class`=*class* if *condition* (with *condition* a conjunction of attribute-value tests) such that each example is classified correctly. To classify an example, one collects all rules which fire (i.e., all rules that cover the example[2]) and predict the class by a simple probabilistic method to resolve clashes. For the moment, we will concentrate on the task to induce a set of rules for one class c: the set of positive examples P are all the examples belonging to the class c, the set of negative examples N are all the others (so $E = P \cup N$).

Reconsider the classification problem in Figure 10.1. CN2 might learn the following hypothesis for class *positive* (as an unordered set of rules):

`class`=*positive* if `color-left`=*white* and `color-right`=*black*
`class`=*positive* if `size-right`=*small* and `shape-right`=*circle*

[1] Formally, an extended example is the minimal Herbrand model of the example and the background theory.

[2] A rule covers an example if the condition of the rule is true for the example.

To learn a theory for one class, CN2 performs a covering approach on the positive examples: it repeatedly finds a single rule that is considered *best* (that is maximizes the positive examples covered and minimizes the negative examples covered). The *best* rule is then added to the hypothesis H and all examples of P that are covered by the rule are removed from P. This process terminates when no *best* rule can be found or when no more positives have to be covered. The algorithm can be found in Table 10.2.

To find a *best* rule, CN2 has to search through the space of rules. The structure of this search space is implied by the subset test. Refining a rule is simply done by adding a new attribute test to the body of the rule (also called condition). CN2 starts with the most general rule of the search space (usually the rule with an empty body: *class* :- *true*). It then performs a beam search. At each step/level, all refinements of the rules in the current *beam* are evaluated. If the rule is statistically significant and better than the current best, it becomes the current best rule. From all the refinements, the $MaxBeamSize$ best rules are kept in the new *beam*. This search repeats until no more rules are in the *beam*. The algorithm for finding a *best* rule can be found in Table 10.3.

10.4 Upgrading CN2

In Section 10.2, we have motivated the need for relational representations and we have introduced a first-order representation for examples. Now, we can focus on the methodology for upgrading propositional learners.

The methodology is summarized in Table 10.4 and discussed in detail below through a case study with CN2 [10.14, 10.13] and ICL [10.23]. The final section will briefly review a number of other cases with the methodology.

Table 10.2. Learn a theory for one class.

Learn-For-One-Class(*Examples*, *Class*) **return** *Hypothesis*;

1. **let** P := $\{e \in \mathit{Examples} \mid$ example e is of class $\mathit{Class}\}$;
2. **let** N := $\{e \in \mathit{Examples} \mid$ example e is not of class $\mathit{Class}\}$;
3. **let** $H := \emptyset$;
4. **repeat**
   a) *BestRule* = **Find-Best-Rule**(P, N);
   b) **if** *BestRule* found **then**
      i. add *BestRule* to H;
      ii. remove from P all examples e covered by *BestRule*;
   until *BestRule* not found **or** P is empty;
5. **return** H;

Table 10.3. Beam search algorithm to find the *best* rule.

Find-Best-Rule(P,N);

1. **let** mgr := most general rule in the search space;
2. **let** $Beam$:= mgr;
3. **let** $BestRule$:= $\emptyset$;
4. **while** $Beam$ is not empty **do**
   a) **let** $NewBeam$:= $\emptyset$;
   b) **for** each rule R in $Beam$ **do**
      for each refinement Ref of R **do**
      i. **if** (Ref is better than $BestRule$ **and** Ref is statistically significant)
         then let $BestRule$:= Ref;
      ii. **if** Ref is not to be pruned
         then
         – **add** Ref to $NewBeam$;
         – **if** size of $NewBeam$ > $MaxBeamSize$
           then remove worst rule from $NewBeam$;
   c) **let** $Beam$:= $NewBeam$;
5. **return** $BestRule$;

10.4.1 The propositional task and algorithm

Suppose that we are asked to design a learning system for Bongard type problems. Machine learning researchers would observe that Bongard problems are classification problems (another popular task is that of descriptive learning, for example discovering association rules [10.2, 10.1, 10.64]). So, the range of possible propositional learning algorithms to consider includes AQ [10.45], TDIDT [10.55] (like C4.5 [10.54]), and CN2 [10.14, 10.13]. Suppose we fancy the latter algorithm because it combines the advantages of AQ and TDIDT, i.e., it produces understandable rules, it is efficient and can cope with noisy data. So, we decide to base our first-order learner on CN2. Then we have also accomplished the first step of the methodology:

Step 1 : *Identify the propositional learner that best matches the learning task.*

Given the goal of upgrading CN2 to first-order logic, the question is how to realize this. At this point, the reader may notice that CN2 will not work on the Bongard problem because:

– the representation of the examples is propositional
– the representation of the rules is propositional
– the search operators are propositional

We will now discuss these issues in detail.

Table 10.4. An overview of the methodology for upgrading propositional learners to first-order logic.

Step 1:	*Identify the propositional learner that best matches the learning task.*
Step 2:	*Use interpretations to represent examples.*
Step 3:	*Upgrade the representation of propositional hypotheses by replacing attribute-value tests by first-order literals and modify the coverage test accordingly.*
Step 4:	*Use θ-subsumption as the framework for generality.*
Step 5:	*Use an operator under θ-subsumption. Use that one that corresponds closely to the propositional operator.*
Step 6:	*Use a declarative bias mechanism to limit the search-space.*
Step 7:	*Implement.*
Step 8:	*Evaluate your (first-order) implementation on propositional and relational data.*
Step 9:	*Add interesting extra features.*

10.4.2 Examples are interpretations

The propositional representation of examples should be upgraded to a first-order one. We propose to use interpretations for this (see Section 10.2) as it is a natural representation for examples and there is a clear relation with attribute-value learning [10.19, 10.18]. This will alleviate the first problem. Also, if desired, background knowledge can be formulated in Prolog as in Section 10.2.3.

Step 2 : *Use interpretations to represent examples.*

10.4.3 First-order hypotheses

As the expressiveness of the examples (inputs) has been extended, we should also extend the expressiveness of the hypotheses (outputs).

Let us have a closer look at the concept representation in CN2. Recall from Section 10.3 that CN2 learns an unordered set of rules of the form `class`=*class* if *condition*, with *condition* a conjunction of attribute-value tests (e.g., `color-left` =*white*). An attribute-value test can be seen as a special case of a literal. For example, `color-left`=*white* can be mapped to `color-left`(*white*) (cf. also the mapping in Section 10.2.2). So if we allow literals (with possibly more than one argument, and with variables or terms as arguments) instead of just attribute-value tests, the hypothesis will be a kind of first-order expression. When using rule sets with literals (the variables in the literals are existentially quantified), we can learn the following rule for the Bongard problem in Figure 10.2:

class=⊕ if ∃T,C: shape(T, *triangle*) and shape(C, *circle*) and in(T, C).

which states that there exists a triangle and a circle (thus an instantiation for the variables T and C) such that the triangle is inside the circle. At this point, the condition of the rule corresponds to a Prolog query. Furthermore, instead of the 'if' notation in rule-based approaches it is common in logic programming and Prolog to write ':-', yielding a typical Prolog clause:

class(⊕) :- shape(T, *triangle*), shape(C, *circle*), in(T, C).

As a result, a first-order upgrade of unordered rule sets for CN2 is of the form:

$$\texttt{class}(class) \text{ :- } l_{1,1}, ..., l_{1,n_1}$$
$$.....$$
$$\texttt{class}(class) \text{ :- } l_{k,1}, ..., l_{k,n_k}$$

where all $l_{i,j}$ are literals and all the variables appearing in the literals are existentially quantified. Note that the variables are independent between different rules.

So far, we have only sketched a syntactic adaptation. We also need to modify the semantics of hypotheses. This boils down to specifying when an example is covered by a hypothesis. An example e will be covered by a hypothesis H (a set of rules for class(c)) if $H \wedge e \models$ class(c). Thus to test coverage, one asserts the hypothesis H (resp. a rule) and the example e in a Prolog knowledge base (one could also use a relational database system) and runs the query ?-class(c). If this query succeeds, the example is covered; otherwise it is not.

Note that this coverage test is more complex in both time and space than its propositional counterpart (which is a simple subset test).

The reader familiar with CN2 may observe that CN2 also uses a simple probabilistic method to resolve conflicts/clashes when an example is covered by rules belonging to multiple classes. These probabilities can straightforwardly be used here too. It is merely the test for coverage that needs to be changed.

The third step of the methodology can be summarized as follows:

Step 3 : *Upgrade the representation of propositional hypotheses by replacing attribute-value tests by first-order literals and modify the coverage test accordingly.*

This third step alleviates the second problem concerning CN2, mentioned earlier.

Note that this step also works for a wide range of propositional hypothesis/concept representations, like ordered or unordered rule sets, decision trees, regression trees, association rules,.... Indeed, all these concept descriptions have one thing in common: they are all based on attribute-value tests. For instance, in a decision tree each branch is based on an attribute-value test, and an association rule is a set of attribute-value tests.

For example, TILDE [10.9, 10.7] introduces first-order logical decision trees - FOLDT (of which binary trees are a special case). A FOLDT is a binary decision tree in which the nodes of the tree contain a conjunction of literals. Moreover, different nodes may share variables, under the restriction that a variable that is introduced in a node (meaning that it does not occur in higher nodes) does not occur in the right branch of that node[3]. An example of a logical decision tree is shown in Figure 10.3. Note the sharing of the variable T in both literals.

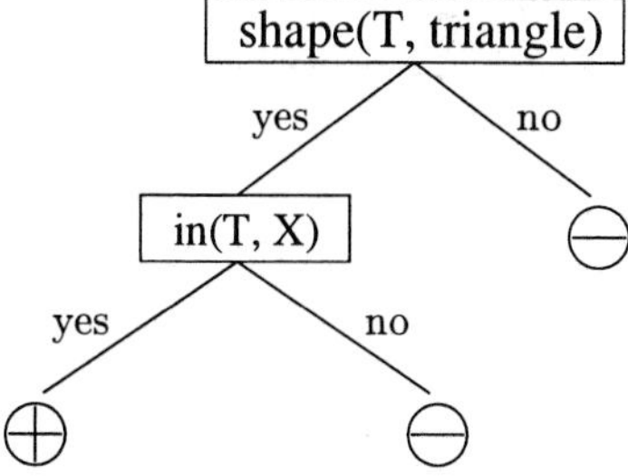

Fig. 10.3. An example of a first-order logical decision tree that discriminates between the classes ⊕ and ⊖ for the Bongard problem in Figure 10.2.

10.4.4 Structuring the search-space

Nearly all symbolic machine learning systems structure the search-space by means of the *is more general than* relation. When working with propositional representations this relation is often quite simple. For instance, in the CN2 algorithm one rule is more general than another rule if all literals (i.e., attribute-value tests) occurring in the first rule are a subset of those occurring in the second rule.

On the other hand, when working with first-order representations the frameworks for generality become rather complex. Various frameworks have been proposed, including θ-subsumption (from Plotkin [10.52]), inverse implication, inverse resolution and inverse entailment (cf. [10.48] for an overview).

[3] The need for this restriction follows from the semantics of FOLDT. A variable X in a literal is existentially quantified within the conjunction of that node. As the the right subtree is only relevant when the conjunction fails (thus saying *there is no such X*), further references to X are meaningless in the right branch.

However, in practice, the large majority of ILP systems (including FOIL [10.56], GOLEM [10.49], PROGOL[4] [10.47], CLAUDIEN [10.20], and TILDE [10.9, 10.7],...) uses θ-subsumption. This is due to the excellent computational properties of θ-subsumption (as compared to inverse resolution and inverse implication, which are both computationally intractable and less understood yet). Another important property of θ-subsumption is that it works at the level of single clauses instead of sets of clauses (as inverse resolution). This is similar to propositional systems which also structure the space at the level of individual rules. On the other hand, working at the level of single clauses only may cause problems when learning recursive clauses or multiple predicates (cf. [10.22, 10.4]). However, in our opinion, recursion is not essential for most real-life applications of relational machine learning and data mining. Indeed, the authors are only familiar of a few recursive rules in the mesh-design [10.27]. Most other current real-life applications of inductive logic programming do not involve recursive regularities (see [10.31, 10.11] for overviews). So, in most applications θ-subsumption is the right framework for generality when upgrading propositional systems to first-order logic.

Step 4 : *Use θ-subsumption as the framework for generality.*

Before showing in the next section how to adapt the operators, we provide a brief review of θ-subsumption and its properties.

Let us first introduce the definition:

$$\text{Clause } C_1 \ \theta\text{-subsumes clause } C_2 \text{ iff } \exists\theta\text{: } C_1\theta \subseteq C_2.$$

A clauses (rule) is a set of literals and a variable substitution θ (=$\{V_1/t_1, \ldots, V_n/t_n\}$) maps each variable V_i to its corresponding term t_i. For instance: C_1 = `father`(X, Y) :- `parent`(X, Y), `male`(X). is more general than clause C_2 = `father`(*jef*, *wim*) :- `parent`(*jef*, *wim*), `parent`(*jef*, *ann*), `male`(*jef*), `female`(*ann*). because $C_1\theta \subseteq C_2$ with $\theta = \{X/jef, Y/wim\}$.

Note that the propositional ordering on the search space is a special case of θ-subsumption. This is an important property in the light of the upgrading procedure. For instance: the clause `class`(*positive*) :- `color-right`(*black*) is more general than `class`(*positive*) :- `color-left`(*white*), `color-right`(*black*).

At this point it is important to realize that θ-subsumption generalizes the well-known *turning constants into variables* introduced by [10.46]. For example, p :- q(X,Y), q(Y,X) is more general than p :- q(a,a) under θ-subsumption, but would not be regarded a generalization using the turning constants into variables. A second point where θ-subsumption generalizes Michalski's framework is that it also works for structured terms. E.g., p :- q(f(a)) is a specialization of p :- q(X).

---

[4] In PROGOL, the θ-subsumption lattice is searched top-down but is bounded from below by a clause computed using inverse entailment.

Some more theoretical properties of θ-subsumption include (for more details see [10.52, 10.51, 10.48, 10.66]):

- it induces a quasi-order (reflexive and transitive) on the space of first-order rules
- if c_1 θ-subsumes c_2 then $c_1 \models c_2$, i.e., c_1 logically entails c_2.
- there exist clauses $c_1 \neq c_2$ that are equivalent under θ-subsumption, e.g., p :- q(X,Y) and p :- q(X,Y),q(X,Z).
- the quasi-order can be turned into a partial order (also anti-symmetric) by defining equivalence classes in the usual way. There is then also a unique (up to variable renaming) representative of each equivalence class, which is called the *reduced* clause. The reduced clause r of a clause c is defined as the smallest subset of literals in c such that r is equivalent under θ-subsumption with c. E.g., p :- q(X,Y) is the reduced clause of p :- q(X,Y),q(X,Z).
- at the level of equivalence classes, one obtains a complete lattice, i.e., any two equivalence classes have a unique least upper bound (also called the least general generalization, the lgg) and a unique greatest lower bound.

10.4.5 Adapting the search operators

Now that we have chosen a framework for generality, we still need to define operators for searching the corresponding rule space. Given the advice of step 4, we will limit the discussion here to operators under θ-subsumption only.

Let us consider three typical operators used by concept learners. A specialization, resp. a generalization, operator that operates on a single clause, and a generalization operator that computes the least general generalization of two clauses.

The typical propositional specialization operator will basically add a condition to a rule. Using clauses, a condition can be added in two manners: either by adding a literal or by applying a substitution to the given clause. E.g., the clauses p :- q(X,Y),r(X), p :- q(X,X) and p :- q(X,a). are specializations of the clause p :- q(X,Y).

This yields the so-called refinement operators (cf. [10.51, 10.66]). There are some additional complications when using refinement operators wrt propositional systems:

- when simply adding literals, one might stay within the same equivalence class, and there might be infinite chains of such refinements, e.g., when refining p :- q(X,Y) to p :- q(X,Y), q(X,Z) and then to p :- q(X,Y), q(X,Z), q(X,W) ...
- it could be that even some proper refinements of a clause do not affect the coverage of the examples, this is known as the *determinacy problem* [10.57]. E.g., refining `class`(*pos*) :- `atom`(X) to `class`(*pos*) :- `atom`(X), `bond`(X,Y). As any atom will have bonds to other atoms, merely adding `bond`(X,Y) will not modify the coverage of the clause. This may misguide the heuristics of the learning engine.

Both difficulties can be alleviated by using a declarative (language) bias that will be discussed in the next section.

The typical propositional generalization operator will delete (or relax) a condition in a rule. Under θ-subsumption there are two ways of relaxing a clause: either delete a literal, or apply an inverse substitution to the clause. The first case is the easy one: e.g., generalize p :- q(X,Y), r(Y) towards p :- r(Y). The second case is more complicated and generalizes the turning constants into variables rule. If the constant (or term) to be generalized occurs only once in the clause, there is no problem: it can merely be generalized into a variable not yet occurring in the clause. E.g., p :- q(a,b) into p :- q(X,b). However, if the constant or term to be generalized occurs multiple times, generalization can be quite complex. Indeed, consider p :- q(a,a). This can be generalized into p :- q(X,X) or p :- q(a,X),q(X,a). One problem is the existence of infinite chains. I.e., the clause positive :- q(a,a) has the following generalizations: positive :- q(X,X), positive :- q(X,Y), q(Y,X), ... The most specific generalization is the infinite rule positive :- q(X_1,X_2), q(X_2,X_3),...,q(X_i,X_{i+1}),...).

This problem and the existence of a *lgg* operator (cf. below) explains why plain generalization operators are less popular in ILP than refinement operators. The problems with generalization operators can again be reduced by an appropriate declarative bias mechanism.

A third popular operator is the (generalization) operator that computes the least general generalization (*lgg*) of two clauses.

A clause g is a *least general generalization* (*lgg*) of the clauses C_1 and C_2 if and only if g θ-subsumes C_1 and g θ-subsumes C_2, and for every clause g' such that g' θ-subsumesC_1 and g' θ-subsumes C_2, g' also θ-subsumes g.

Plotkin has given a procedure to compute the *lgg* of two clauses: The *lgg* of two identical terms is the term itself ($lgg(t,t) = t$). The *lgg* of the terms $f(s_1,...,s_n)$ and $f(t_1,...,t_n)$ is $f(lgg(s_1,t_1),...,lgg(s_n,t_n))$. The *lgg* of the terms $f(s_1,...,s_n)$ and $g(t_1,...,t_m)$ where $f \neq g$ is the variable v where v represents this pair of terms throughout. The *lgg* of two atoms/literals $p(s_1,...,s_n)$ and $p(t_1,...,t_n)$ is $p(lgg(s_1,t_1),...,lgg(s_n,t_n))$, the *lgg* being undefined when the sign or the predicate symbols are different. Finally, the *lgg* of two clauses C_1 and C_2 is then

$$\bigvee_{l_1 \in set(C_1), l_2 \in set(C_2)} lgg(l_1, l_2).$$

Some examples are given below. The *lgg* of

`father`(*luc*,*soetkin*) :- `parent`(*luc*,*soetkin*),`male`(*luc*),`female`(*soetkin*)

and

`father`(*jef*,*wim*) :- `parent`(*jef*,*wim*),`male`(*jef*),`male`(*wim*)

is

```
father(X,Y) :- parent(X,Y), male(X),male(Z).
```

The *lgg* is used in specific to general inductive logic programming systems like GOLEM [10.49]. The problem with the lgg operator is that the complexity of the lgg (i.e., the number of literals) may grow exponentially with the number of examples in the worst case.

Step 5 : *Use an operator under θ-subsumption. Use that one that corresponds closely to the propositional operator.*

In the ICL system, we choose a specialization operator under θ-subsumption. Due to the problems sketched above, we will embed it within a declarative bias mechanism.

10.4.6 The need for bias

In the previous section, several problems with pure θ-subsumption operators were mentioned. These problems are mainly due to the combinatorics of the search-space, the fact that the space is infinite rather than finite (as in the propositional case) and the determinacy problem. To make the search tractable and efficient, it is thus necessary to constrain the search space in some way. In ILP, this is solved using syntactical or semantical declarative bias mechanisms. Various formalisms exist (see [10.50] for an overview), but the overall idea is to limit the number of clauses considered. The most straightforward methods merely employ some bounds on the number of variables, or literals in clauses and make the search-space finite. Other methods will specify syntactic limitations on the clauses considered (from which an operator can be derived). E.g., using a number of schemata to enforce that clauses satisfy certain patterns, e.g., the pattern P(X,Y) :- Q(X), R(X,Y), where P, Q and R are 'predicate' variables (see [10.40]). Other methods use a kind of grammar construction to explicitly declare the range of acceptable clauses [10.16]. A third class of techniques uses so-called mode declarations to state how clauses can be refined, like in PROGOL [10.47], TILDE [10.9, 10.7] and WARMR [10.26].

Step 6: *Use a declarative bias mechanism to limit the search-space.*

In ICL we selected the $\mathcal{D}$LAB declarative bias formalism of [10.20], which encodes a kind of grammar[5]. An example is given in Table 10.5. Min-Max:List means that at least Min and at most Max literals of List are allowed (`len` is the length of `List`). Note that shape(Object, 1-1:[circle,triangle]) is a shorthand for 1-1:[shape(Object,circle), shape(Object,triangle)]. Recursion is allowed. There is also a notion of dlab_variable (not used in the example)

[5] We could have used as well the mode declarations as in TILDE and WARMR.

Table 10.5. A $\mathcal{D}$LAB bias for the Bongard problem.

```
dlab_template('
  1-len:[shape(Object1, 1-1:[circle,triangle]),
         size(Object1, 1-1:[small,large]),
         shape(Object2, 1-1:[circle,triangle]),
         size(Object2, 1-1:[small,large]),
         1-1:[in(Object1, Object2), left_of(Object1,Object2)]
        ]').
```

that allows the user to define shortcuts for frequently occurring parts (like 1-1:[circle,triangle]).

Given a $\mathcal{D}$LAB expression, a refinement operator can be used to traverse the (restricted) search space. A complete refinement operator for $\mathcal{D}$LAB is given in [10.20]. For example, based on the $\mathcal{D}$LAB expression in Table 10.5, the top rule of the search space is
`class`(⊕) :- *true.*
The refinement operator will generate the following refinements for this rule:
`class`(⊕) :- `shape`(Object1, *circle*).
`class`(⊕) :- `shape`(Object1, *triangle*).
.....
`class`(⊕) :- `left_of`(Object1, Object2).
These rules again can be refined further on. For the first rule:
`class`(⊕) :- `shape`(Object1, *circle*), `size`(Object1, *small*).
.....
`class`(⊕) :- `shape`(Object1, *circle*), `shape`(Object2, *circle*).
.....
Note that `class`(⊕) :- `shape`(Object1, *circle*), `shape`(Object1, *triangle*). is not a valid refinement.

The advantage of $\mathcal{D}$LAB is its expressive power. It allows the user to strongly bias the learning system ICL. On the other hand, a $\mathcal{D}$LAB expression can become very complex. Writing a $\mathcal{D}$LAB (and bias in general) is an iterative process and not always straightforward.
Note that some kind of lookahead can be performed by $\mathcal{D}$LAB to overcome the determinacy problem. Indeed, when using len-len:[List] in the template, all the literals in List must be added in one step as a refinement.

10.4.7 Implementing the algorithm

By now, we are ready to implement our first-order learner. All basic modifications needed have been sketched. In this step, it is important that as many features of the original algorithm as possible are preserved, like search strategy, heuristics, noise-handling, pruning, parameters,... For example, ICL uses the same heuristics (Laplace estimate) as its propositional counterpart CN2.

Some advanced and specific features of propositional learning algorithms may need further changes. For example, discretization on numerical attributes cannot be mapped directly towards our relational representation (see Section 10.4.9).

Step 7: *Implement.*

Currently, ICL is implemented in MasterProLog (formerly ProLog by BIM) and freely available for academic use (runtime version for Solaris 2.5 is available on request).

10.4.8 Evaluation of ILP system

A first evaluation is testing the implementation on a propositional problem, for example the problem in Figure 10.1. The results should be compatible with the results of the corresponding propositional learner. In the ideal case these should be the same, but in reality some minor differences might occur. Many reasons exist: small differences in implementation, lack of some features (like handling of unknown values), a slightly different hypothesis space for the propositional and relational system,...

Next is learning on relational data that one is unable to learn on with a propositional learner. A good starting point is some artificial problem, like the Bongard problem in Figure 10.2, where one has already a solution (obtained by hand or by some other relational learner). Experiments with these data can give a good insight in the system (the parameters, the output, the behavior of learning, some problems,...).

Then one can start the *real* work and run the system on real-like problems. A well-known application area is in molecular biology.

Step 8: *Evaluate your (first-order) implementation on propositional and relational data.*

In Section 10.5, some experimental results with ICL will be discussed.

10.4.9 Extensions to the basic system

Many propositional systems have been extended with extra features or optimizations. These can also be incorporated in its relational counterpart, if possible. Some extensions can be plugged in as they are, others will need some adaptation/upgrade similar to the other steps in the sketched methodology. In this way, ILP learners can reuse results from research on propositional learners.

Note that also ideas from other ILP systems can be incorporated. As such, results from ILP can be reused by propositional learners, so both communities can learn from each other.

Step 9: *Add interesting extra features.*

The system ICL has many extensions/optimizations w.r.t. the basic system described up to now (see [10.68, 10.67] for more details). To handle numerical data, we upgraded the discretization method of Fayyad and Irani (see [10.36, 10.29]) towards ICL. To handle multiple classes, we extended the CN2 method with a Bayes approach (inspired by [10.53]). This result can be integrated in CN2 without any problem (illustrating that results in the context of relational learners can be mapped back to propositional learners). Other extensions/optimizations of ICL include: learning both DNF and CNF theories, using the m-estimate (as in [10.32]) instead of the Laplace estimate as heuristic, extra pruning of the search space,... While CN2 has a specific handling of unknown values (* and ?), ICL just assumes the closed world assumption.

10.5 Some experimental results with ICL

To illustrate the utility of the method and the effectiveness of ICL, we will give an overview of some experiments performed with ICL.

10.5.1 Experimental settings

The experiments have been performed with ICL version 4.2, implemented in MasterProLog (formerly ProLog by BIM). We used a Sun Ultra 2 with two 167 Mhz UltraSPARC processors running Solaris 2.6, and a SUN Ultra 10 with a 333 Mhz UltraSPARCII processor running Solaris 7.

Unless stated otherwise, we used the default settings of ICL. The most important ones: significance level is 90%, heuristic is m_estimate (with parameter m the number of classes), the size of the beam is 5, and classes are *pos* and *neg*.

10.5.2 Propositional data

One of the nice properties of our methodology is its *backward compatibility*, meaning that the upgraded relational system behaves similar as its propositional predecessor on propositional data. However, some deflections occur due to differences in implementation.

To simulate CN2 with ICL, we can use the following simple $\mathcal{D}$LAB expression: 1-len:[att_1 = 1-1:[$v_{1,1}$, ..., v_{1,i_1}],..., att_k = 1-1:[$v_{k,1}$, ..., v_{k,i_k}]], with $v_{i,j}$ the values of the attribute att_i.

We have run ICL on a few propositional data sets used in [10.13]: voting-records, breast-cancer, lymphography and primary-tumor. Some information on the data sets is given in Table 10.6. We have chosen these data sets because

Table 10.6. Details of the propositional domains used in the experiments. We did the same data conversions as documented in [10.13].

Domain	Number of			Unknown	Numerical
	Exs	Atts	Classes	values	values
voting-records	435	16	2	yes	no
breast-cancer	286	9	2	few	few
lymphography	148	18	4	no	few
primary-tumor	330	17	15	yes	no

they have no (or few) numerical values and only few unknown values. So this allows a close comparison.

We performed a similar experimental procedure as in [10.13]. The accuracies have been estimated by averaging the results over 20 runs (for each run, 2/3 of the data is selected randomly for training and the remainder for testing). The results are shown in Table 10.7. We have run ICL with the same language bias and the same settings as in [10.13]: beam=20 and heuristic=laplace. The last column gives the result of the default ICL performance (with default parameters: beam=5, significance level=90% and heuristic=m_estimate).

Table 10.7. Comparison of ICL and CN2 on accuracy (with standard deviation) and rule set size (number of attribute tests/literals). Results for CN2 are taken from [10.13], Appendix 1.

Accuracy	ICL		CN2 (unordered)		ICL
Sign. Threshold	0%	99.5%	0%	99.5%	default settings
voting-records	94.1±1.5	92.5±2.0	94.8±1.8	93.3±2.1	94.1±1.9
breast-cancer	69.7±3.3	71.8±3.7	73.0±4.5	70.8±3.5	69.4±4.1
lymphography	80.3±4.0	76.2±6.3	81.7±4.3	76.5±5.3	81.9±5.8
primary-tumor	41.7±4.9	42.0±4.8	45.8±3.6	41.4±5.8	41.4±5.5

Rule set size	ICL		CN2 (unordered)		ICL
Sign. Threshold	0%	99.5%	0%	99.5%	default settings
voting-records	43.5	14.9	64.8	19.9	49.7
breast-cancer	158.5	17.4	100.5	18.0	136.1
lymphography	38.5	14.9	40.4	13.5	45.5
primary-tumor	267	115.35	351.0	131.4	253.6

When we look at the accuracy and rule set size, we can conclude that ICL's performance is similar to CN2's, what we expected. Differences can be accounted by the small differences between CN2 and ICL w.r.t. options and implementation details There are however 2 boundary cases: for breast-cancer and primary-tumor with significance threshold 0% ICL performs less than

CN2 w.r.t. accuracy. In both cases, the theories sizes also differ significantly. We haven't found any explanation for this[6].

10.5.3 Relational data

ICL has been used in many experiments with (real-life) relational data sets. We will give some results, and refer to the literature for more details.

One of the most used data set in ILP is the *mutagenesis* one (see [10.63]). The data consists of 188 molecules, of which 125 are active (thus mutagenic) and 63 are inactive. A molecule is described by its atoms `atom(AtomID, Element, Type, Charge)` (the number of atoms differs between molecules, ranging from 15 to 35) and the bonds `bond(Atom1, Atom2, BondType)` between these atoms. Four different sets of background have been used (same as in [10.63]). Background 1 uses only the information on atoms and bonds, background 2 allows tests on the *charge* of an atom, background 3 adds 2 specific measures w.r.t. the molecule (e.g., $\log P$ and ϵ_{LUMO}) and background 4 consists of descriptions of higher-level structures that appear in the molecule (like aromatic rings).

Experiments with ICL on this data set can be found in [10.67]. Results with ICL version 4.2 are given in Table 10.8. We manually discretized the numerical values (i.e., $\log P$, ϵ_{LUMO} and the *Charge* of the atoms). It seems that the multi-class theory is always better than the separate (DNF) theory for each class. This is not so surprising as the multi-class theory combines the two separate theories for each class, and resolves clashes between the two. The (multi-class) accuracy of ICL is significantly better than FOIL for background 1 and 2, and marginally better for background 3 and 4. ICL is also marginally better than PROGOL and TILDE for background 1. For Background 2, 3 and 4 however, the performance of ICL, PROGOL and TILDE are similar. Note that the accuracy increases as more background is added.

Results on the *biodegradability* domain can be found in [10.67] (preliminary results) and [10.34] (more recent results). The task is to predict the half-life in water for aerobic aqueous biodegradation of a compound from its chemical structure. The biodegradation time has been discretized into 4 classes: fast, moderate, slow and resistant. The structure of a compound is represented by facts about atoms and bonds, much like in the mutagenesis domain. In [10.34] experiments on the relational data (denoted $R1$) and 2 propositional versions of the data (denoted $P1$ and $P2$) has been performed with the propositional classification systems C4.5 and RIPPER [10.15], and the relational learners FFOIL [10.58], S-CART [10.44], ICL and TILDE. A short overview of the results can be found in Table 10.9. Accuracy is classifi-

[6] In other papers we have found similar results for CN2 as ours. For example in [10.28], the result for breast-cancer is 70.0±1.4 accuracy with a theory size of 114.5, and for primary-tumor the accuracy is 39.9±1.0 with a theory size of 302.8. These are similar to our results. The experiments in that paper used a beam of size 5 instead of 20. The other settings are the same.

Table 10.8. Accuracies for the four different backgrounds of the mutagenesis data (estimated by a 10-fold cross-validation). The first three columns are results for ICL (with the *default settings*, except for maxbody=8, and without discretization). Pos and Neg are the two classes and for each of them a DNF theory is learned and evaluated. Multi merges the 2 theories into a multi-class theory. The results for PROGOL, FOIL and TILDE have been taken from [10.8].

			Accuracies (%)			
	Neg	Pos	Multi	PROGOL	FOIL	TILDE
BG1	79.3±8.2	67.6±5.1	80.9±7.4	76	61	75
BG2	80.3±8.9	74.5±6.9	82.4±7.4	81	61	79
BG3	85.1±8.7	83.5±5.9	86.7±10.0	83	83	85
BG4	85.1±7.7	86.2±7.6	88.3±8.0	88	82	86

Table 10.9. Accuracies of machine learning systems predicting Biodegradability. Results are taken from [10.34]. We have left out the results of the regression systems.

System	Representation	Accuracy	Accuracy (+/-)
C4.5	P1	55.2	86.2
C4.5	P2	56.9	82.4
RIPPER	P1	52.6	89.8
RIPPER	P2	57.6	93.9
FFOIL	R1'	53.0	88.7
ICL	R1	55.7	92.6
S-CART-C	P1 + R1	55.0	90.0
TILDE-C	R1	51.0	88.6
TILDE-C	P1 + R1	52.0	89.0

cation accuracy and Accuracy (+/-1) is the accuracy where only misclassification by more than one class counts as an error (e.g., slow as fast, moderate as resistant,...). ICL has only been applied to the relational representation R1. Of all the relational learners using R1, ICL achieves the highest Accuracy and Accuracy(+/-1). Compared to the propositional systems, ICL is better than all systems in term of Accuracy(+/-1), except for RIPPER on P2. For more specific results and discussions we refer to the paper.

ICL also participated in the *PTE-2 challenge* of which the results have been published in [10.61, 10.62]. The challenge was to make carcinogenesis predictions for 30 compounds, based on models constructed by Machine Learning programs. There were 9 (legal) submissions using ILP systems (TILDE, WARMR/MACCENT, ICL, P-PROGOL) and combinations of propositional systems (like C4.5) and ILP (like rules from WARMR). ICL and the other ILP systems perform unexpectedly well on scales of quantitative performance. ICL itself is in the top 3 of ILP systems (with 78% accuracy). ILP assisted models appear to be better than expert assisted ones (w.r.t. the PTE-2 data). Interesting results are obtained with propositional prediction methods using results from ILP systems (for example C4.5 using rules/substructures generated by WARMR).

Other successful experiments with ICL include: *finite element mesh design* by [10.67]; automated acquisition of knowledge on *traffic problem detection* by [10.30, 10.33]; and the problem of *diterpene structure elucidation* from ^{13}C NMR spectra by [10.30].

To conclude, ICL performs as well as other well-known ILP systems, and thus can be said to be a successful upgrade.

10.6 Related work and conclusions

There are plenty of other inductive logic programming systems whose development more or less fits in with the proposed methodology: FOIL [10.56], RIBL [10.35], S-CART [10.44], TILDE [10.9, 10.7], WARMR [10.26, 10.25], MACCENT [10.24], jk-CT learner [10.21], CLAUDIEN [10.20], Probabilistic Relational Models [10.43], Cohen's Flipper (in [10.17]), [10.59] and RDBC [10.42].

E.g., Quinlan's FOIL can also be considered an upgrade of either Michalski's AQ (1983) or CN2, RIBL upgrades the classical k-nearest neighbor algorithm (using a first-order distance due to [10.6]), S-CART and TILDE upgrade the well-known decision (and regression) tree paradigm incorporated in CART [10.12] and C4.5 [10.54, 10.55], WARMR upgrades APRIORI [10.2, 10.1], MACCENT upgrades the Maximum Entropy approach in [10.5], De Raedt and Dzeroski's PAC-learning results (as well as its incorporation in the CLAUDIEN system) for jk-CT are derived from results in [10.65] for k-CNF, Reddy and Tadepalli's results are based on the well-known results on learning horn-sentences by [10.3], Flipper upgrades Cohen's earlier RIPPER [10.15], Koller's probabilistic relational models upgrade (propositional) Bayesian networks, and Kirste and Wrobel's cluster system upgrades bottom-up agglomerative clustering algorithms to first-order logic.

Hence, it is clear that the methodology we presented is not really new. It has been applied - implicitly - several times before to obtain effective inductive logic programming systems. One might even argue that it has been applied in some of the pre-ILP work on relational learning (e.g., [10.69, 10.46]). The most important contribution of our work therefore is to describe the underlying recipe *explicitly* and to show through a case study that it can be used to obtain novel inductive logic programming techniques. By no mean we wish to imply that this recipe is the *only* way to obtain inductive logic programming systems. Certainly, some systems, of which perhaps PROGOL [10.47] and MIS [10.60] are the best examples of well-known inductive logic programming systems that have been derived from logical principles (without our recipe). Yet, we hope that our work gives new insights into the field of inductive logic programming and its relation to propositional machine learning.

Acknowledgments

At the time of writing this chapter, Wim Van Laer and Luc De Raedt were supported by the ALADIN Esprit Project 28.623. The authors would like to thank the machine learning group of Leuven, and more specifically Hendrik Blockeel, Luc Dehaspe and Maurice Bruynooghe, for their feedback, comments and interesting discussions on this topic.

The breast cancer, lymphography and primary-tumor domain were obtained from the University Medical Centre, Institute of Oncology, Ljubljana, Slovenia. Thanks go to M. Zwitter and M. Soklič for providing the data. The Mutagenesis dataset is made public by King and Srinivasan [10.63] and is available at `http://web.comlab.ox.ac.uk/oucl/research/areas/machlearn/mutagenesis.html`.

More info on ICL can be found online:
`http://www.cs.kuleuven.ac.be/~wimv/ICL/main.html`.

References

10.1 R. Agrawal, H. Mannila, R. Srikant, H. Toivonen, and A. Verkamo. Fast discovery of association rules. In U. Fayyad, G. Piatetsky-Shapiro, P. Smyth, and R. Uthurusamy, editors, *Advances in Knowledge Discovery and Data Mining*, pages 307–328. MIT Press, Cambridge, MA, 1996.

10.2 R. Agrawal, T. Imielinski, and A. Swami. Mining association rules between sets of items in large databases. In *Proceedings of the ACM SIGMOD Conference on Management of Data*, pages 207–216. ACM Press, New York, 1993.

10.3 D. Angluin, M. Frazier, and L. Pitt. Learning conjunctions of Horn clauses. *Machine Learning*, 9: 147–162, 1992.

10.4 F. Bergadano and D. Gunetti. An interactive system to learn functional logic programs. In *Proceedings of the Thirteenth International Joint Conference on Artificial Intelligence*, pages 1044–1049. Morgan Kaufmann, San Mateo, CA, 1993.

10.5 A. Berger, V. Della Pietra, and S. Della Pietra. A maximum entropy approach to natural language processing. *Computational Linguistics*, 22(1): 39–71, 1996.

10.6 G. Bisson. Conceptual clustering in a first order logic representation. In *Proceedings of the Tenth European Conference on Artificial Intelligence*, pages 458–462. John Wiley and Sons, 1992.

10.7 H. Blockeel. *Top-down induction of first order logical decision trees.* PhD thesis, Department of Computer Science, Katholieke Universiteit Leuven, 1998. `http://www.cs.kuleuven.ac.be/~ml/PS/blockeel98:phd.ps.gz`.

10.8 H. Blockeel and L. De Raedt. Experiments with top-down induction of logical decision trees. Technical Report CW 247, Department of Computer Science, Katholieke Universiteit Leuven, Belgium, 1997. `http://www.cs.kuleuven.ac.be/publicaties/rapporten/CW1997.html`.

10.9 H. Blockeel and L. De Raedt. Top-down induction of first order logical decision trees. *Artificial Intelligence*, 101(1-2): 285–297, 1998.

10.10 M. Bongard. *Pattern Recognition.* Spartan Books, 1970.

10.11 I. Bratko and S. Muggleton. Applications of inductive logic programming. *Communications of the ACM*, 38(11): 65–70, 1995.
10.12 L. Breiman, J. H. Friedman, R. A. Olshen, and C. J. Stone. *Classification and Regression Trees.* Wadsworth, Belmont, 1984.
10.13 P. Clark and R. Boswell. Rule induction with CN2: Some recent improvements. In *Proceedings of the Fifth European Working Session on Learning*, pages 151–163. Springer, Berlin, 1991.
10.14 P. Clark and T. Niblett. The CN2 algorithm. *Machine Learning*, 3(4): 261–284, 1989.
10.15 W. W. Cohen. Fast effective rule induction. In *Proceedings of the Twelfth International Conference on Machine Learning*, pages 115–123. Morgan Kaufmann, San Mateo, CA, 1995.
10.16 W. Cohen. Grammatically biased learning: learning logic programs using an explicit antecedent description language. *Artificial Intelligence*, 68: 303–366, 1994.
10.17 L. De Raedt, editor. *Advances in Inductive Logic Programming.* IOS Press, Amsterdam, 1996.
10.18 L. De Raedt. Logical settings for concept learning. *Artificial Intelligence*, 95: 187–201, 1997.
10.19 L. De Raedt. Attribute-value learning versus inductive logic programming: The missing links (extended abstract). In *Proceedings of the Eighth International Conference on Inductive Logic Programming*, pages 1–8. Springer, Berlin, 1998.
10.20 L. De Raedt and L. Dehaspe. Clausal discovery. *Machine Learning*, 26: 99–146, 1997.
10.21 L. De Raedt and S. Džeroski. First order *jk*-clausal theories are PAC-learnable. *Artificial Intelligence*, 70: 375–392, 1994.
10.22 L. De Raedt, N. Lavrač, and S. Džeroski. Multiple predicate learning. In *Proceedings of the Thirteenth International Joint Conference on Artificial Intelligence*, pages 1037–1042. Morgan Kaufmann, San Mateo, CA, 1993.
10.23 L. De Raedt and W. Van Laer. Inductive constraint logic. In *Proceedings of the Sixth International Workshop on Algorithmic Learning Theory*, pages 80–94. Springer, Berlin, 1995.
10.24 L. Dehaspe. Maximum entropy modeling with clausal constraints. In *Proceedings of the Seventh International Workshop on Inductive Logic Programming*, pages 109–124. Springer, Berlin, 1997.
10.25 L. Dehaspe. *Frequent Pattern Discovery in First-Order Logic.* PhD thesis, Department of Computer Science, Katholieke Universiteit Leuven, Belgium, 1998. `http://www.cs.kuleuven.ac.be/~ldh/`.
10.26 L. Dehaspe and L. De Raedt. Mining association rules in multiple relations. In *Proceedings of the Seventh International Workshop on Inductive Logic Programming*, pages 125–132. Springer, Berlin, 1997.
10.27 B. Dolšak, I. Bratko, and A. Jezernik. Finite element mesh design: An engineering domain for ilp application. In *Proceedings of the Fourth International Workshop on Inductive Logic Programming*, pages 305-320. GMD, Sankt Augustin, Germany, 1994.
10.28 P. Domingos. A process-oriented heuristic for model selection. In *Proceedings of the Fifteenth International Conference on Machine Learning*, pages 127–135. Morgan Kaufmann, San Francisco, CA, 1998.
10.29 J. Dougherty, R. Kohavi, and M. Sahami. Supervised and unsupervised discretization of continuous features. In *Proceedings of the Twelfth International Conference on Machine Learning.* Morgan Kaufmann, San Mateo, CA, 1995.

10.30 S. Džeroski, S. Schulze-Kremer, et al. Diterpene structure elucidation from 13C NMR spectra with inductive logic programming. *Applied Artificial Intelligence*, 12(5): 363–384, 1998.
10.31 S. Džeroski and I. Bratko. Applications of inductive logic programming. In L. De Raedt, editor, *Advances in inductive logic programming*, pages 65–81. IOS Press, Amsterdam, 1996.
10.32 S. Džeroski, B. Cestnik, and I. Petrovski. Using the m-estimate in rule induction. *Journal of Computing and Information Technology*, 1(1): 37–46, 1993.
10.33 S. Džeroski, N. Jacobs, M. Molina, and C. Moure. ILP experiments in detecting traffic problems. In *Proceedings of the Tenth European Conference on Machine Learning*, pages 61–66. Springer, Berlin, 1998.
10.34 S. Džeroski, H. Blockeel, B. Kompare, S. Kramer, B. Pfahringer, and W. Van Laer. Experiments in Predicting Biodegradability. In *Proceedings of the Ninth International Workshop on Inductive Logic Programming*, pages 80–91. Springer, Berlin, 1999.
10.35 W. Emde and D. Wettschereck. Relational instance-based learning. In *Proceedings of the Thirteenth International Conference on Machine Learning*, pages 122–130. Morgan Kaufmann, San Mateo, CA, 1996.
10.36 U. Fayyad and K. Irani. Multi-interval discretization of continuous-valued attributes for classification learning. In *Proceedings of the Thirteenth International Joint Conference on Artificial Intelligence*, pages 1022–1027. Morgan Kaufmann, San Mateo, CA, 1993.
10.37 P. Flach. Strongly typed inductive concept learning. In *Proceedings of the Eighth International Conference on Inductive Logic Programming*, pages 185–194. Springer, Berlin, 1998.
10.38 J. Ganascia and Y. Kodratoff. Improving the generalization step in learning. In R. Michalski, J. Carbonell, and T. Mitchell, editors, *Machine Learning: An Artificial Intelligence Approach*, pages 215–241. Morgan Kaufmann, San Mateo, CA, 1986.
10.39 F. Hayes-Roth and J. McDermott. An interference matching technique for inducing abstractions. *Communications of the ACM*, 21: 401–410, 1978.
10.40 J.-U. Kietz and S. Wrobel. Controlling the complexity of learning in logic through syntactic and task-oriented models. In S. Muggleton, editor, *Inductive logic programming*, pages 335–359. Academic Press, London, 1992.
10.41 R. D. King, M. J. E. Sternberg, A. Srinivasan, and S. Muggleton. Relating chemical activity to structure: an examination of ILP successes. *New Generation Computing*, 13(3-4): 411–434, 1995.
10.42 M. Kirsten and S. Wrobel. Relational distance-based clustering. In *Proceedings of the Eighth International Conference on Inductive Logic Programming*, pages 261–270. Springer, Berlin, 1998.
10.43 D. Koller. Probabilistic relational models. In *Proceedings of the Ninth International Workshop on Inductive Logic Programming*, pages 3–13. Springer, Berlin, 1999.
10.44 S. Kramer. Structural regression trees. In *Proceedings of the Thirteenth National Conference on Artificial Intelligence*, pages 812–819. AAAI Press, Menlo Park, CA, 1996.
10.45 R. Michalski, I. Mozetič, J. Hong, and N. Lavrač. The multi-purpose incremental learning system AQ15 and its testing application on three medical domains. In *Proceedings of the Fifth National Conference on Artificial Inteligence*, pages 1041–1045. Morgan Kaufmann, San Mateo, CA, 1986.
10.46 R. S. Michalski. A theory and methodology of inductive learning. In R. S. Michalski, J. Carbonell, and T. Mitchell, editors, *Machine Learning:*

An Artificial Intelligence Approach, pages 83–134. Morgan Kaufmann, San Mateo, CA, 1983.

10.47 S. Muggleton. Inverse entailment and Progol. *New Generation Computing*, 13(3-4): 245–286, 1995.

10.48 S. Muggleton and L. De Raedt. Inductive logic programming : Theory and methods. *Journal of Logic Programming*, 19, 20: 629–679, 1994.

10.49 S. Muggleton and C. Feng. Efficient induction of logic programs. In *Proceedings of the First Conference on Algorithmic Learning Theory*, pages 368–381. Ohmsma, Tokyo, Japan, 1990.

10.50 C. Nédellec, H. Adé, F. Bergadano, and B. Tausend. Declarative bias in ILP. In L. De Raedt, editor, *Advances in Inductive Logic Programming*, pages 82–103. IOS Press, 1996.

10.51 S.-H. Nienhuys-Cheng and R. Wolf. *Foundations of inductive logic programming.* Springer, Berlin, 1997.

10.52 G. Plotkin. A note on inductive generalization. In B. Meltzer and D. Michie, editors, *Machine Intelligence*, pages 153–163. Edinburgh University Press, Edinburgh, 1970.

10.53 U. Pompe and I. Kononenko. Probabilistic first-order classification. In *Proceedings of the Seventh International Workshop on Inductive Logic Programming.* Springer, Berlin, 1997.

10.54 J. R. Quinlan. *C4.5: Programs for Machine Learning.* Morgan Kaufmann, San Mateo, CA, 1993.

10.55 J. R. Quinlan. Induction of decision trees. *Machine Learning*, 1: 81–106, 1986.

10.56 J. R. Quinlan. Learning logical definitions from relations. *Machine Learning*, 5: 239–266, 1990.

10.57 J. R. Quinlan. Determinate Literals in Inductive Logic Programming. In *Proceedings of the Eighth International Workshop on Machine Learning*, pages 442–446. Morgan Kaufmann, San Mateo, CA 1991.

10.58 J. R. Quinlan. Learning first-order definitions of functions. *Journal of Artificial Intelligence Research*, 5: 139–161, 1996.

10.59 C. Reddy and P. Tadepalli. Learning first-order acyclic Horn programs from entailment. In *Proceedings of the Eighth International Conference on Inductive Logic Programming*, pages 23–37. Springer, Berlin, 1998.

10.60 E. Shapiro. An algorithm that infers theories from facts. In *Proceedings of the Seventh International Joint Conference on Artificial Intelligence*, pages 446–452. Morgan Kaufmann, San Mateo, CA, 1981.

10.61 A. Srinivasan, R. D. King, S. Muggleton, and M. J. E. Sternberg. The predictive toxicology evaluation challenge. In *Proceedings of the Fifteenth International Joint Conference on Artificial Intelligence*, pages 1–6. Morgan Kaufmann, San Mateo, CA, 1997.

10.62 A. Srinivasan, R. King, and D. Bristol. An assessment of ILP-assisted models for toxicology and the PTE-3 experiment. In *Proceedings of the Ninth International Workshop on Inductive Logic Programming*, pages 291–302. Springer, Berlin, 1999.

10.63 A. Srinivasan, S. Muggleton, M. J. E. Sternberg, and R. D. King. Theories for mutagenicity: A study in first-order and feature-based induction. *Artificial Intelligence*, 85(1,2): 277–299, 1996.

10.64 H. Toivonen, M. Klemettinen, P. Ronkainen, K. Hätönen, and H. Mannila. Pruning and grouping discovered association rules. In *Proceedings of the ML-net Familiarization Workshop on Statistics, Machine Learning and Knowledge Discovery in Databases*, pages 47–52. FORTH, Heraklion, Greece, 1995.

10.65 L. Valiant. A theory of the learnable. *Communications of the ACM*, 27: 1134–1142, 1984.

10.66 P. R. J. van der Laag and S.-H. Nienhuys-Cheng. Completeness and properness of refinement operators in inductive logic programming. *Journal of Logic Programming*, 34(3): 201–225, 1998.

10.67 W. Van Laer, L. De Raedt, and S. Džeroski. On multi-class problems and discretization in inductive logic programming. In *Proceedings of the Tenth International Symposium on Methodologies for Intelligent Systems*, pages 277–286. Springer, Berlin, 1997.

10.68 W. Van Laer, S. Džeroski, and L. De Raedt. Multi-class problems and discretization in ICL. In *Proceedings of the MLnet Familiarization Workshop on Data Mining with Inductive Logic Programming*, pages 53–60. FORTH, Heraklion, Greece, 1995.

10.69 S. Vere. Induction of concepts in the predicate calculus. In *Proceedings of the Fourth International Joint Conference on Artificial Intelligence*, pages 282–287. Morgan Kaufmann, San Mateo, CA, 1975.

10.70 C. Vrain. Ogust: A system that learns using domain properties expressed as theorems. In Y. Kodratoff and R. S. Michalski, editors, *Machine Learning: An Artificial Intelligence Approach*, pages 360–381. Morgan Kaufmann, San Mateo, CA, 1990.

10.71 P. Winston. Learning structural descriptions from examples. In P. Winston, editor, *Psychology of Computer Vision*. MIT Press, Cambridge, MA, 1975.

11. Propositionalization Approaches to Relational Data Mining

Stefan Kramer[1], Nada Lavrač[2], and Peter Flach[3]

[1] Machine Learning and Natural Language Processing Lab
Institute for Computer Science, Albert-Ludwigs University Freiburg
Am Flughafen 17, D-79110 Freiburg i. Br., Germany

[2] Jožef Stefan Institute, Jamova 39, 1000 Ljubljana, Slovenia

[3] Department of Computer Science, University of Bristol
The Merchant Venturers Building, Woodland Road
Bristol BS8 1UB, United Kingdom

Abstract

This chapter surveys methods that transform a relational representation of a learning problem into a propositional (feature-based, attribute-value) representation. This kind of representation change is known as propositionalization. Taking such an approach, feature construction can be decoupled from model construction. It has been shown that in many relational data mining applications this can be done without loss of predictive performance. After reviewing both general-purpose and domain-dependent propositionalization approaches from the literature, an extension to the LINUS propositionalization method that overcomes the system's earlier inability to deal with non-determinate local variables is described.

11.1 Introduction

Given learning problems represented relationally, one can either develop learning algorithms to deal with these problems directly in their original relational representation, or transform these problems into a format suitable for propositional (feature-based, attribute-value) learning algorithms. This latter transformation requires the construction of features that capture relational properties of the learning examples. Such a transformation of a relational learning problem into a propositional one is called *propositionalization.* This chapter gives a survey of propositionalization methods and presents one of the approaches in sufficient detail for practical use in relational data mining tasks.

The chapter is concerned only with prediction tasks, defined as follows.

Given some evidence E (examples, given extensionally either as a set of ground facts or tuples representing a predicate/relation whose intensional definition is to be learned),
and an initial theory B (background knowledge, given either extensionally as a set of ground facts, relational tuples or sets of clauses over the set of background predicates/relations)

Find a theory H (hypothesis, in the form of a set of logical clauses) that together with B *explains* some properties of E.

In most cases the hypothesis H has to satisfy certain restrictions, which we shall refer to as the *bias*. Bias is needed to reduce the number of candidate hypotheses. It consists of the language bias L, determining the hypothesis space, and the search bias which restricts the search of the space of possible hypotheses.

The background knowledge used to construct hypotheses is a distinctive feature of ILP. It is well-known that relevant background knowledge may substantially improve the results of learning in terms of accuracy, efficiency and the explanatory potential of the induced knowledge. On the other hand, irrelevant background knowledge will have just the opposite effect. Consequently, much of the art of inductive logic programming lies in the appropriate selection and formulation of background knowledge to be used by the selected ILP learner.

This chapter shows that by devoting enough effort to the construction of features, which encode the background knowledge in learning, even complex relational learning tasks can be solved by simple propositional rule learning systems. Generally, representation changes for learning are known under the term *constructive induction*. In propositional learning, constructive induction involves the construction of new features based on existing ones [11.43, 11.54, 11.42, 11.27]. A first-order counterpart of feature construction is *predicate invention* (see, for instance a paper by Stahl [11.50] for an overview of predicate invention in ILP). This chapter takes the middle ground by performing a simple form of predicate invention through first-order feature construction, and using the constructed features for propositional learning. In this way, feature construction and model construction can be decoupled.

It has been shown that in many practical relational data mining tasks feature construction and model construction can be decoupled without loss of performance. Two recent successful applications of propositionalization are described by Srinivasan *et al.* [11.48] and Džeroski *et al.* [11.16]. In their summary of the results from the Predictive Toxicology Evaluation (PTE) challenge, an International competition in carcinogenicity prediction [11.48], one of the authors' major observations is that feature construction methods played a significant role in the top-ranked results. In another application that deals with the prediction of biodegradation rates [11.16], the authors applied a large number of propositional and relational learning algorithms to both classification and regression variants of the learning problem. Similar to the PTE results, propositionalization (feature construction) methods turned out to perform best.

So what are features, and how are they constructed? A natural choice – to which we will restrict ourselves in this chapter – is to define a feature as a conjunction of (possibly negated) literals. In ILP, these literals share local variables referring, for instance, to parts of the individual to be classified, such as atoms within molecules.[1] Features may describe subsets of the training set that are for some reason unusual or interesting. For instance, the class distribution among the instances described by a feature may be different from the class distribution over the complete training set in a statistically significant way. Alternatively, a feature may simply be shared by a sufficiently large fraction of the training set. In the first case, the feature is said to describe an *interesting subgroup* of the data, and several propositional and first-order systems exist that can discover such subgroups (e.g., EXPLORA [11.25], MIDOS [11.55], and TERTIUS [11.21]). In the second case, the feature is said to describe a *frequent itemset*, and again several algorithms and systems exist to discover frequent itemsets (e.g., APRIORI [11.1] and WARMR [11.12]). Notice that the systems just mentioned are all discovery systems, which perform descriptive rather than predictive induction. Indeed, feature construction is a discovery task rather than a classification task.

The review of propositionalization approaches in this chapter considers approaches to feature construction from relational background knowledge and structural properties of individuals. It excludes other forms of constructive induction and predicate invention, which are not used to construct features from relational background predicates. Moreover, the review excludes transformation approaches which result in a table where several rows correspond to a single example. Such transformations do not involve a representation change to a propositional form but rather transform a relational problem to a multiple-instance problem [11.15] that can not be solved directly by an arbitrary propositional learner.

It can be shown that the presented approaches to propositionalization through first-order feature construction can mostly be applied in so-called *individual-centered* domains, where there is a clear notion of individual [11.18] (e.g., molecules or trains) and learning occurs on the level of individuals only. Usually, individuals are represented by a single variable, and the predicates whose definition is to be learned are either unary predicates concerning Boolean properties of individuals, or binary predicates assigning an attribute-value or a class-value to each individual. Individual-centered representations have the advantage of a strong language bias, because of the restrictions on local (existentially quantified) variables. However, not all domains are amenable to the approach we present in this chapter – in particular, we cannot learn recursive clauses, and we cannot deal with domains where there is not a clear notion of an individual (e.g., many program synthesis problems).

This chapter is organized as follows. Section 11.2 explains the background and introduces basic terminology. In Section 11.3 we give an example of a simple propositionalization of a relational learning problem. Sections 11.4 and 11.5 review the literature on general-purpose methods and special-purpose

[1] *Local* or *existential* variables are variables not occurring in the head of a clause. In contrast, variables in the head are called *global* variables. Since *global* variables mostly denote individuals in this chapter, they are also called *individual* variables here.

approaches, respectively. In Section 11.6 we discuss other related work. Finally, in Section 11.7 we pick one of the methods (based on LINUS [11.32]), and describe it in some detail.

11.2 Background and definition of terms

Representation languages in Machine Learning greatly differ with respect to their expressiveness. The main distinction is drawn between propositional, feature-based representations and relational, graph-based or so-called first-order representations.

For *propositional representations*, we assume that examples are represented as feature-vectors of fixed size. In other words, they all can be described using the same set of features. For a fixed number of features, a feature-vector lists the values that a specific observation (or example) takes. Hence it is of the form $f_1 = v_1 \wedge ... \wedge f_n = v_n$, where f_i are the features and the v_i are the values. The problem with feature-based representations is that they are unable to encode certain types of examples (such as labeled graphs) directly. Indeed, to model e.g., labeled graphs with feature-based representations, the knowledge engineer has to select structural features of interest before performing experiments. The problem is that it is impossible to know in advance which features will be needed to solve a particular problem.

In bio-chemical domains, where the learning examples are molecules, features could be encodings of structural properties (such as counts of fragments or substructures) or 'global' physico-chemical properties of chemical compounds (such as the molecular weight). An example of a propositional representation is shown in Table 11.1.

Table 11.1. Example of a simple propositional representation of chemical compounds, where *methyl* denotes the number of methyl groups, *six_rings*, the number of rings of size 6, *logP* measures the lipophilicity, *mweight* is the molecular weight and the dependent variable to be predicted is the *degradation rate*.

		Attributes				*Prediction*
methyl	...	*six_ring*	...	*logP*	*mweight*	*degr. rate*
3	...	1	...	2.62	138.20	6.04
1	...	2	...	1.33	232.24	6.51
...	...	...	...	...	...	...

In *relational representations*, each example is given as a set of tuples in several relations. Thus, it is possible to represent structural information *as is*, without having to encode it in features. Table 11.2 shows an example of a relational representation. Both examples have been taken from the biodegradability domain [11.16].

Table 11.2. Example of a relational representation of chemical compounds. *atoms* and *bonds* are represented as tuples of their respective relations. Each *atom* and each *bond* can have their own properties. For instance, atoms are of a certain element, and bonds are of a particular bond type (or bond order).

```
% atom(cas, atom_id, element).
atom('78-59-1','78-59-1_1', o).
atom('78-59-1','78-59-1_2', c).
...

% bond(cas, atom_id, atom_id, bond_type).
bond('78-59-1','78-59-1_1','78-59-1_2', 2).
bond('78-59-1','78-59-1_2','78-59-1_3', 1).
...

% <functional_group>(cas, member_atoms, connected_list).
methyl('78-59-1', ['78-59-1_7',...,'78-59-1_7_3'],
                  ['78-59-1_6']).
methyl('78-59-1', ['78-59-1_8',...,'78-59-1_8_3'],
                  ['78-59-1_6']).
methyl('78-59-1', ['78-59-1_9',...,'78-59-1_9_3'],
                  ['78-59-1_4']).
...

six_ring('78-59-1', ['78-59-1_10',...,'78-59-1_6'],
                    ['78-59-1_1',...,'78-59-1_9']).
...

logP('78-59-1', 2.61). mweight('78-59-1', 138.20).

degradation_rate('78-59-1', 6.04).
```

Propositionalization is defined as a representation change from a relational representation to a propositional one. It involves the construction of features from relational background knowledge and structural properties of individuals. One nice feature of this approach is that any propositional learner can be applied after propositionalization. Propositionalization is a special case of constructive induction and feature construction.

Propositionalizations can be either complete or partial (heuristic). In the former case, no information is lost in the process; in the latter, information is lost and the representation change is incomplete: we do not aim at obtaining all possible features that can be derived from the relational representation, but only their subset. So, our goal is to automatically generate a small but relevant set of structural features.

De Raedt [11.14] showed that in the simplest first-order learning setting (learning from interpretations [11.13]), a *complete transformation* into a propositional representation is possible, but results in a feature space that is exponential in the number of parameters of the original learning problem. In addition, the subsumption relation between the features (the lattice of clauses) is no longer represented explicitly, so that it cannot be used or exploited in search in a straightforward manner.

Several authors have proposed partial or heuristic approaches to propositionalization [11.30, 11.47, 11.12]. In partial propositionalization, one is looking for a set of features, where each feature is defined in terms of a corresponding program clause. If the number of features is m, then a propositionalization of the relational learning problem is simply a set of clauses:

$$\begin{aligned} f_1(X) &:- Lit_{1,1}, \ldots, Lit_{1,n_1}. \\ f_2(X) &:- Lit_{2,1}, \ldots, Lit_{2,n_2}. \\ &\ldots \\ f_m(X) &:- Lit_{m,1}, \ldots, Lit_{m,n_m}. \end{aligned} \tag{11.1}$$

where each clause defines a feature f_i. Clause body $Lit_{i,1}, ..., Lit_{i,n_i}$ is said to be the definition of feature f_i; these literals are derived from the relational background knowledge. In the clause head $f_i(X)$, argument X refers to an individual (e.g., a molecule, a train, ...), as introduced in the discussion on individual-centered representations in Section 11.1. If such a clause is called for a particular individual (i.e., if X is bound to some example identifier) and this call succeeds at least once, the corresponding Boolean feature is defined to be "true" for the given example; otherwise, it is defined to be "false".[2]

Notice that propositionalization could also be defined more broadly, involving any representation change that would allow relational learning problems to be transformed into propositional ones, not only those involving the construction of features from relational background knowledge and structural properties of individuals. Some of these approaches will be reviewed in the section on related work. It is also worth pointing out that most propositionalization approaches decouple feature construction from model construction (i.e., learning). In a non-propositionalization approach, features are constructed during learning, rather than more or less exhaustively as a pre-processing step. This means that hte applied propositional learner method must be able to cope with many features, and that it must have mechanisms to avoid overfitting.

11.3 An example illustrating a simple propositionalization

To illustrate propositionalization, consider a simple relational learning problem of learning illegal positions on a chess board with only two pieces, white

[2] Some features can be non-Boolean, e.g., 'the lowest charge of an atom in the molecule'. Such features would require a second variable in the head of the clause defining the feature, returning the value of the feature. Non-Boolean features often occur when the local variables in the feature are all determinate, i.e., take on a unique value given a specific individual.

king and black king. In this simplified example, a position P is a four-tuple P = <WKf,WKr,BKf,BKr>, where WKf,WKr and BKf,BKr stand for the file and rank of white king and black king, respectively. Let illegal(WKf,WKr,BKf,BKr) denote that the position in which the white king is at (WKf,WKr) and the black king at (BKf,BKr) is illegal. Arguments WKf and BKf are of type file (with values a to h), while WKr and BKr are of type rank (with values 1 to 8).

Suppose that the learner is given one positive example (illegal chess endgame position) p1 = <a,6,a,7>, and two negative examples (legal endgame positions): p2 = <f,5,c,4> and p3 = <b,7,b,3>, and the learning task is to induce the hypothesis defining predicate illegal(P) from the given labeled examples:

```
illegal(a,6,a,7).  ⊕
illegal(f,5,c,4).  ⊖
illegal(b,7,b,3).  ⊖
```

Consider the background knowledge about chess positions, reflecting the common knowledge that for checking whether a position is illegal it is worthwhile checking if the two kings are next to each other. The background knowledge may thus involve the definition of whether two files/ranks are identical or adjacent. So, as indicated above, the background knowledge for our knowledge is represented by predicates adjF(X,Y) and adjR(X,Y), which can be applied on arguments of type file and rank, respectively, expressing that X and Y are adjacent. These predicates are defined extensionally by simple facts of the following form:

```
adjF(a,b)    adjR(1,2)
adjF(b,c)    adjR(2,3)
   ...          ...
   ...          ...
adjF(g,h)    adjR(7,8)
```

We assume that the learner knows that the two predicates are symmetric, i.e., adjF(X,Y) = adjF(Y,X) and adjR(X,Y) = adjR(Y,X). Suppose that a built-in symmetric equality predicate eq(X,Y), which works on arguments of the same type, is also provided as part of the background knowledge. As opposed to predicates adjF/2 and adjR/2, predicate equal/2 is defined intensionally as eq(X,Y):- X = Y, for X and Y of the same type.[3]

The learning problem to be propositionalized is to learn a hypothesis that defines the predicate illegal(WKf,WKr,BKf,BKr) in terms of background predicates adjF(X,Y), adjR(X,Y), and eq(X,Y).

How does propositionalization work? In its simplest form, propositionalization will consider only simple features (involving only one literal). Given the background knowledge, considering typing and symmetry of background

[3] Checking that the two arguments are of the same type must be implemented in the system, since this is not automatically provided by Prolog.

predicates (allowing for only one of the two symmetric variants of a literal to be considered), only four simple features f_i are taken into account: `eq(WKf,BKf)`, `eq(WKr,BKr)`, `adjF(WKf,BKf)` and `adjR(WKr,BKr)`.

The given examples can now be transformed into four-tuples of the form:

⟨`eq(WKf,BKf)`, `eq(WKr,BKr)`, `adjF(WKf,BKf)`, `adjR(WKr,BKr)`⟩

The actual propositional tuples are then constructed by evaluating the truth value of features for the given argument values of the ground facts of predicate `illegal`. More formally if $e \in E$ is an example, and σ_j is a substitution such that `illegal(P)`$\sigma_j \in E$, the truth value of features is obtained by testing whether $f_i\sigma_j$ is true for the given background theory B ($f_i\sigma_j$ is true if $B \models f_i\sigma_j$). There are three such substitutions, one for each training example: σ_1 = `P/<a,6,a,7>`, σ_2 = `P/<f,5,c,4>` and σ_3 = `P/<b,7,b,3>`.

The propositional table, obtained as a result of such a propositionalization, is shown in Table 11.3. Notice that the truth value tuples in Table 11.3 are generalizations (relative to the given background knowledge) of the individual facts about the target predicate `illegal/4`.

Table 11.3. Propositionalized form of the `illegal` position learning problem.

ID	*Position* <WKf,WKr, BKf,BKr>	*Features* eq(WKf,BKf)	eq(WKr,BKr)	adjF(WKf,BKf)	adjR(WKr,BKr)	*Class*
p1	<a,6,a,7>	true	false	false	true	⊕
p2	<f,5,c,4>	false	false	false	true	⊖
p3	<b,7,b,3>	true	false	false	false	⊖

Given the above representation change, by which a relational learning problem was transformed into a propositional one, resulting in a tabular representation of examples, an arbitrary propositional learner can now be used to induce a hypothesis. Suppose that a rule learner is used, inducing the following *if-then* rule:

`Class` = $\oplus$ **if** `(eq(WKf,BKf))` = `true` $\wedge$ `adjR(WKr,BKr)` = `true`

This *if-then* rule corresponds to a clause:

```
illegal(WKf,WKr,BKf,BKr) :- eq(WKf,BKf), adjR(WKr,BKr).
```

This simple example concentrates on the use of relational background knowledge. The examples themselves are already in the single-relation attribute-value format (notice that we could also have used those attributes, such as "the file of the white king", in learning). The individuals here are chess positions (i.e., 4-tuples of white king file and rank and black king file and rank), and the concept to be learned is a property of such positions. The features are properties of individuals as well, e.g., the feature "white king and black king have the same file" would be defined as

```
f1(WKf,WKr,BKf,BKr) :- eq(WKf,BKf).
```

Notice that this does not correspond to the definition of features given in Equation 11.1, where we assumed that the individual would be represented by a single variable. To solve this, a board position could be represented by a functor p with four arguments, from which the files and ranks could be extracted by projection functions:

```
pos2wkf(p(WKf,_  ,_  ,_  ),WKf).
pos2wkr(p(_  ,WKr,_  ,_  ),WKr).
pos2bkf(p(_  ,_  ,BKf,_  ),BKf).
pos2bkr(p(_  ,_  ,_  ,BKr),BKr).

f1(P) :- pos2wkf(P,WKf), pos2bkf(P,BKf), eq(WKf,BKf).
f2(P) :- pos2wkr(P,WKr), pos2bkr(P,BKr), adjR(WKr,BKr).

illegal(P) :- f1(P), f2(P).
```

This representation clearly illustrates the propositional nature of the learned rule: it has no interactions between local variables, which are all hidden in the features. Notice that the features consist of literals *introducing* local variables and literals *consuming* local variables. In the bias used in Section 11.7, the former are called *structural* predicates and the latter *utility* predicates. In this simple example, all structural predicates are determinate (i.e., in each endgame position P there is a unique value for WKf, WKr, etc.), but in the general case they could be non-determinate. Also, note that we can avoid the use of function symbols and use position identifiers instead, in which case the structural predicates would be extensionally defined rather than intensionally (see Example 11.7.1).

Not all domains are inherently individual-centered. For instance, consider a family domain, where the system may need to learn the definition `daughter(X,Y) :- female(X), parent(Y,X)` from features `female(X)` and `parent(Y,X)`. On the one hand, we could choose persons as individuals, but then we could only learn properties of persons, and not the `daughter` predicate. Taking pairs of individuals would solve this problem, but then the pair `(Y,X)` would be different from the pair `(X,Y)` and would need to be constructed from the latter by a background predicate. The third alternative, taking whole family trees as individuals, suffers from the same problem as the first, namely that `daughter` is not a property of the individual. This is not to say that this problem would not be amenable to propositionalization (on the contrary, it can easily be solved due to the lack of local variables), but rather that this would be a non-typical problem for most of the propositionalization approaches discussed in this chapter.[4]

[4] The family domain is more closely related to *program synthesis*, which typically involves some calculation to determine the value of one argument of the target

Finally, the transformation shown in this section is an example of the LINUS system [11.32, 11.31] at work. In the following section, we will present this classical propositionalization system in some detail.

11.4 Feature construction for general-purpose propositionalization

This and the next section review some of the existing approaches to propositionalization. The review concentrates on the feature construction step. We distinguish between general-purpose approaches (this section) and special-purpose approaches which are either domain-dependent, assume a strong declarative bias, or are applicable to a limited problem class (Section 11.5).

11.4.1 LINUS

LINUS [11.32, 11.31] is a descendant of the learning algorithm used in QUMAS (Qualitative Model Acquisition System) which was used to learn functions of components of a qualitative model of the heart in the KARDIO expert system for diagnosing cardiac arrhythmias [11.3]. The QUMAS learner and LINUS were the first systems to transform a relational representation into a propositional representation. The hypothesis language of LINUS is restricted to function-free constrained DHDB (deductive hierarchical database) clauses. This implies that no recursion is allowed, and that no new variables may be introduced. A clause is constrained if all variables in the body also occur in the head. An example of such a clause is `daughter(X,Y) :- female(X), parent(Y,X)`. It should be clear that one can transform this kind of hypotheses into a propositional form, if `female(X)` and `parent(Y,X)` are defined as features.

After propositionalization, users of the LINUS system can choose among a number of propositional learning algorithms. These include the decision tree induction system ASSISTANT [11.5], and two rule induction systems: an ancestor of AQ15, named NEWGEM [11.37], and CN2 [11.8, 11.7]. Recently, LINUS has been upgraded also with an interface to MLC++ [11.26]. The hypotheses induced by either of these systems are translated back into first-order logic to facilitate their interpretation.

DINUS [11.31] weakens the language bias of LINUS so that the system can learn clauses with a restricted form of new variables, introduced by determinate literals. A literal is determinate if each of its variables that does not appear in preceding literals has only one possible binding given the bindings

predicate, given the others. It can be trivially transformed into a Boolean classification problem, by ignoring the input-output relations and viewing the program as a predicate rather than a function, but this ignores crucial information and makes the task even harder.

of its variables that appear in preceding literals. An example of such clauses is:

```
grandmother(X,Y) :- father(Z,Y), mother(X,Z).
grandmother(X,Y) :- mother(U,Y), mother(X,U).
```

This restriction allows for a similar transformation approach as the one taken in LINUS.

11.4.2 Stochastic propositionalization

Kramer et al. [11.30] developed a method that finds sets of features that together possess good discriminatory power. Stochastic search is used to find clauses of arbitrary length, where each clause corresponds to a binary feature. The search strategy employed is similar to random mutation hill-climbing. Over a predefined number of episodes, the algorithm develops a set of m clauses. An elementary step in this search consists of the replacement of a number of clauses in the current generation. The removal of clauses as well the addition of new ones is done probabilistically, i.e., with a probability proportional to the fitness of individual clauses. The fitness function of individual clauses/features is based on the Minimum Description Length (MDL) principle. Also, we impose constraints on desirable features, for instance, on their generality and specificity. For the final selection of one generation of features, we evaluate sets of features/clauses by the resubstitution estimate (i.e., the training set error) of the decision table constructed from all features in the set.

An example for a feature found by this method is:

```
f1(A) :- atm(A, B, _, 27, _),
         bond(A, B, C, _),
         atm(A, C, _, 29, _).
```

This feature is “true” if in compound A there exists a bond between an atom of type 27 (according to the biomolecular modeling package QUANTA) and an atom of type 29. While this is still a relatively simple feature, in some domains the average length was about 4 and the standard deviation was up to 3. So, the method often is able to search deeper than other methods due to its flexible “search horizon”, but cannot provide guarantees for the optimality or completeness of its propositionalizations. This method also differs from other methods in that it works in a class-sensitive manner and that it attempts to find features that work well together (not just features that are useful individually).

11.4.3 PROGOL

Srinivasan and King [11.47] presented a method for feature construction based on hypotheses returned by PROGOL [11.38]. For each clause, each input-output connected subset of literals is used to define a feature. For instance, if PROGOL returns a theory

```
inactive(A) :- has_rings(A, [R1, R2]),
               hydrophobic(A, H),
               H > 1.0.
```

then this approach would come up with the following feature definitions:

```
f1(A) :- has_rings(A, [R1, R2]).
f2(A) :- hydrophobic(A, H).
f3(A) :- hydrophobic(A, H), H  > 1.0.
f4(A) :- has_rings(A, [R1, R2]), hydrophobic(A, H).
f5(A) :- has_rings(A, [R1, R2]), hydrophobic(A, H), H > 1.0.
```

Since f2 would be true for all examples, the system would discard it in a subsequent step. The authors used these "ILP-features" in multiple linear regression.

This method works for all types of background knowledge. Besides, it could be applied to any ILP algorithm.

11.4.4 WARMR

WARMR [11.12] is an algorithm for finding association rules over multiple relations. The main step of the algorithm consists of detecting frequently succeeding Datalog queries. Queries returned by WARMR have been successfully used to construct features from relational background knowledge.

Datalog is a Prolog-like language used for deductive databases that excludes function symbols. A Datalog query is an expression of the form $? - A_1, \ldots, A_n$, where $A_1, \ldots, A_n$ are logical atoms. Such a query denotes a conjunction of conditions. The WARMR system automatically generates queries up to a certain length, and returns those that succeed for a sufficient number of cases. A query is frequently succeeding, if the conditions in it hold for a sufficient number of examples in the dataset.

Given a representation in Datalog, this approach can be applied to the problem of detecting commonly occurring substructures in chemical compounds. One simply asks for frequently succeeding queries, where $A_1, \ldots, A_n$ denote structural properties of compounds. An example for such a query is

$$? - six\_ring(C, S), atomel(C, A_1, h), atomel(C, A_2, c), bond(C, A_1, A_2, X).$$

If such a query is known to succeed for a sufficient number examples C, it is turned into a feature:

```
f1(C) :- six_ring(C,S),  atomel(C,A1, h),
         atomel(C,A2,c), bond(C,A1,A2,X).
```

WARMR features have been used with several propositional learning algorithms, e.g., C4.5[11.44].

11.5 Special-purpose feature construction

In this section, we review some of the existing approaches to propositionalization that proved to be successful in solving specific learning tasks, either due to assuming a very strong language bias (graph-based approaches) or due to being designed only for very specialized tasks. Despite their specificity, reviewing of these approaches may serve the reader as a source of ideas for designing special-purpose feature construction methods for new learning problems to be solved.

11.5.1 Turney's approach applied in the East-West challenge

In the East-West challenge aimed at discovering low size-complexity Prolog programs for classifying trains as eastbound or westbound [11.35] (see also Section 11.7.1), Turney's RL-ICET algorithm achieved one of the best results [11.51]. Its success was due to exhaustive feature construction, introducing new propositional features for all combinations of up to three Prolog literals. This led to new features, as illustrated in Table 11.4 below. To minimize size complexity, costs of features were computed and taken into account by a cost-sensitive decision tree induction algorithm ICET [11.51].

Table 11.4. Examples of features and their costs.

Feature	*Prolog Fragment*	*Cost (Complexity)*
ellipse	*has_car(T, C), ellipse(C).*	5
short_closed	*has_car(T, C), short(C), closed(C).*	7
train_4	*len1(T, 4).*	3
train_hexagon	*has_load1(T, hexagon).*	3
ellipse_peaked_roof	*has_car(T, C), ellipse(C), arg(5, C, peaked).*	9
u_shaped_no_load	*has_car(T, C), u_shaped(C), has_load(C, 0).*	8
rectangle_load_infront jagged_roof	*infront(T, C1, C2), has_load0(C1, rectangle), arg(5, C2, jagged).*	11

In summary, the program constructs 28 features applicable to a single car, and features of the type
$<property\ of\ car_i>\_infront\_<property\ of\ car_{i+1}>$ for any possible property of a single car, e.g., *u_shaped_in_front_peaked_roof*. This type of feature construction from relational background knowledge only works for "linear sequences" such as the trains in the East-West challenge.

11.5.2 Cohen's set-valued features

Cohen [11.10] introduced the notion of "set-valued features", which can be used to transform certain types of background knowledge. A value of a set-valued feature is allowed to be a set of strings. This type of feature can easily be incorporated in existing propositional learning algorithms. Some first-order learning problems (e.g., text categorization) can be propositionalized in this way.

11.5.3 Geibel and Wysotzki's graph-based approach

Geibel and Wysotzki [11.22] proposed a method for feature construction in a graph-based representation. The features are obtained through fixed-length paths in the neighborhood of a node in the graph. This approach distinguishes between "context-dependent node attributes of depth n" and "context-dependent edge attributes of depth n". A context-dependent node attribute $A^n(G)[i,i]$ is defined as follows. For each node i in graph G, we define the context of i (of depth n) as all length n paths from node i and to node i. Each such context is used to define a feature. The feature value for an example graph is the number of occurrences of the corresponding context in it. Analogously, a context-dependent edge attribute $A^n(G)[i,j]$ is defined in terms of all length n paths from node i to node j in graph G.

11.5.4 SUBDUE

SUBDUE [11.11] is an MDL-based algorithm for substructure discovery in graphs. SUBDUE finds substructures that compress the original data and represent structural concepts in the data. After SUBDUE has identified a compressive substructure, it replaces all occurrences in the graph and continues search in the transformed graph. In multiple passes of this kind, SUBDUE produces a hierarchical description of the structural regularities in the data. It has to be emphasized that SUBDUE is capable of identifying subgraphs, not just linearly connected fragments.

11.5.5 CASE/MULTICASE

The CASE/MULTICASE systems [11.23, 11.24] represent remarkable propositionalization approaches in the computational chemistry literature.

The development of a CASE [11.23] model starts with a training set, containing chemical structures and biological activities. Each molecule is split into all possible linear subunits with two to ten non-hydrogen atoms (the size of fragment varies in CASE related publications). All fragments belonging to an active molecule are labeled active, while those belonging to an inactive molecule are labeled inactive. To identify relevant substructures, CASE assumes a binomial distribution and takes a statistically significant deviation from random distribution as indication that the fragment is relevant for biological activity. In the prediction step, the probability of activity is estimated using Bayesian statistics based on the presence of the significant fragments. MULTICASE differs from CASE in that it employs separate-and-conquer (as in rule induction) to select one relevant fragment after the other. CASE/MULTICASE is quite similar to the approach described in the following section. However, since no technical details are known about fragment generation in CASE/MULTICASE, it is hard to determine the similarities and differences between these approaches.

11.5.6 Kramer and Frank's propositionalization methods for graph-based domains

Kramer and Frank [11.29] present a method for propositionalization that is tailored for domains, where the examples are given as labeled graphs. The method is designed to discover all linearly connected vertices of some type (i.e., labeled in some way) that occur frequently in the examples (up to some user-specified length).

The authors applied the method to bio-chemical databases, where the examples are 2-D descriptions of chemical compounds. In these domains, the goal is to discover frequently occurring "fragments", i.e., chains of atoms. One example for such a fragment would be "a carbon atom with a single bond to an oxygen atom with a single bond to a carbon atom". It can easily be transformed into a new feature:

```
f1(X) :- dif(A1,A2), dif(A2,A3), dif(A1,A3),
         atom(X,A1), element(A1,c),
         connected(A1,A2,B1), bondtype(B1,1), element(A2,o),
         connected(A2,A3,B2), bondtype(B2,1), element(A3,c).
```

The `dif`'s in the beginning of the clause make sure that the variables `A1`, `A2` and `A3` are bound to different individuals. In this way, we avoid cycles as well as the incorrect recognition of substructures.

Kramer and Frank took a *bottom-up approach* to discover all such frequently occurring fragments. Similar to other bottom-up approaches (e.g., PROGOL), the idea is to generate only those fragments that really occur in examples. Basically, the algorithm *generates* all fragments occurring in all molecules, but makes efficient use of frequency-based pruning to make the problem tractable. Once a fragment is generated, it is checked whether its coverage is already known. If it is already known to be infrequent, search can be stopped in this branch of the search space. If it is not known to be infrequent, and it has not yet been evaluated, it is *evaluated* on the full dataset. Evaluations are memorized in order to avoid redundant computations. Using this method, it is practical to search for fragments up to length 15 (which takes a few hours computation time on a Linux PC with a Pentium II processor).

After propositionalization, an algorithm for support vector machines [11.52, 11.53, 11.4] is applied to the transformed problem. Support vector machines seem particularly useful in the context of propositionalization, as they can deal with a large number of moderately significant features.

Preliminary experiments in the domain of carcinogenicity prediction showed that bottom-up propositionalization is a promising approach to feature construction from relational data: frequent fragments are found in a reasonable time and the predictive accuracy obtained using these fragments is very good.

11.6 Related transformation approaches

In its broadest sense, the related work involves various different approaches to constructive induction [11.43, 11.54, 11.42, 11.27] and predicate invention (some of which are presented in an overview paper by Stahl [11.50]).

More closely related are transformation approaches, usually restricted to some particular form of background knowledge, which do not necessarily result in a table where one row corresponds to a single example. Some of these result in a table with multiple rows corresponding to a single training example; these representations are known as multiple-instance learning problems [11.15].

11.6.1 Zucker and Ganascia's transformation approach

Zucker and Ganascia [11.56, 11.57] proposed to decompose structured examples into several learning examples, which are descriptions of parts of what they call the "natural example". The transformation results in a table with multiple rows corresponding to a single example. For instance, any combination of two cars of a train would be defined to be an example. Note that a problem which is reformulated in this way is not equivalent to the original problem. Although it is not clear whether this approach could successfully be applied to, e.g., arbitrary graph structures, their system REMO solves the problem of learning structurally indeterminate clauses, much in line with the extended LINUS propositionalization approach presented in Section 11.7. Their definition of structural clauses whose bodies contain exclusively structural literals, and their algorithm for learning structurally indeterminate clauses can be seen as one of the predecessors of the extended LINUS approach. Recent work by Chevaleyre and Zucker [11.6] further elaborates on the issue of transformed representations for multiple-instance data.

11.6.2 Transformation approach by Fensel et al.

Fensel et al. [11.17] achieve the transformation from the first-order representation to the propositional representation by ground substitutions over a given, user-defined alphabet which transform clauses to ground clauses. As a consequence, they introduce a new definition of positive and negative examples: instead of ground facts they regard ground substitutions as examples. For every possible ground substitution (depending on the number of variables and the alphabet), there is one example in the transformed problem representation. Each background predicate, together with the variable it uses, defines a binary attribute. Like in the previous approach, an example is not described by a single row in the resulting table, but by several rows.

11.6.3 STILL

STILL [11.46] is an algorithm that performs "stochastic matching" in the test whether a hypothesis covers an example. It operates in the so-called Disjunctive Version Space framework, where for each positive example E, one is interested in the space of all hypotheses covering E and excluding all negative examples F_i.

In order to transfer the idea of disjunctive version spaces to first-order logic, STILL reformulates first-order examples in a propositional form. Like in the work by Fensel et al., substitutions are handled as attribute-value examples. In contrast to LINUS and REMO, this reformulation is bottom-up rather than top-down. It is not performed for the complete dataset, but only for one seed example E and for counter-examples F_i. The reformulation is one-to-one (one example, one row) for the seed example, and one-to-many (one example, several rows) for the counter-examples. Since it would be intractable to use all possible substitutions for the counter examples, STILL stochastically samples a subset of these. STILL uses a representation which effectively yields black-box classifiers instead of intelligible features.

11.6.4 PROPAL

PROPAL [11.2] is a recent propositionalization approach which does not propositionalize according to a fixed set of features, but rather extends each hypothesis (Boolean vector) when it is found to be overly general (covering a negative example). PROPAL works in function-free, non-recursive Horn logic. A seed example is chosen and maximally variabilized, serving as the pattern against which other examples are propositionalized. The pattern consists both of non-ground literals and the equality constraints which hold between variables in the seed example. The actual learning algorithm is an AQ-like algorithm.

11.7 A sample propositionalization method: Extending LINUS to handle non-determinate literals

In this section, we show that LINUS can be extended to learning of non-determinate clauses, provided that the problem domain is individual-centered. The approach can not be extended to program synthesis or tasks which do not have a clear notion of an individual. To illustrate this propositionalization approach, we first present a running example that will be used throughout the remainder of the section.

11.7.1 The East-West challenge example

In the running example, the learning task is to discover low size-complexity Prolog programs for classifying trains as eastbound or westbound [11.35]. The problem is illustrated in Figure 11.7.1. Each train consists of two to four cars; the cars have attributes like shape (rectangular, oval, u-shaped, ...), length (long, short), number of wheels (2, 3), type of roof (none, peaked, jagged, ...), shape of load (circle, triangle, rectangle, ...), and number of loads (1-3). A possible rule distinguishing between eastbound and westbound trains is 'a train is eastbound if it contains a short closed car, and westbound otherwise'.

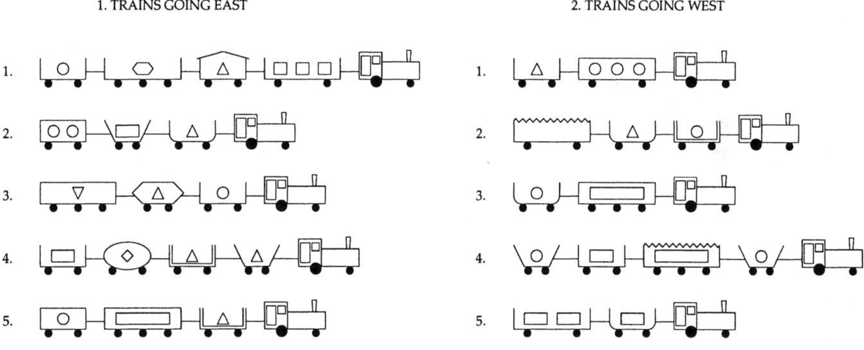

Fig. 11.1. The ten train East-West challenge.

Example 11.7.1 (East-West challenge). Using a flattened (i.e., function-free) representation, the first train in Figure 11.7.1 can be represented as in Table 11.5.

Essentially, a train is represented as a list of cars; a car is represented as a six-tuple indicating its shape, length, whether it has a double wall or not,

Table 11.5. Representation of the first train as facts.

```
east(t1).

hasCar(t1,c11).              hasCar(t1,c12).
cshape(c11,rect).            cshape(c12,rect).
clength(c11,short).          clength(c12,long).
cwall(c11,single).           cwall(c12,single).
croof(c11,no).               croof(c12,no).
cwheels(c11,2).              cwheels(c12,3).
hasLoad(c11,l11).            hasLoad(c12,l12).
lshape(l11,circ).            lshape(l12,hexa).
lnumber(l11,1).              lnumber(l12,1).

hasCar(t1,c13).              hasCar(t1,c14).
cshape(c13,rect).            cshape(c14,rect).
clength(c13,short).          clength(c14,long).
cwall(c13,single).           cwall(c14,single).
croof(c13,peak).             croof(c14,no).
cwheels(c13,2).              cwheels(c14,2).
hasLoad(c13,l13).            hasLoad(c14,l14).
lshape(l13,tria).            lshape(l14,rect).
lnumber(l13,1).              lnumber(l14,3).
```

its roof shape, number of wheels, and its load; finally, a load is represented by a pair indicating the shape of the load and the number of objects. The six-tuples are deconstructed (flattened) to get the six predicates and two predicates for the load.

A possible inductive hypothesis, stating that a train is eastbound if it has a short open car, is as follows:

```
east(T):-hasCar(T,C),clength(C,short),croof(C,no)
```

□

The key idea of first-order features is to restrict all interaction between local variables to literals occurring in the same feature. This is not really a restriction, as in some cases the whole body may constitute a single feature. However, often the body of a rule can be partitioned into separate parts which only share global variables.

Example 11.7.2. Consider the following Prolog rule, stating that a train is eastbound if it contains a short car and a closed car:

```
east(T) :- hasCar(T,C1),clength(C1,short),
           hasCar(T,C2),not croof(C2,no).
```

The body of this clause consists of two features: `hasCar(T,C1)`, `clength(C1,short)` or 'has a short car' and `hasCar(T,C2)`, `not croof(C2,no)` or 'has a closed car'. In contrast, the following rule

```
east(T):-hasCar(T,C),clength(C,short),not croof(C,no)
```

contains a single feature expressing the property 'has a short closed car'. □

11.7.2 The extended LINUS propositionalization approach: Propositionalization through first-order feature construction

In this section, we propose a general-purpose propositionalization method, complete with respect to a given language bias, provided that the problem under consideration can be represented in an individual-centered representation. We start by providing a declarative bias for first-order feature construction, and continue by giving an experimental evaluation of the proposed approach.

The declarative feature bias is based on the term-based individual-centered representations introduced by Flach *et al.* [11.19] and further developed by Flach and Lachiche [11.20]. Such representations collect all information about one individual in a single term, e.g., a list of 6-tuples. Rules are formed by stating conditions on the whole term, e.g., `length(T,4)`, or by referring to one or more subterms and stating conditions on those subterms. Predicates which refer to subterms are called *structural predicates*: they come with the type of the term, e.g., list membership comes with lists, projections (n different ones) come with n-tuples, etc. Notice that projections are determinate, while list membership is not. In fact, the only place where non-determinacy can occur in individual-centered representations is in structural predicates.

Individual-centered representations can also occur in flattened form. In this case each of the individuals and most of its parts are named by constants, as in Example 11.7.1. It is still helpful to think of the flattened representation to be obtained from the term-based representation. Thus, `hasCar` corresponds to list/set membership and is non-determinate (i.e., one-to-many), while `hasLoad` corresponds to projection onto the sixth component of a tuple and thus is determinate (i.e., one-to-one). Keeping this correspondence in mind, the following definitions for the non-flattened case can be easily translated to the flattened case.

We assume a given type structure defining the type of the individual. Let τ be a given type signature, defining a single top-level type in terms of subtypes. A *structural predicate* is a binary predicate associated with a complex type in τ representing the mapping between that type and one of its subtypes. A *functional* structural predicate, or *structural function*, maps to a unique subterm, while a *non-determinate* structural predicate is non-functional.[5]

[5] In other words, the predicate maps to sets of terms, i.e., it does not return a unique value.

In general, we have a structural predicate or function associated with each non-atomic subtype in τ. In addition, we have *utility predicates* as in LINUS [11.32, 11.31] (called *properties* by Flach and Lachiche [11.20]) associated with each atomic subtype, and possibly also with non-atomic subtypes and with the top-level type (e.g., the class predicate). Utility predicates differ from structural predicates in that they do not introduce new variables.

Example 11.7.3. For the East-West challenge we use the following type signature. `train` is declared as the top-level set type representing an individual. The structural predicate `hasCar` non-deterministically selects a car from a train. `car` is defined as a 6-tuple. The first 5 components of each 6-tuple are atomic values, while the last component is a 2-tuple representing the load, selected by the structural function `hasLoad`. In addition, the type signature defines the following utility predicates (properties): `east` is a property of the top-level type `train`, the following utility predicates are used on the subtype `car`: `cshape`, `clength`, `cwall`, `croof` and `cwheels`, whereas `lshape` and `lnumber` are properties of the sub-subtype `load`.
□

To summarize, structural predicates refer to parts of individuals (these are binary predicates representing a link between a complex type and one of its components; they are used to introduce new local variables into rules), whereas utility predicates present properties of individuals or their parts, represented by variables introduced so far (they do not introduce new variables). The language bias expressed by mode declarations used in other ILP learners such as PROGOL [11.38] or WARMR [11.12]) partly achieves the same goal by indicating which of the predicate arguments are input (denoting a variable already occurring in the hypothesis currently being constructed) and which are output arguments, possibly introducing a new local variable. However, mode declarations constitute a bias for the body rather than a bias for features.

Declarations of types, structural predicates and utility predicates define the bias for features. The actual first-order feature construction will be restricted by parameters that define the maximum number of literals constituting a feature, maximal number of variables, and the number of occurrences of individual predicates.

A first-order feature can now be defined as follows. A *first-order feature* of an individual is constructed as a conjunction of structural predicates and utility predicates which is well-typed according to τ. Furthermore:

1. there is exactly one *individual* variable with type τ, which is free (i.e., not quantified) and will play the role of the global variable in rules;
2. each structural predicate introduces a new existentially quantified local variable, and uses either the global variable or one of the local variables introduced by other structural predicates;
3. utility predicates do not introduce new variables (this typically means that one of their arguments is required to be instantiated);

4. all variables are used either by a structural predicate or a utility predicate.

The following first-order feature could be constructed in the above feature bias, allowing for 4 literals and 3 variables:

```
hasCar(T,C),hasLoad(C,L),lshape(L,tria)
```

Now that we have given precise definitions of features in first-order languages such as Prolog, we show the usefulness of this approach by solving two non-determinate ILP tasks with the transformation-based rule learner LINUS [11.32, 11.31]. Since first-order features bound the scope of local variables, constructing bodies from features is essentially a propositional process that can be solved by a propositional rule learner such as CN2.

We provide LINUS with features defining background predicates. For instance, in the trains example we add clauses of the following form to the background knowledge:

```
train42(T):-
  hasCar(T,C),hasLoad(C,L),lshape(L,tria)
```

LINUS would then use the literal `train42(T)` in its hypotheses. Such literals represent propositional properties of the individual.

To formally define propositionalization as understood in the extended LINUS approach consider the following proposition. Let R be a Prolog rule, and let R' be constructed as follows. Replace each feature F in the body of R by a literal L consisting of a new predicate with R's global variable(s) as argument(s), and add a rule $L : -F$. R' together with the newly constructed rules is equivalent to R, in the sense that they have the same success set.

Example 11.7.4. Consider again the following Prolog rule R, stating that a train is eastbound if it contains a short car and a closed car:

```
east(T):-
   hasCar(T,C1), clength(C1,short),
   hasCar(T,C2), not croof(C2,no)
```

By introducing L' as `hasShortCar(T):-hasCar(T,C),clength(C,short)` and L'' as `hasClosedCar(T):-hasCar(T,C),not croof(C,no)`, then rule R is equivalent to rule R' defined as follows:

```
east(T):-hasShortCar(T),hasClosedCar(T)
```

□

Notice that R' contains only global variables, and therefore is essentially a propositional rule. R's first-order features have been confined to the background theory. Thus, provided we have a way to construct the necessary first-order features, an ILP problem can be transformed into a propositional learning problem.

11.7.3 Experimental results

In the two experiments reported in this section we simply provide LINUS with all features that can be generated within a given feature bias (recall that such a feature bias includes bounds on the number of literals and first-order features).

Experiment 1 (LINUS **applied to the East-West challenge**) *We ran* LINUS *on the ten trains in Figure 11.7.1, using a non-determinate background theory consisting of all 190 first-order features with up to two utility predicates and up to two local variables. Using* CN2*, the following rules were found:*

```
east(T):-
  hasCar(T,C1),hasLoad(C1,L1),lshape(L1,tria),lnumber(L1,1),
  not (hasCar(T,C2),clength(C2,long),
       croof(C2,jagged)),
  not (hasCar(T,C3),hasLoad(C3,L3),
       clength(C3,long),lshape(L3,circ)).
west(T):-
  not (hasCar(T,C1),cshape(C1,ellipse)),
  not (hasCar(T,C2),clength(C2,short),croof(C2,flat)),
  not (hasCar(T,C3),croof(C3,peak),cwheels(C3,2)).
```

If negation is allowed within features, the following simple rules are induced:

```
east(T):-
  hasCar(T,C),clength(C,short),not croof(C,no).
west(T):-
  not (hasCar(T,C),clength(C,short),not croof(C,no)).
```

That is, a train is eastbound if and only if it has a short closed car.

This example demonstrates that one should draw a distinction between the feature bias and the rule bias. The first set of rules was found without negation in the features, but allowing it within rules, i.e., over features (a feature of CN2). The second, simpler rule set was found by also allowing negation within features.

The mutagenesis learning task [11.49, 11.41] concerns predicting which molecular compounds cause DNA mutations. The mutagenesis dataset consists of 230 classified molecules; 188 of these have been found to be amenable to regression modeling, and the remaining 42, to which we restrict attention here, as 'regression-unfriendly'. The dataset furthermore includes two handcrafted indicator attributes I_1 and I_a to introduce some degree of structural detail into the regression equation; following some experiments by Muggleton *et al.* [11.41], we did not include these indicators.

Experiment 2 (**L**INUS **applied to mutagenesis**) *We ran* LINUS *on the 42 regression-unfriendly molecules, using a non-determinate background theory consisting of all 57 first-order features with one utility literal concerning atoms (i.e., discarding bond information). Using* CN2*, the following rules were found:*

```
mutag(M,false) :- not (has_atom(M,A),atom_type(A,21)),
                  logP(M,L),between(1.99,L,5.64).
mutag(M,false) :- not (has_atom(M,A),atom_type(A,195)),
                  lumo(M,Lu),between(-1.74,Lu,-0.83),
                  logP(M,L),L>1.81.
mutag(M,false) :- lumo(M,Lu),Lu>-0.77.

mutag(M,true)  :- has_atom(M,A),atom_type(A,21),
                  lumo(M,Lu),Lu<-1.21.
mutag(M,true)  :- logP(M,L),between(5.64,L,6.36).
mutag(M,true)  :- lumo(M,Lu),Lu>-0.95,
                  logP(M,L),L<2.21.
```

Three out of 6 clauses contain first-order features. Notice how two of these concern the same first-order feature 'having an atom of type 21' – incidentally, such an atom also features in the (single) rule found by PROGOL on the same dataset. Running CN2 with only the lumo and logP attributes produced 8 rules; thus, this experiment suggests that first-order features can enhance the understandability of learned rules. Furthermore, we also achieved higher predictive accuracy: 83% with first-order features (as opposed to 76% using only lumo and logP). This accuracy is the same as achieved by PROGOL, having access to bond information and further structural background knowledge.

11.7.4 Related and future work

The approach discussed in this section is a fairly straightforward exhaustive feature construction approach. Its most interesting aspect is the use of a strong individual-centered declarative bias, originating from the 1BC first-order Bayesian classifier [11.20]. This bias is related to the one used by Zucker and Ganascia [11.56, 11.57], although their transformation approach results in a multiple-instance learning problem. It is also interesting to note that Cohen proved that non-determinate clauses whose features do not exceed k literals for some constant k are PAC-learnable [11.9].

One way of continuing this research is by applying a relevancy filter [11.33] to eliminate irrelevant features prior to learning. Alternatively, we can use a descriptive learner such as TERTIUS [11.21], MIDOS [11.55] or WARMR [11.12] to generate only features that correlate sufficiently with the class attribute, or that are frequent.

11.8 Concluding remarks

In this chapter, we gave an overview of work in propositionalization, and presented one of the methods in detail. To summarize the scope of the chapter:

- We focused on induction in the learning from interpretations framework [11.14]. This means, we restricted ourselves to cases where an "ordinary" relational database is given for learning. Thus, we exclude all settings for learning in first-order logic that are more complicated (e.g., dealing with structured terms and recursion).
- De Raedt [11.14] showed that for the above mentioned learning setting, complete propositionalizations are possible, but result in learning problems that are, in size, exponential in a number of parameters of the original learning problem. This chapter presented work on heuristic, partial propositionalizations, which generate only a subset of the features derivable from the relational background knowledge within a given language bias.
- The work presented here focusses on individual-centered representations, i.e., representations that build upon the notion of individuals (here, described in terms of relations). For the given set of individuals, we compute a relatively small, but relevant (sub-)set of structural properties. This type of representation excludes applications such as program synthesis and learning of relations between/among individuals. (Note that this restriction is related to the first point in this list).
- All approaches described in this chapter have in common that they only construct features described in terms of conjunctions of literals. It would be conceivable to construct disjunctive features as well, but obviously the search space would be vast.

In this chapter, we have seen general-purpose propositionalization methods as well as special-purpose ones. While the first group of methods is readily available and applicable, it cannot exploit characteristics of the application domains to improve efficiency. Clearly, the advantages and disadvantages of these two groups of methods are complementary.

The work described in this chapter is also related to applications where ILP is used to handle learning examples that are not handled successfully by propositional systems. For instance, the original paper on the well-known mutagenicity application [11.49] shows that PROGOL performs well on the set of 42 regression-unfriendly examples, i.e., those examples, that are not well handled by stepwise multiple linear regression. Another work in this direction was the submission made by Srinivasan *et al.* to the PTE-2 challenge [11.48], where errors made by C4.5 on the training set were corrected by PROGOL. The combined model turned out to be optimal under certain misclassification costs and/or class distributions according to ROC curve analysis. An interesting direction of further work would be to use full-fledged relational learning to correct errors made by a propositional learner applied to a propositionalization.

It should be stressed that feature construction can play an important role with respect to the reusability of constructed features. Individual problem areas have particular characteristics and requirements, and researchers aim at developing specialized learners for such applications. This may involve also the development of libraries, background knowledge and previously learned hypotheses to be stored for further learning in selected problem areas. In the context of this chapter, we wish to emphasize that feature construction may result in components that are worth to be stored in background knowledge libraries and reused for similar types of applications. Notice that such libraries are now being established for selected problem areas in molecular biology. One should be aware, however, that an increased volume of background knowledge may have also undesirable properties: not only that learning will become less efficient, but with possibly irrelevant information being stored in background knowledge libraries, results of learning will be less accurate. Therefore it is crucial to employ criteria for evaluating the relevance of constructed features before they are allowed to become part of a library of background knowledge for a specific application area. Work in irrelevant feature elimination may help to filter out features irrelevant for the task at hand [11.33].

Propositionalizations have been shown to work successfully in a number of domains. However, it has to be emphasized that this is no argument against ILP, because ILP-type techniques are required for propositionalizations as well. Work on propositionalization has to be considered as a part of ILP research. (Actually, it was there quite early on [11.32, 11.31].) While learning directly in relational representations is still an open research topic (despite recent successes), learning with propositionalizations is a practical option with the nice property that progress in propositional learning can immediately be utilized (see, e.g., the recent work by one of the authors applying Support Vector Machines [11.52, 11.53, 11.4] after propositionalization).

Acknowledgments

We are grateful to Marko Grobelnik and Sašo Džeroski for joint work on LINUS, and Nicolas Lachiche for the joint work on individual-centered representation resulting and implementing feature generation within the 1BC system which we used for the extended LINUS propositionalization method. This work has been supported by the Slovenian Ministry of Science and Technology, the EU-funded project Data Mining and Decision Support for Business Competitiveness: A European Virtual Enterprise (IST-1999-11495), and the British Council (ALIS link 69).

References

11.1 R. Agrawal, H. Mannila, R. Srikant, H. Toivonen, and A.I. Verkamo. Fast discovery of association rules. In U. Fayyad, G. Piatetsky-Shapiro, P. Smyth, and R. Uthurusamy, editors, *Advances in Knowledge Discovery and Data Mining*, pages 307–328. MIT press, Cambridge, MA, 1996.

11.2 E. Alphonse and C. Rouveirol. Lazy propositionalisation for relational learning. *Proceedings of the Fourteenth European Conference on Artificial Intelligence*, pages 256–260. IOS Press, Amsterdam, 2000.

11.3 I. Bratko, I. Mozetič, and N. Lavrač. *KARDIO: A Study in Deep and Qualitative Knowledge for Expert Systems.* MIT Press, Cambridge, MA, 1989.

11.4 C.J.C. Burges. A tutorial on support vector machines for pattern recognition. *Data Mining and Knowledge Discovery*, 2(2), pages 121–167, 1998.

11.5 B. Cestnik, I. Kononenko, and I. Bratko. ASSISTANT 86: A knowledge elicitation tool for sophisticated users. In *Proceedings of the Second European Working Session on Learning*, pages 31–44. Sigma Press, Wilmslow, UK, 1987.

11.6 Y. Chevaleyre and J-D.Zucker. Noise-tolerant rule induction from multi-instance data. *Proceedings of the ICML-2000 workshop on Attribute-Value and Relational Learning: Crossing the Boundaries*, pages 1–11. Stanford University, Stanford, CA, 2000.

11.7 P. Clark and R. Boswell. Rule induction with CN2: Some recent improvements. In *Proceedings Fifth European Working Session on Learning*, pages 151–163. Springer, Berlin, 1991.

11.8 P. Clark and T. Niblett. The CN2 induction algorithm. *Machine Learning*, 3(4):261–283, 1989.

11.9 W.W. Cohen. PAC-learning nondeterminate clauses. In *Proceedings of the Twelfth National Conference on Artificial Intelligence*, pages 676–681. AAAI Press, Menlo Park, CA, 1994.

11.10 W.W. Cohen. Learning trees and rules with set-valued features. In *Proceedings of the Thirteenth National Conference on Artificial Intelligence*, pages 709–716. AAAI Press, Menlo Park, CA, 1996.

11.11 D.J. Cook and L.B. Holder. Substructure discovery using minimum description length and background knowledge. *Journal of Artificial Intelligence Research*, 1:231–255, 1994.

11.12 L. Dehaspe and H. Toivonen. Discovery of frequent Datalog patterns. *Data Mining and Knowledge Discovery*, 3(1):7–36, 1999.

11.13 L. De Raedt. Logical settings for concept learning. *Artificial Intelligence*, 95:187–201, 1997.

11.14 L. De Raedt. Attribute-value learning versus inductive logic programming: The missing links (extended abstract). In *Proceedings of the Eighth International Conference on Inductive Logic Programming*, pages 1–8. Springer, Berlin, 1998.

11.15 T.G. Dietterich, R.H. Lathrop and T. Lozano-Pérez. Solving the multiple-instance problem with axis-parallel rectangles. *Artificial Intelligence* 89(1–2): 31–71, 1997.

11.16 S. Džeroski, H. Blockeel, B. Kompare, S. Kramer, B. Pfahringer, and W. Van Laer. Experiments in Predicting Biodegradability. In *Proceedings of the Ninth International Workshop on Inductive Logic Programming*, pages 80–91. Springer, Berlin, 1999.

11.17 D. Fensel, M. Zickwolff, and M. Wiese. Are substitutions the better examples? Learning complete sets of clauses with Frog. In *Proceedings of the Fifth International Workshop on Inductive Logic Programming*, pages 453–474. Department of Computer Science, Katholieke Universiteit Leuven, 1995.

11.18 P. Flach. Knowledge representation for inductive learning. In *Proceedings of the European Conference on Symbolic and Quantitative Approaches to Reasoning and Uncertainty*, pages 160–167. Springer, Berlin, 1999.

11.19 P. Flach, C. Giraud-Carrier, and J.W. Lloyd. Strongly typed inductive concept learning. In *Proceedings of the Eighth International Conference on Inductive Logic Programming*, pages 185–194. Springer, Berlin, 1998.
11.20 P. Flach and N. Lachiche. 1BC: A first-order Bayesian classifier. In *Proceedings of the Ninth International Workshop on Inductive Logic Programming*, pages 92–103. Springer, Berlin, 1999.
11.21 P. Flach and N. Lachiche. Confirmation-guided discovery of first-order rules with Tertius. *Machine Learning*, 42(1-2): 61–95, 2001.
11.22 P. Geibel and F. Wysotzki. Relational learning with decision trees. In *Proceedings Twelfth European Conference on Artificial Intelligence*, pages 428–432. IOS Press, Amsterdam, 1996.
11.23 G. Klopman. Artificial intelligence approach to structure-activity studies: computer automated structure evaluation of biological activity of organic molecules. *Journal of the American Chemical Society*, 106:7315–7321, 1984.
11.24 G. Klopman. MultiCASE: A hierarchical computer automated structure evaluation program. *Quantitative Structure Activity Relationships*, 11:176–184, 1992.
11.25 W. Klösgen. EXPLORA: A multipattern and multistrategy discovery assistant. In U. Fayyad, G. Piatetsky-Shapiro, P. Smyth, and R. Uthurusamy, editors, *Advances in Knowledge Discovery and Data Mining*, pages 249–271. AAAI Press, Menlo Park, CA, 1996.
11.26 R. Kohavi, D. Sommerfield, and J. Dougherty. Data mining using MLC++: A machine learning library in C++. In *Proceedings of the Eighth IEEE International Conference on Tools for Artificial Intelligence*, pages 234-245. IEEE Computer Society Press, Los Alamitos, CA, 1996. http://www.sgi.com/Technology/mlc.
11.27 D. Koller and M. Sahami. Toward optimal feature selection. In *Proceedings of the Thirteenth International Conference on Machine Learning*, pages 284–292. Morgan Kaufmann, San Francisco, CA, 1996.
11.28 S. Kramer. Structural regression trees. In *Proceedings of the Thirteenth National Conference on Artificial Intelligence*, pages 812–810. AAAI Press, Menlo Park, CA, 1996.
11.29 S. Kramer and E. Frank. Bottom-Up propositionalization. In *Proceedings of the ILP-2000 Work-In-Progress Track*, pages 156–162. Imperial College, London, 2000.
11.30 S. Kramer, B. Pfahringer, and C. Helma. Stochastic propositionalization of non-determinate background knowledge. In *Proceedings of the Eighth International Conference on Inductive Logic Programming*, pages 80–94. Springer, Berlin, 1998.
11.31 N. Lavrač and S. Džeroski. *Inductive Logic Programming: Techniques and Applications*. Ellis Horwood, Chichester, 1994. Freely available at http://www-ai.ijs.si/SasoDzeroski/ILPBook/.
11.32 N. Lavrač, S. Džeroski, and M. Grobelnik. Learning nonrecursive definitions of relations with LINUS. In *Proceedings of the Fifth European Working Session on Learning*, pages 265–281. Springer-Verlag, Berlin, 1991.
11.33 N. Lavrač, D. Gamberger, P. Turney. A relevancy filter for constructive induction. *IEEE Intelligent Systems*, 13: 50–56, 1998.
11.34 H. Mannila and H. Toivonen. Levelwise search and borders of theories in knowledge discovery. *Data Mining and Knowledge Discovery*, 1:241–258, 1997.
11.35 D. Michie, S. Muggleton, D. Page, and A. Srinivasan. To the international computing community: A new East-West challenge. Technical report, Oxford University Computing laboratory, Oxford, UK, 1994.

11.36 F. Mizoguchi, H. Ohwada, M. Daidoji, and S. Shirato. Learning rules that classify ocular fundus images for glaucoma diagnosis. In *Proceedings of the Sixth International Workshop on Inductive Logic Programming*, pages 146–162. Springer-Verlag, Berlin, 1996.

11.37 I. Mozetič. NEWGEM: Program for learning from examples, technical documentation and user's guide. Reports of Intelligent Systems Group UIUCDCS-F-85-949, Department of Computer Science, University of Illinois, Urbana Champaign, IL, 1985.

11.38 S. Muggleton. Inverse entailment and Progol. *New Generation Computing*, 13: 245–286, 1995.

11.39 S. Muggleton and C. Feng. Efficient induction of logic programs. In S. Muggleton, editor, *Inductive Logic Programming*, pages 281–298. Academic Press, London, 1992.

11.40 S. Muggleton, R.D. King, and M.J.E Sternberg. Protein secondary structure prediction using logic. In *Proceedings of the Second International Workshop on Inductive Logic Programming*, pages 228–259. TM-1182, ICOT, Tokyo, 1992.

11.41 S. Muggleton, A. Srinivasan, R. King, and M. Sternberg. Biochemical knowledge discovery using Inductive Logic Programming. In *Proceedings of the First Conference on Discovery Science*, pages 326–341. Springer, Berlin, 1998.

11.42 A.L. Oliveira and A. Sangiovanni-Vincentelli. Constructive induction using a non-greedy strategy for feature selection. In *Proceedings of the Ninth International Workshop on Machine Learning*, pages 354–360. Morgan Kaufmann, San Francisco, CA, 1992.

11.43 G. Pagallo and D. Haussler. Boolean feature discovery in empirical learning. *Machine Learning*, 5:71–99, 1990.

11.44 J. R. Quinlan. *C4.5: Programs for Machine Learning.* Morgan Kaufmann, San Mateo, CA, 1993.

11.45 B.L. Richards and R.J. Mooney. Learning relations by pathfinding. In *Proceedings of the Tenth National Conference on Artificial Intelligence*, pages 50–55. AAAI Press, Menlo Park, CA, 1992.

11.46 M. Sebag and C. Rouveirol. Tractable induction and classification in first order logic via stochastic matching. In *Proceedings of the Fifteenth International Joint Conference on Artificial Intelligence*, pages 888–893. Morgan Kaufmann, San Francisco, CA, 1997.

11.47 A. Srinivasan and R. King. Feature construction with inductive logic programming: a study of quantitative predictions of biological activity aided by structural attributes. *Data Mining and Knowledge Discovery*, 3(1):37–57, 1999.

11.48 A. Srinivasan, R. King and D.W. Bristol, An assessment of submissions made to the Predictive Toxicology Evaluation Challenge *Proceedings of the Sixteenth International Joint Conference on Artificial Intelligence*, 270–275. Morgan Kaufmann, San Francisco, CA, 1999.

11.49 A. Srinivasan, S. Muggleton, R.D. King and M. Sternberg. Theories for mutagenicity: a study of first-order and feature based induction. *Artificial Intelligence*, 85(1-2):277–299, 1996.

11.50 I. Stahl. Predicate invention in inductive logic programming. In L. De Raedt, editor, *Advances in Inductive Logic Programming*, pages 34–47. IOS Press, Amsterdam, 1996.

11.51 P. Turney. Low size-complexity inductive logic programming: The East-West challenge considered as a problem in cost-sensitive classification. In

L. De Raedt, editor, *Advances in Inductive Logic Programming*, pages 308–321. IOS Press, Amsterdam, 1996.

11.52 V. Vapnik. *Estimation of Dependencies Based on Empirical Data.* Springer Verlag, Berlin, 1982.

11.53 V. Vapnik. *The Nature of Statistical Learning Theory.* Springer Verlag, Berlin, 1995.

11.54 J. Wnek and R.S. Michalski. Hypothesis-driven constructive induction in AQ17: A method and experiments. In *Proceedings of IJCAI-91 Workshop on Evaluating and Changing Representations in Machine Learning*, pages 13–22. Sydney, Australia, 1991.

11.55 S. Wrobel. An algorithm for multi-relational discovery of subgroups. In *Proceedings of the First European Symposium on Principles of Data Mining and Knowledge Discovery*, pages 78–87. Springer, Berlin, 1997.

11.56 J-D. Zucker and J-G. Ganascia. Representation changes for efficient learning in structural domains. In *Proceedings of the Thirteenth International Conference on Machine Learning*, pages 543–551. Morgan Kaufmann, San Francisco, CA, 1996.

11.57 J-D. Zucker and J-G. Ganascia. Learning structurally indeterminate clauses. In *Proceedings of the Eighth International Conference on Inductive Logic Programming*, pages 235–244. Springer, Berlin, 1998.

12. Relational Learning and Boosting

Ross Quinlan

School of Computer Science and Engineering, University of New South Wales
Sydney 2052, Australia

Abstract

Boosting, a methodology for constructing and combining multiple classifiers, has been found to lead to substantial improvements in predictive accuracy. Although boosting was formulated in a propositional learning context, the same ideas can be applied to first-order learning (also known as inductive logic programming). Boosting is used here with a system that learns relational definitions of functions. Results show that the magnitude of the improvement, the additional computational cost, and the occasional negative impact of boosting all resemble the corresponding observations for propositional learning.

12.1 Introduction

Boosting [12.15] is a new approach to improving the predictive accuracy achievable with classifier-learning systems. Although its origins lie in theoretical machine learning, especially the Winnow procedure of Littlestone and Warmuth [12.19] and Schapire's weak learning framework [12.34], the practical benefits of boosting have been demonstrated in numerous trials [12.5, 12.11, 12.15, 12.30].

Many authors have found that improved predictions can be obtained by combining multiple classifiers; Dietterich [12.10] gives an excellent overview. Common methods for generating several classifiers from one dataset can be divided into two families. The first, characterized as *multistrategy learning* [12.20], varies factors such as the learning method or abstraction level to produce different classifiers. Boosting belongs to the second family in which the learning algorithm remains constant and other means are employed to ensure diversity. For example, Buntine's *option trees* [12.6] approximate the collection of trees consistent with the dataset. More recently, Breiman [12.4] has used bootstrap samples from the data in a process that he calls *bagging*. Boosting introduces a crucial innovation – the data from which a classifier is learned is biased towards the training instances handled poorly by the previous classifier. It is interesting to observe that my earlier *windowing* procedure [12.26] has a similar bias, although it discards all but the most recent classifier rather than combining all classifiers and is much less effective than boosting.

In this chapter, boosting is moved from its usual classification context to first-order relational learning, sometimes known as *inductive logic programming*. The beneficiary is a system, FFoil [12.31], that learns Prolog-like definitions of functional relations. The boosting paradigm is a flexible one and this transition proves to be relatively uncomplicated.

The following sections of the chapter introduce boosting, FOIL and FFOIL, then Section 5 describes the way in which FFOIL is boosted. Results of experiments with nine relational datasets are presented in Section 6. The final section compares these with earlier boosting results from a more familiar classification setting, showing that both have a similar profile.

12.2 Boosting

The boosting procedure used in this chapter generally follows Freund and Schapire's *AdaBoost.M1* [12.15]. We assume a set of training instances $i = 1, 2, ..., N$. The classifier-learning procedure is invoked repeatedly in a series of *trials* $t = 1, 2, ..., T$, with trial t generating classifier C^t. Each instance i is assigned a weight w_i^t for trial t so that the sum of the instance weights is equal to 1 for each trial. The instances initially have uniform weights $1/N$ and these are adjusted after each trial so as to influence the performance of the classifier learning system at the subsequent trial.

The process at trial t can be summarized as follows:

- The learning system constructs classifier C^t from the training instances using the weights w_i^t.
- The error rate ϵ^t of this classifier on the training data is determined as the sum of the weights w_i^t of the misclassified instances.
- If $\epsilon^t = 0$, the process terminates before the limit of T trials is reached. Boosting also stops prematurely if $\epsilon^t \geq 1/2$, in which case the classifier C^t is discarded because its error rate is too high. Otherwise the weights w_i^{t+1} for the next trial are set so that the error rate of the current classifier C^t under weights w_i^{t+1} is exactly $1/2$. Freund and Schapire accomplish this by multiplying the weight of each correctly classified instance by $\epsilon^t/(1-\epsilon^t)$ and renormalizing. This is equivalent to setting the weight w_i^{t+1} of instance i at trial $t+1$ to

$$w_i^{t+1} = \begin{cases} w_i^t/2\epsilon^t & \text{if } C^t \text{ misclassifies instance } i \\ w_i^t/2(1-\epsilon^t) & \text{otherwise.} \end{cases}$$

A *composite classifier* C^* is constructed from this sequence of classifiers $C^1, C^2, ..., C^T$ by aggregating weighted votes of the individual classifiers. If C^t classifies some instance x as belonging to class k, the total vote for k is incremented by $\log((1-\epsilon^t)/\epsilon^t)$ where, as above, ϵ^t is the error of classifier C^t on the weighted training instances from which it was constructed. C^* then assigns x to the class with the greatest total vote.

Provided that the learning system, when given a training set of weighted instances, can reliably generate a hypothesis that has less than 50% error on those instances, a sequence of "weak" classifiers $\{C^t\}$ can be boosted to a "strong" classifier C^* that is at least as accurate as, and usually more accurate than, the best weak classifier. Freund and Schapire prove that the error rate of C^* on the unweighted training instances approaches zero exponentially quickly as T increases.

Some learning systems cannot readily employ weighted training instances. For these, the training set for trial $t = 2, 3, ..., T$ can be created by drawing from the instances a resample of size N, using w_i^t as the probability of selecting instance i at each draw. This is equivalent to approximating the weights w_i^t by another weight vector, all of whose components are multiples of $1/N$.

12.3 FOIL

We turn now to relational learning and an early system, FOIL [12.27, 12.32], that is the direct ancestor of FFOIL.

As with most first-order learning, the task addressed by FOIL is to learn a Horn clause definition of a *target relation* R in terms of itself and other *background relations.* Input to FOIL consists of relations defined extensionally as sets of tuples of constants. FOIL also needs examples of tuples that do not belong to the target relation R in order to rule out a simple definition of R that includes all tuples.

Functions are not permitted within terms, so every function

$$f(V_1, V_2, .., V_k)$$

is represented using a relation

$$F(V_1, V_2, ..., V_k, V_{k+1})$$

where a tuple $\langle c_1, c_2, ..., c_k, c_{k+1}\rangle$ from this relation F specifies the value c_{k+1} of $f(c_1, c_2, ..., c_k)$. Rouveirol [12.33] proves that *flattening* functions in this way does not affect the expressive power of Horn clause definitions.

A *binding* of a clause

$$R(V_1, V_2, .., V_x) \leftarrow L_1,\ L_2,\ ...,\ L_y$$

consists of a mapping of each variable in the clause to a ground term (here a constant) so that all literals in the clause body are satisfied. If the variables in the clause are

$$V_1, V_2, ..., V_z$$

a binding is conveniently represented as a z-tuple of constants.

The basic approach used by FOIL is *covering.* Initially, all tuples in the target relation R have to be explained by clauses. A clause is constructed as described below so that all its bindings also satisfy the head of the clause. For each such binding

$$\langle c_1, c_2, ..., c_x, c_{x+1}, ..., c_z\rangle$$

the tuple

$$\langle c_1, c_2, ..., c_x \rangle$$

in R has now been explained by this clause. All such tuples are removed from the set of tuples still to be explained and a further clause is constructed to explain some of the remainder. This iterative process continues until all tuples in the target relation have been explained by at least one clause in the definition.

A clause is found by a general-to-specific search, starting from the head

$$R(V_1, V_2, ..., V_x) \leftarrow$$

where the V_is are unique variables. This clause matches all tuples in R, but also all tuples that are known not to belong to R. In order to exclude the latter, literals are added to the clause body to restrict applicability of the clause, guided by the bindings of the partial clause. A binding is labeled $\oplus$ if the assignment of values to variables satisfies the head of the clause, and $\ominus$ if it does not. Since the goal is to construct a clause, all of whose bindings satisfy the head of the clause (and so are labeled $\oplus$), FOIL seeks to add literals that increase the concentration of $\oplus$ bindings, and continues adding literals until the clause has no $\ominus$ bindings.

As an example, consider the task of learning a definition for the relation *member*(E, L), indicating that atom E occurs in the flat list L. Atoms will be represented by digits, which ground lists will be denoted by constants such as [123] that stands for the list containing atoms 1, 2 and 3. The empty list is thus written as [ ].

The first requirement is for examples of tuples that belong to relation *member*, such as

⟨1,[1]⟩, ⟨1,[31]⟩, ⟨2,[23]⟩, ⟨2,[232]⟩, ...

and some that do not, such as

⟨2,[ ]⟩, ⟨1,[23]⟩, ⟨2,[31]⟩, ...

For this task, we will use a single background relation *components*(L, H, T) indicating that list L has head H and tail T. FOIL expects this relation also to be specified extensionally by the tuples

⟨[1],1,[ ]⟩, ⟨[31],3,[1]⟩, ⟨[23],2,[3]⟩, ...

and so on.

The search for the first clause defining *member* proceeds as follows. The initial clause

$$member(A, B) \leftarrow$$

has bindings that contain all the tuples known to be in or not in the target relation, i.e.,:

$\langle 1,[1]\rangle \oplus$
$\langle 1,[31]\rangle \oplus$
$\langle 2,[23]\rangle \oplus$
$\langle 2,[232]\rangle \oplus$
. . . . .
$\langle 2,[\,]\rangle \ominus$
$\langle 1,[23]\rangle \ominus$
$\langle 2,[31]\rangle \ominus$
. . . . .

Notice that the bindings are labeled $\oplus$ or $\ominus$ according to whether or not they satisfy the clause head.

If the clause is further developed to

$$member(A, B) \leftarrow components(B, C, D)$$

by the addition of a literal to the body of the clause, all bindings will now give values for the four variables A, B, C, and D. Each previous binding (except those in which B is equal to the null list) will provide one new binding, since there are unique values of C and D that specify the head and tail of B. The bindings at this stage are

$\langle 1,[1],1,[\,]\rangle \oplus$
$\langle 1,[31],3,[1]\rangle \oplus$
$\langle 2,[23],2,[3]\rangle \oplus$
$\langle 2,[232],2,[32]\rangle \oplus$
.. . . . .
$\langle 1,[23],2,[3]\rangle \ominus$
$\langle 2,[31],3,[1]\rangle \ominus$
.. . . . .

Extending the clause further to

$$member(A, B) \leftarrow components(B, C, D),\ member(A, D)$$

excludes most of the previous bindings because they do not satisfy

$$member(A, D).$$

The bindings that remain, such as

$\langle 1,[31],3,[1]\rangle \oplus$
$\langle 2,[232],2,[32]\rangle \oplus$
.. . . . .

all satisfy the head of the clause. Since the clause now has only bindings labeled $\oplus$, it is known to be complete and becomes the first clause of the definition. The tuples

$\langle 1,[31]\rangle \oplus$
$\langle 2,[232]\rangle \oplus$
.. . . . .

in the target relation that are explained by this clause are dropped from the initial bindings for subsequent clauses.

A second clause

$$member(A, B) \leftarrow components(B, A, C)$$

accounts for all outstanding tuples from the target relation member, so the complete definition found is

$$member(A, B) \leftarrow components(B, C, D),\ member(A, D).$$
$$member(A, B) \leftarrow components(B, A, C).$$

This account has ignored important issues such as how to identify useful literals to be added to the clause body and how to ensure that recursive definitions are well-behaved. Such matters lie well beyond the scope of this chapter, but are discussed in [12.32, 12.7].

12.4 Overview of FFOIL

First-order learning systems are of two types. General systems such as FOCL [12.25], FOIL [12.27, 12.32], GRENDEL [12.8] and PROGOL [12.22] do not constrain the nature of the target relation R or the form of clauses in the definition of R. On the other hand, more specialized systems such as DOLPHIN [12.36], FILP [12.3] and GOLEM [12.23] exploit some property of a class of learning tasks in order to improve their efficiency and/or effectiveness on such tasks. Examples of these classes include learning guard clauses for logic programs, learning definitions of functional relations, and learning definitions consisting of determinate clauses.

FFOIL [12.31] belongs in this second group, being an adaptation of FOIL for functional relations (and so strongly related to the work of Mooney and Califf [12.21] and Bergadano and Gunetti [12.3]). A target relation $R(V_1, V_2, ..., V_n)$ is *functional* if, given ground values v_1, v_2, ..., v_{n-1} for V_1, V_2, ..., V_{n-1}, there is a single value v_n for V_n such that $\langle v_1, v_2, ..., v_n \rangle$ belongs to R. The last variable V_n of a functional target relation will be referred to as the *output* variable.

Relations of this kind are common and have several distinguishing characteristics:

- Queries to be answered by the learned definition cannot be assumed to be ground. For example, if $mult(A, B, C)$ denotes the functional relation $A \times B = C$, a sensible query is not "Is $4 \times 3 = 12$?", but rather "What is 4×3?".
- Systems that perform general relational learning do not necessarily bind every variable in the clause head by its appearance in one or more literals in the clause body. With functional relations, however, the output variable must be bound somewhere in the clause body in order to answer queries of the form above.

- Such a query has only one correct answer. If a ground tuple $\langle v_1, v_2, ..., v_n \rangle$ belongs to the target relation R, then changing v_n to any different constant is guaranteed to yield a tuple that does not belong to R. This fact can be used to generate negative examples without them having to be provided explicitly, and without recourse to the closed world assumption [12.2]. Similarly, no further clauses should be investigated after a clause succeeds, i.e., each clause should end with a cut. This implies that a learned definition consists of an ordered sequence of clauses instead of the unordered clauses generated by most general first-order induction systems.
- Some queries might exhaust the list of clauses without finding one whose body is satisfied. It may then be useful to allow for this by providing a final default clause of the form

  $$R(V_1, V_2, ..., k).$$

  for some constant k. Here k is set to the most frequent value of the function.

Table 12.1. Outline of FFOIL.

```
/* Initialization */
definition := null program
remaining := all tuples in target relation R

/* Construct definition */
While remaining is not empty:
  /* Grow a new clause */
  clause := R(V1, V2, ..., Vn) ←
  While clause does not bind output Vn or gives incorrect answers
    /* Specialize clause */
    Find appropriate literal(s) L
      - if unable, exit from outer while loop
    Add L to body of clause
  Remove from remaining tuples in R covered by clause
  Remove unnecessary literals from clause
  Append cut !
  Add clause to definition

Add default clause

/* Global pruning */
While errors do not increase:
  Remove literals from each clause
  Remove clauses from definition
```

The above notwithstanding, the outline structure of FFOIL shown in Table 12.1 is similar to that of FOIL. The main differences are the addition of cuts to clauses, the generation of a default clause, and a new global pruning phase. This last is required because FFOIL tends to produce heavily qualified clauses that overfit the data, and to give up while some tuples of R remain uncovered. The final pruning phase removes literals from clauses and clauses from the definition while this does not increase the number of errors on the training data.

12.5 Boosting FFOIL

Although FFOIL is not a classifier-learning system, there is an intuitive correspondence between attribute-value learning and finding definitions of functional relations. The analogue of a training instance is a tuple of constants belonging to the target relation R, and instead of a classifier we have a Horn clause definition of R. This definition "misclassifies" a tuple $\langle v_1, v_2, ..., v_n \rangle$ if it gives the wrong answer to a query $R(v_1, v_2, ..., v_{n-1}, X)$, i.e., the value returned for X is not equal to v_n.

Table 12.2. Method of boosting FFOIL.

Given:

target relation R containing tuples $P = P_1,...,P_N$
background relations

/* *Initialization* */
For $i := 1$ *to* N *do* $w_i^1 := 1/N$

/* *Form component definitions* */
For $t := 1$ *to* T *do*

If $t = 1$ *then*
training set := P
else
training set := *resample of size* N *from* P *using distribution* w_i^t
Run FFOIL *on training set giving definition* D^t *of* R
Calculate error rate ϵ^t *of* D^t *on* P
if $\epsilon^t > 0.5$, *discard* D^t *and reinitialize weights*
if $\epsilon^t = 0$ *or* $\epsilon^t = 0.5$, *reinitialize weights*
/* *Adjust weights* */
For $i := 1$ *to* N *do*

$$w_i^{t+1} := w_i^t / \begin{cases} 2\epsilon^t & \textit{if } D^t \textit{ incorrect for } P_i \\ 2(1-\epsilon^t) & \textit{otherwise.} \end{cases}$$

To answer a query using the boosted definition:

Answer query with each definition D^t
Choose most common value for output variable

Since FFOIL contains no mechanism for associating weights with training instances, the resampling procedure described at the end of Section 2 is used. The complete boosting scenario for FFOIL, set out in Table 12.2, follows

AdaBoost.M1 [12.15] except for two minor changes that address the use of the resampling regimen:

- For trials after the first, some tuples in the target relation are omitted from the sample that forms the training set. These "unseen" training instances are often handled incorrectly by the learned definition, with the result that the error ϵ^t can be large and highly variable. Besides affecting the next weights, ϵ^t also determines the voting weight of classifier t. It seems unwise to base a definition's voting weight on this erratic value; empirical trials have confirmed that a uniform voting weight of 1 gives better results.
- The original boosting procedure stops if the learning system returns either a perfect classifier or one with error greater than 50% on the weighted training instances. Breiman [12.5] suggests the alternative of reinitializing the weights so that the next training set is sampled from the original data, and this procedure has been adopted here.

12.6 Experiments

The boosted version of FFOIL was tested on several datasets, most of which have appeared in published studies and are available either from the UCI Repository of Machine Learning Databases or the more recent ILPnet Repositories described by Lavrač et al. [12.17] and the Chapter 16 of Todorovski et al. in this volume. The tasks cover a spectrum of difficulties, from those for which very accurate definitions can be learned, to those for which it is hard to do much better than the default clause. For each dataset, the true error rates of the "standard" and boosted versions of FFOIL were estimated either by cross-validation or by repeated runs with different training and test sets, using where possible the same divisions of the data as previously-reported experiments. Unless otherwise noted, a 10-fold cross-validation was employed. Each train/test run was performed twice with the number T of boosting trials set to the values 10 and 20 respectively.

12.6.1 Description of tasks

Satellite faults: The first task uses data from a qualitative simulation of a satellite power supply [12.14]. For 601 discrete times for which there are no faults, and for 20 discrete times at which a single fault is present, 29 background relations give the on/off values of the relevant switches and sensors. There are two further background relations, one that defines the phase of the satellite mission at each time, and another than indicates the sequence of discrete times. The target relation *fault*(A, B) gives, for each time A, a value B of *true* or *false*.

Mutagenicity: This task from Srinivasan et al. [12.35] concerns developing a theory of mutagenicity. The data consist of 188 molecules, 125 of which are mutagenic, described by 18 background relations. Two relations provide global properties of the molecule, but the other 16 relations address the molecule's structure and specify component atoms, bonds between them, and information about the presence and connectivity of rings. The target relation is *active*(A, B) where A identifies one of the molecules and B has the value *true* or *false*.

Protein secondary structure: Experiments with this data are reported in [12.24]. The goal is to learn to predict whether a protein has an alpha-helix at a given position. The target relation is *alpha*(A, B, C) where A identifies a protein, B is a position in that protein, and C is *true* or *false* according to whether there is an alpha-helix at that position. One background relation specifies the amino acid residue at each position in every protein, and three more give patterns of relative positions that are important for this task. A further twenty unary background relations provide information about each of the twenty amino acid residues, and two more relations denote partial orderings of pairs of amino acids. The data cover a total of 2028 positions in 16 proteins. A 16-fold cross-validation was used, at each fold omitting all information about one protein from the training data and testing the learned definition on the positions in this protein.

Document components: This learning task, reported in Esposito et al. [12.13], involves identifying the purposes served by components of single-page letters. Each component consists of a rectangular area of the letter such as might be identified by a scanning device. Components vary from small (positioning marks on the letterhead) to large (paragraphs of text), and the data consist of thirty letters containing a total of 364 components. The target relation used here is *function*(A, B), where A is a component of one of the letters and B is one of *sender, receiver, date, reference, logo* and *other*. Fifty-seven background relations describe properties of components such as their width and height, and inter-relationships such as horizontal and vertical alignment with other components. A 30-fold cross-validation was conducted, at each fold omitting all information about components of one letter, learning a definition for *function*, then testing this definition on all components of the omitted letter.

Past tense of English verbs: Learning how to transform a verb from present to past tense was first studied in the connectionist community, but relational representations of the task have also been explored [12.29, 12.21]. These experiments imitate those reported by Ling [12.18] and use ten training and test sets of verbs taken from a corpus of 1500 phonetic verbs. This formulation uses two target relations, *delete*(A, B) and *add*(A, C), which jointly state that the past tense of verb A is found by removing the string B from its end and appending the string C. A single background relation *split*(X, Y, Z) indicates all ways in which verb X (in present or past tense) can be divided into non-null substrings Y and Z. Every run involves two separate learning tasks, one for each target relation, with the same training and test verbs used in each case. A test verb is judged to be correct only when both target relations give the correct result.

Sensor data from a mobile robot: This is one of a hierarchy of learning tasks investigated by Klingspor et al. [12.16]. The target relation *S-convex*(A, B, C, D, E) identifies a particular trace A, a sensor B, a time interval C to D, and a direction E that is one of *diagonal*, *parallel*, *straight-away* or *straight-towards*. Eleven high-arity background relations provide information on basic perceptual features extracted from sensor data for the 538 tuples in the target relation.

Finite element mesh design: This learning task, first discussed by Dolšak and Muggleton [12.12], deals with the division of edges of an object into an appropriate number of intervals for finite element simulation. Data are provided for five objects containing 277 edges in total, each edge being divided into one to twelve segments. The target relation *mesh*(A, B) gives the number B of segments recommended by an expert for edge A. Thirty background relations cover properties of each edge and its topological relationship to other edges in the same object. A leave-one-out cross-validation was carried out with this data.

Assigning chapters to conference sections: This task was suggested by Cohen [12.9] in the context of applying relational learning to problems in information retrieval, although this experiment uses different data. From the titles of the 110 chapters in five of the largest sections at *IJCAI'95*, the goal is to learn how to associate a chapter with its appropriate section. The target relation, *section*(A, B), identifies a chapter A and its section – one of *action/perception*, *planning*, *learning*, *non-monotonic reasoning*, and *natural language*. There is one background relation $W(A, P)$ for each capitalized word W that occurs in at least three chapter titles, where P is the position at which W occurs in the title of chapter A. Two further background relations *near*(P, Q) and *after*(P, Q) record whether position P is close to position Q (i.e., within three words) and whether P is after Q.

Moves to win in a chess end-game: For King and Rook versus King, Bain [12.1] gives the complete set of more than 28,000 positions that can arise, and either the number of moves to mate under optimal play (0 to 16) or the fact that the position is drawn. These experiments use a sample of 1000 of the positions rather than the entire dataset. The target relation *moves*(A, B, C, D, E, F, G) gives the rank and file of each piece, then the value *drawn*, *0m*, *1m*, ..., or *16m*. Two background relations *rank-dist*(X, Y, D) and *file-dist*(X, Y, D) specify the distance D between pairs of ranks and pairs of files, and two further relations *rank-lt*(X, Y) and *file-lt*(X, Y) record whether one rank or file is less than another.

12.6.2 Results

Table 12.3 presents the results of these experiments; for each dataset, it shows:

- the total number of tests of the learned definitions – for all datasets except *past tense* (where separate training and test sets are used), this is the same as the number of tuples in the target relation;
- the number of possible values of the function;
- the error rate of a definition consisting of just the default clause;
- the number of errors and the error rate from FFOIL without boosting;
- the corresponding values when 10-trial boosting is applied to FFOIL; and
- similar values for 20-trial boosting.

Table 12.3. Errors on unseen data with and without boosting.

Relational dataset	Tests	Values	Default Error	FFOIL		Boosting (10 trials)		Boosting (20 trials)	
Satellite faults	621	2	3.2%	4	0.6%	3	0.5%	3	0.5%
Mutagenicity	188	2	33.6%	25	13.3%	22	11.7%	22	11.7%
Protein structure	2028	2	47.5%	777	38.3%	710	35.0%	671	33.0%
Document components	364	6	48.6%	22	6.0%	11	3.0%	10	2.7%
Past tense	5000	66	53.2%	566	11.3%	559	11.2%	557	11.1%
Robot motion	538	4	60.4%	51	9.5%	28	5.2%	26	4.8%
Mesh design	277	12	73.6%	154	55.6%	186	67.1%	189	68.2%
Conference section	110	5	74.6%	54	49.0%	48	43.6%	50	45.4%
Moves to mate	1000	18	82.3%	641	64.1%	574	57.4%	551	55.1%

The performance of unboosted FFOIL on these datasets ranges from strong (on tasks such as *satellite faults* and *document components* for which learned definitions have high predictive accuracy) to weak (on *mesh design*, for example, where the definitions perform not much above the level of the default). Boosting with 10 trials improves performance on all datasets except *mesh design*, but the benefit is not uniform – there is little improvement on the *past tense* task, but error rate is halved on *document components*. Increasing the number of boosting trials to 20 gives further improvement on five tasks, no change on two tasks, and lower accuracy than 10-trial boosting on two tasks. Even when it proves helpful to increase the number of boosting trials, the magnitude of the further improvement is less than the initial gain from 10-trial boosting.

For each dataset, the effect of boosting can be summarized by the ratio of the error rate of the boosted learner to the error rate with no boosting. A value smaller than 1 indicates that boosting is beneficial, while a value greater than 1 signifies that boosting increases error. These values are shown in Table 12.4 together with the (geometric) means that provide a useful precis of the experiments. The overall effect of 10-trial boosting is to reduce error by 19%. There seems little point going beyond 10-trial boosting, at least for these datasets, as doubling the number of trials yields a further improvement of only 2%.

Table 12.4. Measuring the impact of boosting .

Relational dataset	Boosting (10 trials)	Boosting (20 trials)
Satellite faults	.75	.75
Mutagenicity	.88	.88
Protein structure	.91	.86
Document components	.50	.45
Past tense	.99	.98
Robot motion	.55	.51
Mesh design	1.21	1.23
Conference section	.89	.93
Moves to mate	.89	.86
Geometric mean	0.81	0.79

12.7 Summary

This chapter has shown that boosting, originally formulated in a propositional classification setting, can readily be adapted to the context of first-order relational learning. The particular learning system used here is typical of a class of empirical learning systems, all of which might be expected to benefit from boosting. Experiments with nine relational datasets demonstrate that useful improvements in predictive accuracy can be obtained with as few as 10 boosting trials, representing an increase of one order of magnitude in the computational effort required for learning.

These findings closely resemble results from boosting C4.5, a propositional classifier [12.30]. In those experiments, a variant of 10-trial boosting (using modified voting weights) improved predictive accuracy on 25 of 27 datasets, reducing error on average by 18%. Just as was found here, boosting actually degrades performance on some datasets, although these fortunately seem to be in the minority. It was subsequently observed that the artificial introduction of class noise in the data could cause boosting to become harmful.

A better understanding of its occasional failures is important if we are to use boosting for best effect in both attribute-value and relational learning. Breiman [12.5] observes that such failures are more likely to occur with relatively small datasets. Freund and Schapire [12.15] suggest that it may be due to overfitting, a common problem with noisy data. This is an issue on which further research is clearly needed.

The complexity of boosted classifiers is another problematic aspect of boosting. The experiments reported here used relatively small numbers of trials (10 and 20), but some researchers use hundreds or even thousands. Even with ten trials, the boosted theory is an order of magnitude more complex than a single theory, and the interactions among the component definitions effectively render the boosted definition unintelligible. This problem could be circumvented if methods could be found for producing an approximation to a boosted theory that preserves most of the accuracy gained through boosting, but has about the same complexity as a single constituent theory.

References

12.1 M. Bain. *Learning Logical Exceptions in Chess.* PhD Thesis. University of Strathclyde, Glasgow, 1994

12.2 S. Bell and S. Weber. On the close logical relationship between FOIL and the frameworks of Helft and Plotkin. In *Proceedings of the Third International Workshop on Inductive Logic Programming*, pages 127–147. Jožef Stefan Institute, Ljubljana, Slovenia, 1993.

12.3 F. Bergadano and D. Gunetti. An interactive system to learn functional logic programs. In *Proceedings of the Thirteenth International Joint Conference on Artificial Intelligence*, pages 1044–1049. Morgan Kaufmann, San Mateo, CA, 1993.

12.4 L.Breiman. Bagging predictors. *Machine Learning*, 24: 123–140,1996.

12.5 L. Breiman. *Bias, variance, and arcing classifiers.* Technical Report 460. Statistics Department, University of California, Berkeley, CA, 1996.

12.6 W. Buntine. *A Theory of Learning Classification Rules.* PhD Thesis. University of Technology, Sydney, Australia, 1990.

12.7 R. M. Cameron-Jones and J. R. Quinlan. Avoiding pitfalls when learning recursive theories. *Proceedings of the Thirteenth International Joint Conference on Artificial Intelligence*, pages 1050–1057. Morgan Kaufmann, San Mateo, CA, 1993.

12.8 W. W. Cohen. Recovering software specifications with inductive logic programming. In *Proceedings of the Twelfth National Conference on Artificial Intelligence*, pages 142–148. AAAI Press, Menlo Park, CA, 1994.

12.9 W. W. Cohen. Text categorization and relational learning. In *Proceedings of the Twelfth International Conference on Machine Learning*, pages 124–132. Morgan Kaufmann, San Francisco, CA, 1995.

12.10 T. G. Dietterich. Machine learning research: four current directions. *AI Magazine*, 18: 97–136, 1997.

12.11 T. G. Dietterich, M. J. Kearns, and Y. Mansour. Applying the weak learning framework to understand and improve C4.5. In *Proceedings of the Thirteenth International Conference on Machine Learning*, pages 96–104. Morgan Kaufmann, San Francisco, CA, 1996.

12.12 B. Dolšak and S. Muggleton. The application of inductive logic programming to finite element mesh design. In S. Muggleton, editor, *Inductive Logic Programming*, pages 453–472. Academic Press, London, 1992.

12.13 F. Esposito, D. Malerba, G. Semeraro, and M. Pazzani. A machine learning approach to document understanding. In *Proceedings of the Second International Workshop on Multistrategy Learning*, pages 276–292. George Mason University, Fairfax, VA, 1993.

12.14 C. Feng. Inducing temporal fault diagnostic rules from a qualitative model. In S. Muggleton, editor, *Inductive Logic Programming*, pages 473–493. Academic Press, London, 1992.

12.15 Y. Freund and R. E. Schapire. A decision-theoretic generalization of on-line learning and an application to boosting. *Journal of Computer and System Sciences*, 55: 119–139, 1997.

12.16 V. Klingspor, K. Morik, and A. Rieger. Learning concepts from sensor data of a mobile robot. *Machine Learning*, 23: 305–332, 1996.

12.17 N. Lavrač, I. Weber, D. Zupanič, D. Kazakov, O. Štěpánková, and S. Džeroski. ILPNET repositories on WWW: inductive logic programming systems, datasets and bibliography. *Artificial Intelligence Communications*, 9: 1–50, 1996.

12.18 C. X. Ling. Learning the past tense of English verbs: the symbolic pattern associator versus connectionist models. *Journal of Artificial Intelligence Research*, 1: 209–229, 1994.

12.19 N. Littlestone and M. K. Warmuth. The weighted majority algorithm. *Information and Computation*, 108: 212–261, 1994.

12.20 R. S. Michalski and G. Tecuci, editors. *Machine Learning: A Multistrategy Approach.* Morgan Kaufmann, San Francisco, CA, 1994.

12.21 R. J. Mooney and M. E. Califf. Induction of first-order decision lists: results on learning the past tense of English verbs. *Journal of Artificial Intelligence Research*, 3: 1–24, 1995.

12.22 S. Muggleton. Inverse entailment and Progol. *New Generation Computing*, 13: 245–286, 1995.

12.23 S. Muggleton and C. Feng. Efficient induction of logic programs. In S. Muggleton, editor, *Inductive Logic Programming*, pages 281–298. Academic Press, London, 1992.

12.24 S. Muggleton, R. D. King, and M. J. Sternberg. Protein secondary structure prediction using logic-based machine learning. *Protein Engineering*, 5:6 46–657, 1992.

12.25 M. Pazzani and D. Kibler. The utility of knowledge in inductive learning. *Machine Learning*, 9: 57–94, 1992.

12.26 J. R. Quinlan. Discovering rules by induction from large collections of examples. In D. Michie, editor, *Expert Systems in the Micro Electronic Age.* Edinburgh University Press, Edinburgh, 1979.

12.27 J. R. Quinlan. Learning logical definitions from relations. *Machine Learning*, 5: 239–266, 1990.

12.28 J. R. Quinlan. *C4.5: Programs for Machine Learning.* Morgan Kaufmann, San Mateo, CA, 1993.

12.29 J. R. Quinlan. Past tenses of verbs and first-order learning. In *Proceedings of the Seventh Australian Joint Conference on Artificial Intelligence*, pages 13–20. World Scientific, Singapore, 1994.

12.30 J. R. Quinlan. Boosting, bagging, and C4.5. In *Proceedings of the Fourteenth National Conference on Artificial Intelligence*, pages 725–730. AAAI Press, Menlo Park, CA, 1996.

12.31 J. R. Quinlan. Learning first-order definitions of functions. *Journal of Artificial Intelligence Research*, 5: 139–161, 1996.

12.32 J. R. Quinlan and R. M. Cameron-Jones. Induction of logic programs: FOIL and related systems. *New Generation Computing*, 13: 287–312, 1995.

12.33 C. Rouveirol. Flattening and saturation: two representation changes for generalization. *Machine Learning*, 14: 219–232, 1994.

12.34 R. E. Schapire. The strength of weak learnability. *Machine Learning*, 5: 197–227, 1990.

12.35 A. Srinivasan, S. Muggleton, M. J. E. Sternberg, and R. D. King. Theories for mutagenicity: A study in first-order and feature-based induction. *Artificial Intelligence*, 85: 277–299, 1996.

12.36 J. M. Zelle and R. J. Mooney. Combining FOIL and EBG to speed-up logic programs. In *Proceedings of the Thirteenth International Joint Conference on Artificial Intelligence*, pages 1106–1111. Morgan Kaufmann, San Mateo, CA, 1993.

13. Learning Probabilistic Relational Models

Lise Getoor[1], Nir Friedman[2], Daphne Koller[1], and Avi Pfeffer[3]

[1] Computer Science Department, Stanford University
Stanford, CA 94305-9010, USA

[2] The School of Computer Science & Engineering, Hebrew University
Jerusalem 91904, Israel

[3] Division of Engineering and Applied Sciences, Harvard University
Cambridge, MA 02138, USA

Abstract

Probabilistic relational models (PRMs) are a language for describing statistical models over typed relational domains. A PRM models the uncertainty over the attributes of objects in the domain and uncertainty over the relations between the objects. The model specifies, for each attribute of an object, its (probabilistic) dependence on other attributes of that object and on attributes of related objects. The dependence model is defined at the level of *classes* of objects. The class dependence model is instantiated for any object in the class, as appropriate to the particular context of the object (i.e., the relations between this objects and others). PRMs can also represent uncertainty over the relational structure itself, e.g., by specifying a (class-level) probability that two objects will be related to each other. PRMs provide a foundation for dealing with the noise and uncertainty encountered in most real-world domains. In this chapter, we show that the compact and natural representation of PRMs allows them to be learned directly from an existing relational database using well-founded statistical techniques. We give an introduction to PRMs and an overview of methods for learning them. We show that PRMs provide a new framework for relational data mining, and offer new challenges for the endeavor of learning relational models for real-world domains.

13.1 Introduction

Relational models are the most common representation of structured data. Enterprise business information, marketing and sales data, medical records, and scientific datasets are all stored in relational databases. Efforts to extract knowledge from partially structured (e.g., XML) or even raw text data also aim to extract relational information. Recently, there has been growing interest in extracting interesting statistical patterns from these huge amounts of data. These patterns provide a deeper understanding of the domain and the relationships in it. In addition, extracted patterns can be used for reaching conclusions about important attributes whose values may be unobserved.

Probabilistic graphical models, and particularly Bayesian networks , have been shown to be a useful way of representing statistical patterns in real-world domains. Recent work [13.5, 13.11] develops techniques for learning these models directly from data, and shows that interesting patterns often emerge in this learning process. However, all of these learning techniques apply only to flat-file representations of the data, and not to the richer relational data encountered in many applications. An obvious solution is to take a relational database and "flatten" it, creating a flat file on which standard Bayesian network learning algorithms can be run. As we discuss in Section 13.7, this approach has several important shortcomings.

Probabilistic relational models (PRMs) are a recent development [13.15, 13.20, 13.23] that extend the standard attribute-based Bayesian network representation to incorporate a much richer relational structure. These models allow the specification of a probability model for *classes* of objects rather than simple attributes; they also allow properties of an entity to depend probabilistically on properties of other *related* entities. The probabilistic class model represents a generic dependence, which is then instantiated for specific circumstances, i.e., for particular sets of entities and relations between them.

We have developed methods for learning PRMs directly from structured data such as relational databases. The two key tasks in the construction of any statistical model are model selection and parameter estimation. We have developed algorithms for each of these tasks. These algorithms are based on the same underlying principles that govern the learning of Bayesian networks.

A learned PRM provides a statistical model that can uncover and discover many interesting probabilistic dependencies that hold in a domain. Unlike a set of (probabilistic) rules for classification, PRMs specify a joint distribution over a relational domain. Thus, like Bayesian networks, they can be used for answering queries about any aspect of the domain given any set of observations. Furthermore, rather than trying to predict one particular attribute, the PRM learning algorithm attempts to tease out the most significant direct dependencies in the data. The resulting model thus provides a high-level, qualitative picture of the structure of the domain, in addition to the quantitative information provided by the probability distribution. Thus, PRMs are ideally suited to exploratory analysis of a domain and relational data mining.

This chapter is structured as follows. We begin by briefly surveying some of the foundations on which PRMs are built: In the next two sections, we provide some basic background on probabilistic models and relational models. In Section 13.4, we define PRMs and give their semantics. In Section 13.5 we describe how to learn a PRM from an existing database. In Section 13.6 we describe experimental results for several real-world domains. We conclude with a discussion of related work and briefly mention some promising directions for future work.

13.2 Probabilistic models

The traditional logic-based approach to representing knowledge is to write down a knowledge base in the form of logical axioms about the domain. The knowledge base restricts the set of possible worlds, or models, to those consistent with the axioms. Additional facts — those that are true in all of these possible worlds — are logically entailed by the knowledge base.

In a standard logical framework, we are restricted to representing only facts that are true absolutely. Thus, this framework is unable to represent and reason with uncertain and noisy information. This is a significant gap in the expressive power of the framework, and a major barrier to its use in many real-world applications. Uncertainty is unavoidable in the real world: our information is often inaccurate and always incomplete, and only a few of the "rules" that we use for reasoning are true in all (or even most) of the possible cases.

This limitation, which is critical in many domains (e.g., medical diagnosis), has led over the last decade to the resurgence of probabilistic reasoning in artificial intelligence. Probability theory models uncertainty by assigning a probability to each of the states of the world that an agent considers possible. Most commonly in probabilistic reasoning, these states are the set of possible assignments of values to a set of *attributes* or *random variables*. Consider, for example, a simple model of the performance of a student in a course. There are six random variables: *Intelligence*, *Difficulty* (of the course), *Good Test Taker*, *Understands Material*, *Exam Grade* and *Homework Grade*. Of these variables, *Intelligence*, *Good Test Taker*, and *Understands Material* are boolean variables, *Difficulty* takes values from {*low, medium, high*}, and *Exam Grade* and *Homework Grade* take values from $\{A, B, C, D, F\}$. The possible worlds are all possible assignments of values to these variables, 600 $(2 \times 2 \times 2 \times 3 \times 5 \times 5)$ in this case.

A probabilistic model specifies a joint probability distribution over all possible worlds. Thus, it specifies implicitly the probability of any event, such as an assignment of values to some subset of variables. Unlike many models, such as a set of rules used for predicting some particular attribute, a probabilistic model is not limited to conclusions about a prespecified set of attributes, but rather can be used to answer queries about any variable or subset of variables. Nor does it require that the values of all other variables be given; it applies in the presence of any evidence. For example, a probabilistic model of a student's performance can be used to predict the distribution over the student's exam grade given his intelligence. As new evidence is obtained, e.g., about his homework grade, conditioning can be used to update this probability, so that the probability of a high exam grade will go up if we observe good homework grades. The same model is used to do the predictive and the evidential reasoning.

Furthermore, a probabilistic model can perform *explaining away*, a reasoning pattern that is very common in human reasoning, but very difficult to obtain in other formal frameworks. Explaining away uses evidence supporting one cause to decrease the probability in another, not because the two are incompatible, but simply because the one cause explains away the evidence, removing the support for the other cause. For example, if we observe that the student's exam grade is high, our belief that she is intelligent will go up. If we then hear that the class is known to be easy, that fact provides an alternative explanation for the student's high grade, reducing our belief that she is intelligent. The same probabilistic model supports all of these reasoning patterns, allowing it to be used in many different tasks.

The traditional objection to probabilistic models has been their computational cost. A complete joint probability distribution over a set of random variables must specify a probability for each of the exponentially many different instantiations of the set. Even in our very simple example, we must specify 600 numbers to specify the joint distribution. This type of representation is impractical both from a knowledge engineering perspective, since it is almost impossible for a person to specify an entry in a complex joint distribution, far less to specify an exponential number of them, and from a learning perspective, since the number of samples required to perform robust parameter estimation would be infeasible.

Therefore, a naive representation of the joint distribution is infeasible for all but the simplest domains. *Bayesian networks* [13.21] use the underlying structure of the domain to overcome this problem. The key insight is the locality of influence present in many real-world domains: each variable is directly influenced by only a few others. For example, a student's intelligence induces a better understanding of the material, which in turns leads to a higher homework grade. But the effect of intelligence on homework grade is an indirect one: if the student does not understand the material, her intelligence does not help her get better grades. A Bayesian network captures this insight graphically; it represents the distribution as a directed acyclic graph whose nodes represent the random variables and whose edges represent direct dependencies. Figure 13.1 shows a Bayesian network for our simple student domain.

A Bayesian network has formal semantics in terms of *probabilistic conditional independence.* Formally, the network asserts that each node (or rather the random variable) is conditionally independent of its non-descendants given values for its parents. For example, if we know that the student does not understand the material, our distribution over her grades is no longer influenced by information that we might have about her intelligence.

These conditional independence assumptions allow a very concise representation of the joint probability distribution over these random variables: we associate with each node a *conditional probability distribution* (CPD), which specifies for each node X the probability distribution over the values of X given each combination of values for its parents, denoted $\mathrm{Pa}(X)$. The conditional independence assumptions associated with the Bayesian networks

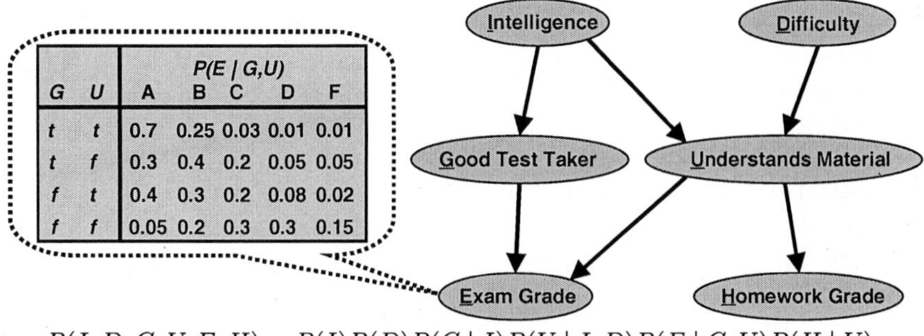

G	U	P(E | G,U) A	B	C	D	F
t	t	0.7	0.25	0.03	0.01	0.01
t	f	0.3	0.4	0.2	0.05	0.05
f	t	0.4	0.3	0.2	0.08	0.02
f	f	0.05	0.2	0.3	0.3	0.15

$$P(I, D, G, U, E, H) = P(I)P(D)P(G \mid I)P(U \mid I, D)P(E \mid G, U)P(H \mid U)$$

Fig. 13.1. A simple Bayesian network for the student performance domain, the decomposition of the joint distribution into a product of CPDs, and the CPD for one of the nodes in the network.

imply that these numbers suffice to uniquely determine the probability distribution over these random variables. More precisely, the joint distribution over all variables can be *factorized* into a product of the CPDs of all the variables via the *Chain Rule for Bayesian Networks*:

$$P(X_1, \ldots, X_n) = \prod_{i=1}^{n} P(X_i \mid \mathrm{Pa}(X_i)). \tag{13.1}$$

Consider the Bayesian network for our student performance domain (Figure 13.1). In this model, the student's performance in tests depends on her intelligence. Her understanding of the material depends both on her intelligence and on the difficulty of the class. Her exam grade depends on whether she is a good test taker, and on her understanding of the material, while her homework grade depends on her understanding of the material. The structure of the network encodes a number of conditional independence assertions. For example, the student's exam grade is conditionally independent of her intelligence given her test taking ability and understanding of the material.

These independence assumptions allow us to factor the joint distribution into a product form as shown in Figure 13.1. Each of the conditional probabilities in this product form is a CPD of one of the variables. In this example, the CPD is simply a table, such as the one shown for $P(E \mid G, U)$ in the figure. This CPD shows that if a student is a good test taker and understands the material, then she has probability 0.7 of getting an A on the exam, whereas if the student is a bad test taker and does not understand the material, her probability of getting an A is only 0.05.

Bayesian networks provide a compact representation of complex joint distributions. By modeling an entire joint distribution, Bayesian networks implicitly specify the answer to any probability query, in particular, any query where we want to find the probability distribution over some variables given evidence about any others. In theory, the problem of doing inference in Bayesian networks is NP-hard [13.4]. However, the dependency structure made explicit by the network representation can be exploited by inference algorithms, allowing for efficient inference in practice, even for very large networks.

The semantics and compact representation of Bayesian networks also allow effective statistical learning from data. Standard statistical parameter estimation techniques can be used for learning the parameters of a given network. For learning the structure of the network, typical approaches use a *score* function (which is typically based on Bayesian considerations) to score how different structures "match" the training data. The learning process then reduces to the task of searching for the highest scoring structure [13.11]. These techniques allow a Bayesian network structure to be discovered from data. The learned structure can often give us insight about the nature of the connections between the variables in the domain. Furthermore, the graph structure can sometimes be interpreted causally [13.24], allowing us to induce cause and effect, which can be very useful for understanding our domain, and to reach conclusions about the consequences of intervening (acting) in the domain. Statistical learning techniques are also robust to the presence of missing data and hidden variables. Techniques such as *EM (Expectation Maximization)* can be used to deal with this issue in the context of parameter estimation [13.16] and have recently been generalized to the harder problem of structure selection [13.7].

Bayesian network learning has been applied successfully to data mining applications. For example, Breese et al. [13.2] show how a Bayesian network can be learned from data describing people's preferences over a variety of items. The learned dependencies correspond to correlations between a person's preference for different items. The resulting Bayesian network can be used for collaborative filtering, and is a better predictor than the standard approaches to this task. In addition to their predictive ability, Bayesian networks have the advantage that they provide a visualization of the most significant direct correlations in the domain, clarifying the domain structure to the user.

13.3 Relational models

Over the last decade, Bayesian networks have been used with great success in a wide variety of real-world and research applications. However, despite their success, Bayesian networks are often inadequate to properly model aspects of complex relational domains. A Bayesian network for a given domain involves a prespecified set of random variables, whose relationship to each other is fixed in advance. Hence, a Bayesian network cannot be used to deal with domains where we might encounter several entities in a variety of configurations. This limitation of Bayesian networks is a direct consequence of the fact that they lack the concept of an "object" (or domain entity). Hence,

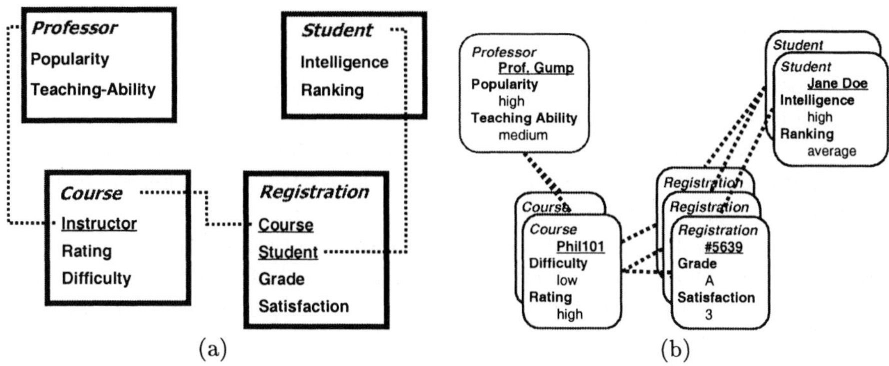

Fig. 13.2. (a) A relational schema for a simple university domain. The underlined attributes are reference slots of the class and the dashed lines indicate the types of objects referenced. (b) An example instance of this schema. Here we do not show the reference slots, we use dashed lines to indicate the relationships that hold between objects.

they cannot represent general principles about multiple similar objects which can then be applied in multiple contexts.

Relational logic, which has traditionally formed the basis for most large-scale knowledge representation systems, addresses these problems. The notions of "individuals", their properties, and the relations between them provide an elegant and expressive framework for reasoning about many diverse domains. The use of quantification allows us to compactly represent general rules, that can be applied in many different situations. For example, when reasoning about genetic transmission of certain properties (e.g., genetically transmitted diseases), we can write down general rules that hold for all people and many properties.[1]

Probabilistic relational models (PRMs) [13.15, 13.22] extend Bayesian networks with the concepts of individuals, their properties, and relations between them. In a way, they are to Bayesian networks as relational logic is to propositional logic. Bayesian networks have a formal semantics in terms of probability distributions over sets of propositional interpretations that are assignments of values to attributes. PRMs have a similar formal semantics in terms of probability distributions over sets of *relational logic* interpretations.

Our relational framework, on which PRMs are based, is derived from the presentation of Friedman et al. [13.8]; it is motivated primarily by the concepts of relational databases, although some of the notation is derived from frame-based and object-oriented systems. However, the framework is a fully general one, and is equivalent to the standard vocabulary and semantics of relational logic.

---

[1] See Chapters 3 and 4 for a more extensive discussion of the advantages of relational representations.

A schema for a relational model describes a set of *classes*, $\mathcal{X} = \{X_1, \ldots, X_n\}$. Each class is associated with a set of *descriptive attributes* and a set of *reference slots*. There is a direct mapping between our representation and that of relational databases. Each class corresponds to a single table. Our descriptive attributes correspond to standard attributes in the table, and our reference slots correspond to attributes that are foreign keys (key attributes of another table).

Figure 13.2(a) shows a schema for a simple domain that we will be using as our main running example. The domain is that of a university, and contains professors, students, courses, and course registrations. The classes in the schema are Professor, Student, Course, and Registration.

The set of descriptive attributes of a class X is denoted $\mathcal{A}(X)$. Attribute A of class X is denoted $X.A$, and its space of values is denoted $\mathcal{V}(X.A)$. We assume here that value spaces are finite. For example, the Student class has the descriptive attributes *Intelligence* and *Ranking*. The value space for Student.*Intelligence* might be $\{$*high, low*$\}$.

The set of reference slots of a class X is denoted $\mathcal{R}(X)$. We use a similar notation, $X.\rho$, to denote the reference slot ρ of X. Each reference slot ρ is typed, i.e., the schema specifies the range type of object that may be referenced. More formally, for each ρ in X, the *domain type* Dom$[\rho]$ is X and the *range type* Range$[\rho]$ is Y for some class Y in $\mathcal{X}$. For example, the class Course has reference slot *Instructor* with range type Professor, and class Registration has reference slots *Course* and *Student.* In Figure 13.2(a) the reference slots are underlined.

For each reference slot ρ, we can define an *inverse slot* ρ^{-1}, which is interpreted as the inverse function of ρ. For example, we can define an inverse slot for the *Student* slot of Registration and call it *Registered-In*. Note that this is not a one-to-one relation, but returns a *set* of Registration objects. Finally, we define the notion of a *slot chain*, which allows us to compose slots, defining functions from objects to other objects to which they are indirectly related. More precisely, we define a *slot chain* $\rho_1, \ldots, \rho_k$ to be a sequence of slots (inverse or otherwise) such that for all i, Range$[\rho_i]$ = Dom$[\rho_{i+1}]$. For example, Student.*Registered-In.Course.Instructor* can be used to denote a student's set of instructors.

An *instance* $\mathcal{I}$ of a schema is simply a standard relational logic interpretation of this vocabulary. It specifies: a set of objects x, partitioned into classes; a value for each attribute $x.A$ (in the appropriate domain); and a value for each reference slot $x.\rho$, which is an object in the appropriate range type. We use $\mathcal{A}(x)$ as a shorthand for $\mathcal{A}(X)$, where x is of class X. For each object x in the instance and each of its attributes A, we use $\mathcal{I}_{x.A}$ to denote the value of $x.A$ in $\mathcal{I}$. For example, Figure 13.2(b) shows an instance of the schema from our running example. In this (simple) instance there is one Professor, two Classes, three Registrations, and two Students. The relations between them show that the Professor is the instructor in both classes, and that one student ("Jane Doe") is registered only for one class ("Phil101"), while the other student is registered for both classes.

13.4 Probabilistic relational models

Probabilistic relational models are a new development that integrates the strengths of probabilistic models and relational logic. Several approaches have been proposed, some based on probabilistic logic programming [13.20, 13.23] and others based on a more object-relational framework [13.15]. Our presentation is based on the ideas presented by Koller and Pfeffer [13.15] and follows the presentation of [13.8]. It also accommodates and generalizes the probabilistic logic programming approaches [13.20, 13.23].

13.4.1 Basic language

PRMs provide a language for specifying a probability distribution over a set of relational interpretations. More precisely, a PRM specifies a distribution over a set of instances of a given schema. One might consider PRMs that specify a distribution over all possible instances of the schema, i.e., all possible databases over that schema. This set of databases is infinitely large, as it includes all the possible variations over the number of objects in each class and the possible relations between them. It is clearly very difficult to place a distribution over this type of space, and it is not obvious that such a general-purpose distribution is useful. On the other hand, unlike in the case of Bayesian networks, we want PRMs to be a general model, that can apply to a wide variety of situations. Hence, a PRM is actually a template: given a set of ground objects, a PRM specifies a probability distribution over a set of interpretations involving these objects (and perhaps other objects). We begin with describing the simplest form of PRMs, where the relational structure of the model — the set of objects and the relations between them — is assumed to be part of the input to the template. Only the attributes of the objects participate in the probabilistic model. In Section 13.8 we discuss how to extend this framework to much richer settings.

A *relational skeleton* σ of a relational schema is a partial specification of an instance of the schema. It specifies the set of objects for each class and the relations that hold between the objects. However, it leaves the values of the attributes unspecified. Figure 13.3(a) shows the relational skeleton of the instance shown in Figure 13.2(b). A PRM specifies a probability distributions over *completions* $\mathcal{I}$ of any given skeleton. As we will show, for any skeleton for the schema, the PRM induces a distribution over instances that complete the skeleton.

A PRM specifies the probability distribution using the same underlying principles used in specifying Bayesian networks. The assumption is that each

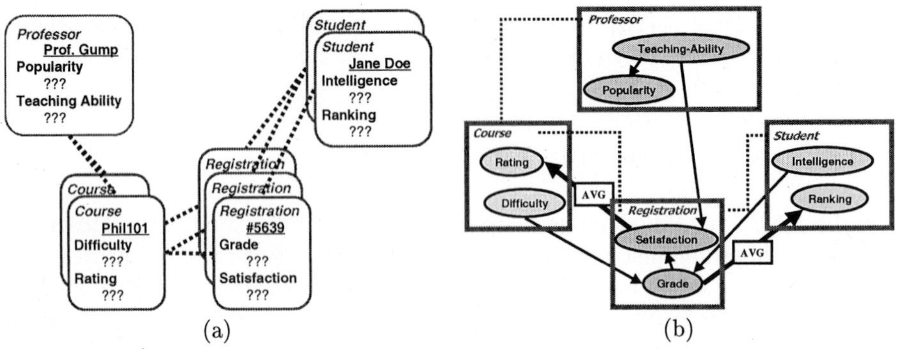

Fig. 13.3. (a) An example relational skeleton for the school domain. Here we know only the objects in our domain and the relationships that hold between them. (b) A PRM structure for the school domain. Edges correspond to probabilistic dependency. Edges from one class to another are routed through slot-chains (chains of references). For clarity these are not explicitly stated in the diagram (see text).

of the random variables in the PRM — in this case the attributes $x.A$ of the individual objects x — is directly influenced by only a few others. The PRM therefore defines for each $x.A$ a set of *parents*, which are the direct influences on it, and a local probabilistic model that specifies the dependence on these parents. However, there are two primary differences between PRMs and Bayesian networks. First, a PRM defines the dependency model at the class level, allowing it to be used for any object in the class. In a sense, the class dependency model is universally quantified and instantiated for every element in the class domain. Second, the PRM explicitly uses the relational structure of the model, in that it allows the probabilistic model of an attribute of an object to depend also on attributes of related objects. The specific set of related objects can vary with the skeleton σ; the PRM specifies the dependency in a generic enough way that it can apply to an arbitrary relational structure.

A PRM consists of two components: the qualitative dependency structure, $\mathcal{S}$, and the set of parameters associated with it, $\theta_{\mathcal{S}}$. Figure 13.3(b) shows an example PRM structure for our school domain. The dependency structure is defined by associating with each attribute $X.A$ a set of *parents* $\mathrm{Pa}(X.A)$. These correspond to *formal* parents; they will be instantiated in different ways for different objects in X. Intuitively, the parents are attributes that are "direct influences" on $X.A$. In Figure 13.3(b), the arrows define the dependency structure.

We distinguish between two types of formal parents. The attribute $X.A$ can depend on another probabilistic attribute B of X. This formal dependence induces a corresponding dependency for individual objects: for any object x in of class X, $x.A$ will depend probabilistically on $x.B$. For example, in Figure 13.3(b), a professor's *Popularity* depends on her *Teaching*

Ability. This dependency model is duplicated for each professor in the skeleton. Thus, we essentially assume that the same probabilistic model applies to all the professors in our domain.

In addition, an attribute $X.A$ can also depend on attributes of related objects $X.\tau.B$, where τ is a slot chain. In Figure 13.3(b), the grade of a student in a course, Registration.*Grade*, depends on Registration.*Student.Intelligence* and Registration.*Course.Difficulty*. The PRM language also allows us to use longer slot chains, for example the dependence of Student.*Satisfaction* on Registration.*Course.Instructor.Teaching-Ability*. Such slot chains are instantiated for each object by following the references that are assigned to it by the skeleton. Thus, for example, for the registration object *#5639*, Registration.*Student.Intelligence* references *Jane-Doe.Intelligence*, and the slot Registration.*Course.Difficulty* references *Phil101.Difficulty*.

Our example PRM also contains a dependence of Student.*Ranking* on Student.*Registered-In.Grade*. Note that a student will typically be registered in several classes; the model specifies a dependence of the student's ranking on the grades that he receives in all of them. In general, $x.\tau$ represents the *set* of objects that are τ-relatives of x. Except in cases where the slot chain is guaranteed to be single-valued, we must specify the probabilistic dependence of $x.A$ on the multiset $\{y.B \; : \; y \in x.\tau\}$. This dependence poses a representational problem, since we need to specify the distribution of $x.A$ given a multiset of values of size 1, 2, 3, and so on. It is clearly impractical to to provide a dependency model for each of the unboundedly many possible multiset sizes.

The notion of *aggregation* from database theory gives us an appropriate tool to address this issue. The dependence of $x.A$ on $x.\tau.B$, is interpreted as a probabilistic dependence of $x.A$ on some (deterministically computed) aggregate property of this multiset. There are many natural and useful notions of aggregation: the mode of the set (most frequently occurring value); mean value of the set (if values are numerical); median, maximum, or minimum (if values are ordered); cardinality of the set; etc. More formally, our language allows a notion of an aggregate γ; γ takes a multiset of values of some ground type, and returns a summary of it. The type of the aggregate can be the same as that of its arguments. However, we allow other types as well, e.g., an aggregate that reports the size of the multiset. More precisely, we allow $X.A$ to have as a parent $\gamma(X.\tau.B)$; the semantics is that for any $x \in X$, $x.A$ will depend on the value of $\gamma(x.\tau.B)$. We define $\mathcal{V}(\gamma(X.\tau.B))$ to be the set of possible values of this aggregate. In our example PRM, there are two aggregate dependencies defined, one that specifies that the ranking of a student depends on the average of his grades and one that specifies that the rating of a course depends on the average satisfaction of students in the course.

As in Bayesian networks, the second component of a PRM is the parameters associated with the qualitative structure. A PRM contains a CPD for

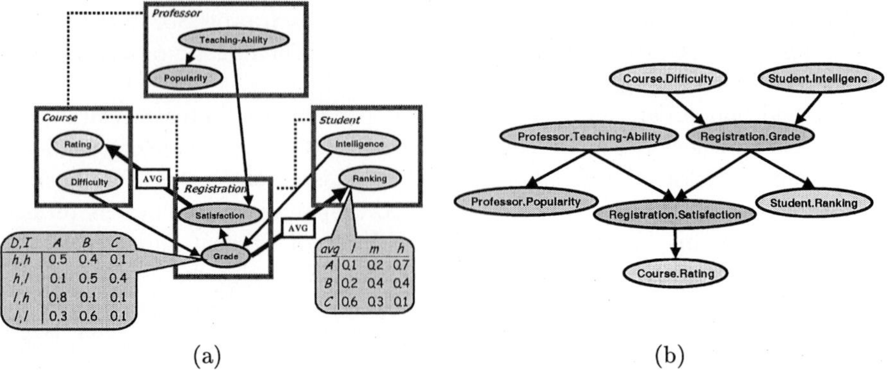

Fig. 13.4. (a) The CPD for Registration.*Grade* and the CPD for an aggregate dependency of Student.*Rank* on Student.*Registered-In*.*Grade*. (b) The dependency graph for the school PRM.

each attribute of each class. As for the dependencies, we assume that the parameters are shared by each object in the class. We associate with each attribute $X.A$ a CPD $P(X.A \mid \text{Pa}(X.A))$. Figure 13.4(a) shows two CPDs, one for a dependency on a single-valued chain and one for an aggregate dependency. More precisely, let $\mathbf{U}$ be the set of parents $\text{Pa}(X.A)$. Each of these parents U_i — whether a simple attribute or an aggregate — has a set of values $\mathcal{V}(U_i)$ in some ground type. For each tuple of values $\mathbf{u} \in \mathcal{V}(\mathbf{U})$, we specify a distribution $P(X.A \mid \mathbf{u})$ over $\mathcal{V}(X.A)$. We use $\theta_{X.A|\mathbf{u}}$ to denote the parameters of this distribution. The entire set of these parameters, for all $X.A$ and all $\mathbf{u}$, comprises $\theta_{\mathcal{S}}$.

Definition 1: A *probabilistic relational model (PRM)* Π for a relational schema $\mathcal{R}$ is defined as follows. For each class $X \in \mathcal{X}$ and each descriptive attribute $A \in \mathcal{A}(X)$, we have:

- a set of *parents* $\text{Pa}(X.A) = \{U_1, \ldots, U_l\}$, where each U_i has the form $X.B$ or $\gamma(X.\tau.B)$, where τ is a slot chain;
- a *conditional probability distribution (CPD)* that represents $P_\Pi(X.A \mid \text{Pa}(X.A))$. ∎

13.4.2 PRM semantics

Given any skeleton, we have a set of random variables of interest: the attributes $x.A$ of the objects in the skeleton. Formally, let $\mathcal{O}^\sigma(X)$ denote the set of objects in skeleton σ whose class is X. The set of random variables for σ is the set of attributes of the form $x.A$ where $x \in \mathcal{O}^\sigma(X_i)$ and $A \in \mathcal{A}(X_i)$ for some class X_i. The PRM specifies a probability distribution over the possible joint assignments of values to these random variables. As with Bayesian networks, the joint distribution over these assignments can be factored. That is, we take the product, over all $x.A$, of the probability in the CPD of the specific value assigned by the instance to the attribute given the values assigned to its parents. Formally, this is written as follows:

$$\begin{aligned} P(\mathcal{I} \mid \sigma, \mathcal{S}, \theta_{\mathcal{S}}) &= \prod_{x \in \sigma} \prod_{A \in \mathcal{A}(x)} P(\mathcal{I}_{x.A} \mid \mathcal{I}_{\mathrm{Pa}(x.A)}) \\ &= \prod_{X_i} \prod_{A \in \mathcal{A}(X_i)} \prod_{x \in \mathcal{O}^{\sigma}(X_i)} P(\mathcal{I}_{x.A} \mid \mathcal{I}_{\mathrm{Pa}(x.A)}) \end{aligned} \tag{13.2}$$

This expression is very similar to the chain rule for Bayesian networks (see Equation (13.1)). There are two primary differences. First, our random variables are the attributes of a set of objects. Second, the set of parents of a random variable can vary according to the relational context of the object — the set of objects to which it is related.

As in any definition of this type, we have to take care that the resulting function from instances to numbers does indeed define a *coherent* probability distribution, i.e., where the sum of the probability of all instances is 1. In Bayesian networks, where the joint probability is also a product of CPDs, this requirement is satisfied if the dependency graph is acyclic: a variable is not an ancestor of itself. A similar condition is sufficient to ensure coherence in PRMs as well. We want to ensure that our probabilistic dependencies are acyclic, so that a random variable does not depend, directly or indirectly, on its own value. To do so, we can consider the graph of dependencies among attributes of objects in the skeleton. Consider the parents of an attribute $X.A$. When $X.B$ is a parent of $X.A$, we define an edge $x.B \rightarrow_{\sigma} x.A$; when $\gamma(X.\tau.B)$ is a parent of $X.A$ and $y \in x.\tau$, we define an edge $y.B \rightarrow_{\sigma} x.A$. We say that a dependency structure $\mathcal{S}$ is *acyclic* relative to a skeleton σ if the directed graph defined by $\rightarrow_{\sigma}$ over the variables $x.A$ is acyclic. In this case, we are guaranteed that the PRM defines a coherent probabilistic model over complete instantiations $\mathcal{I}$ consistent with σ.

This procedure allows us to check whether a dependency structure $\mathcal{S}$ is acyclic relative to a fixed skeleton σ. However, we often want stronger guarantees: we want to ensure that our dependency structure is acyclic for any skeleton that we are likely to encounter. How do we guarantee this property based only on the class-level PRM? To do so, we consider potential dependencies at the class level. More precisely, we define a *class dependency graph*, which reflects these dependencies [13.15, 13.8]. This class dependency graph has an edge from $Y.B$ to $X.A$ if either: $X = Y$ and $X.B$ is a parent of $X.A$; or $\gamma(X.\tau.B)$ is a parent of $X.A$ and $\mathrm{Range}[X.\tau] = Y$. Figure 13.4(b) shows the dependency graph for our school domain.

The most obvious approach for using the class dependency graph is to simply require that it be acyclic. This requirement is equivalent to assuming a stratification among the attributes of the different classes, and requiring that the parents of an attribute precede it in the stratification ordering. It is clear that if the class dependency graph is acyclic, we can never have

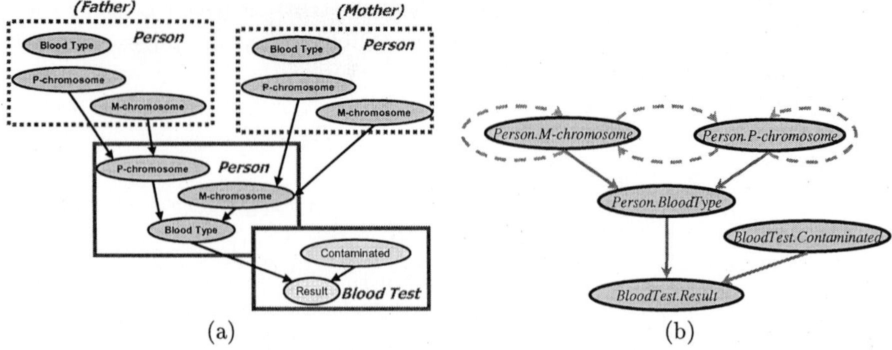

Fig. 13.5. (a) A simple PRM for the genetics domain. (b) The corresponding dependency graph. Dashed edges correspond to "guaranteed acyclic" dependencies.

that $x.A$ depends (directly or indirectly) on itself. For example, if we examine the PRM of Figure 13.3(b), we can easily convince ourselves that we cannot create a cycle in any instance. Indeed, as we saw in Figure 13.4(b), the class dependency graph is acyclic. Note, however, that if we introduce additional dependencies we can create cycles. For example, if we make Professor.*Teaching-Ability* depend on the rating of courses she teaches (e.g., if high teaching ratings increase her motivation), then the resulting class dependency graph is cyclic, and there is no stratification order that is consistent with the PRM structure. An inability to stratify the class dependency graph implies that there are skeletons for which the PRM will induce a distribution with cyclic dependencies. In general, however, a cycle in the class dependency graph does not imply that all skeletons induce cyclic dependencies.

While this simple approach clearly ensures acyclicity, it is too limited to cover many important cases. Consider, for example, a simple genetic model of the inheritance of a single gene that determines a person's blood type, shown in Figure 13.5(a). Each person has two copies of the chromosome containing this gene, one inherited from her mother, and one inherited from her father. There is also a possibly contaminated test that attempts to recognize the person's blood type. Our schema contains two classes Person and BloodTest. Class Person has reference slots *Mother* and *Father* and descriptive attributes *Gender*, *P-Chromosome* (the chromosome inherited from the father), and *M-Chromosome* (inherited from the mother). *BloodTest* has a reference slot *Test-Of* that points to the owner of the test, and descriptive attributes *Contaminated* and *Result.*

In our genetic model, the genotype of a person depends on the genotype of his parents; thus, at the class level, we have *Person.P-Chromosome* depending directly on *Person.P-Chromosome*. As we can see in Figure 13.5(b), this dependency results in a cycle that clearly violates the requirements of our simple approach. However, it is clear to us that the dependencies in this model are not actually cyclic for any skeleton that we will encounter in this domain. The reason is that, in "legitimate" skeletons for this schema, a person cannot be his own ancestor, which disallows the situation of the person's genotype depending (directly or indirectly) on itself. In other words, although the model appears to be cyclic at the class level, we know that this cyclicity is always resolved at the level of individual objects.

Our ability to guarantee that the cyclicity is resolved relies on some prior knowledge that we have about the domain. We want to allow the user to give us information such as this, so that we can make stronger guarantees about acyclicity. The user can specify that certain slots are *guaranteed acyclic.* In our genetics example, *Father* and *Mother* are guaranteed acyclic; cycles involving these attributes may in fact be legal. In fact, they are mutually guaranteed acyclic, so that compositions of the slots are also guaranteed acyclic. Figure 13.5(b) shows the class dependency graph for the genetics domain, with guaranteed acyclic edges shown as dashed edges. It turns out that because all of the cycles in this graph contain mutually guaranteed acyclic relations, the structure is legal. In [13.8], we give an algorithm for checking the legality of structures that contain guaranteed acyclic slots and slot chains.

13.5 Learning PRMs

In the previous sections, we defined the PRM language and its semantics. We now move to the task of learning a PRM from data. In the learning problem, our input contains a relational schema, that specifies the basic vocabulary in the domain — the set of classes, the attributes associated with the different classes, and the possible types of relations between objects in the different classes (which simply specifies the mapping between a foreign key in one table and the associated primary key). Our training data consists of a fully specified instance of that schema. We assume that this instance is given in the form of a relational database. Although our approach would also work with other representations (e.g., a set of ground facts completed using the closed world assumption), the efficient querying ability of relational databases is particularly helpful in our framework, and makes it possible to apply our algorithms to large datasets.

There are two variants of the learning task: parameter estimation and structure learning. In the parameter estimation task, we assume that the qualitative dependency structure of the PRM is known; i.e., the input consists of the schema and training database (as above), as well as a qualitative dependency structure $\mathcal{S}$. The learning task is only to fill in the parameters that define the CPDs of the attributes. In the structure learning task, there is no additional required input (although the user can, if available, provide prior knowledge about the structure, e.g., in the form of constraints). The goal is to extract an entire PRM, structure as well as parameters, from the training database alone. We discuss each of these problems in turn.

13.5.1 Parameter estimation

We begin with the parameter estimation task for a PRM where the dependency structure is known. In other words, we are given the structure $\mathcal{S}$ that determines the set of parents for each attribute, and our task is to learn the parameters $\theta_{\mathcal{S}}$ that define the CPDs for this structure. While this task is relatively straightforward, it is of interest in and of itself. Experience in the setting of Bayesian networks shows that the qualitative dependency structure can be fairly easy to elicit from human experts, in cases where such experts are available. In addition, the parameter estimation task is a crucial component in the structure learning algorithm described in the next section.

The key ingredient in parameter estimation is the *likelihood function*, the probability of the data given the model. This function measures the extent to which the parameters provide a good explanation of the data. Intuitively, the higher the probability of the data given the model, the better the ability of the model to predict the data. The likelihood of a parameter set is defined to be the probability of the data given the model: $L(\theta_{\mathcal{S}} \mid \mathcal{I}, \sigma, \mathcal{S}) = P(\mathcal{I} \mid \sigma, \mathcal{S}, \theta_{\mathcal{S}})$. As in many cases, it is more convenient to work with the logarithm of this function:

$$
\begin{aligned}
l(\theta_{\mathcal{S}} \mid \mathcal{I}, \sigma, \mathcal{S}) &= \log P(\mathcal{I} \mid \sigma, \mathcal{S}, \theta_{\mathcal{S}}) \\
&= \sum_{X_i} \sum_{A \in \mathcal{A}(X_i)} \left[\sum_{x \in \mathcal{O}^{\sigma}(X_i)} \log P(\mathcal{I}_{x.A} \mid \mathcal{I}_{\mathrm{Pa}(x.A)}) \right] \qquad (13.3)
\end{aligned}
$$

The key insight is that this equation is very similar to the log-likelihood of data given a Bayesian network [13.11]. In fact, it is the likelihood function of the Bayesian network induced by the structure given the skeleton: the network with a random variable for each attribute of each object $x.A$, and the dependency model induced by $\mathcal{S}$ and σ, as discussed in Section 13.4.2. The only difference from standard Bayesian network parameter estimation is that parameters for different nodes in the network — those corresponding to the $x.A$ for different objects x from the same class — are forced to be identical. This similarity allows us to use the well-understood theory of learning from Bayesian networks.

Consider the task of performing *maximum likelihood* parameter estimation. Here, our goal is to find the parameter setting $\theta_{\mathcal{S}}$ that maximizes the likelihood $L(\theta_{\mathcal{S}} \mid \mathcal{I}, \sigma, \mathcal{S})$ for a given $\mathcal{I}$, σ and $\mathcal{S}$. Thus, the maximum likelihood model is the model that best predicts the training data. This estimation is simplified by the *decomposition* of log-likelihood function into a summation of terms corresponding to the various attributes of the different classes. Each of the terms in the square brackets in (13.3) can be maximized independently of the rest. Hence, maximum likelihood estimation reduces to independent maximization problems, one for each CPD. In fact, a little further work reduces Equation (13.3) even further, to a sum of terms, one for each multinomial distribution $\theta_{X.A|\mathbf{u}}$. Furthermore, there is a closed form solution for the parameter estimates. In addition, while we do not describe the details here, we can take a *Bayesian approach* to parameter estimation by incorporating parameter priors. For an appropriate form of the prior and by making standard assumptions, we can also get a closed form solution for the estimates.

13.5.2 Structure learning

We now move to the more challenging problem of learning a dependency structure automatically, as opposed to having it given by the user. The main problem here is finding a good dependency structure among the potentially infinitely many possible ones. As in most learning algorithms, there are three important issues that need to be addressed in this setting:

- **hypothesis space:** specifies which structures are candidate hypotheses that our learning algorithm can return;
- **scoring function:** evaluates the "goodness" of different candidate hypotheses relative to the data;
- **search algorithm:** a procedure that searches the hypothesis space for a structure with a high score.

We discuss each of these in turn.

Hypothesis space. Fundamentally, our hypothesis space is determined by our representation language: a hypothesis specifies a set of parents for each attribute $X.A$. Note that this hypothesis space is infinite. Even in a very simple schema, there may be infinitely many possible structures. In our genetics example, a person's genotype can depend on the genotype of his parents, or of his grandparents, or of his great-grandparents, etc. While we could impose a bound on the maximal length of the slot chain in the model, this solution is quite brittle, and one that is very limiting in domains where we do not have much prior knowledge. Rather, we choose to leave open the possibility of arbitrarily long slot chains, leaving the search algorithm to decide how far to follow each one.

We must, however, restrict our hypothesis space to ensure that the structure we are learning is a legal one. Recall that we are learning our model based on one training database, but would like to apply it in other settings, with potentially very different relational structure. We want to ensure that the structure we are learning will generate a consistent probability model for any skeleton we are likely to see. As we discussed in Section 13.4.2, we can test this condition using the class dependency graph for the candidate PRM. It is straightforward to maintain the graph during learning, and consider only models whose dependency structure passes the appropriate test.

Scoring structures. The second key component is the ability to evaluate different structures in order to pick one that fits the data well. We adapt Bayesian *model selection* methods to our framework. Bayesian model selection utilizes a probabilistic scoring function. In line with the Bayesian philosophy, it ascribes a prior probability distribution over any aspect of the model about which we are uncertain. In this case, we have a prior $P(\mathcal{S})$ over structures, and a prior $P(\theta_{\mathcal{S}} \mid \mathcal{S})$ over the parameters given each possible structure. The *Bayesian score* of a structure $\mathcal{S}$ is defined as the *posterior* probability of the structure given the data $\mathcal{I}$. Formally, using Bayes rule, we have that:

$$P(\mathcal{S} \mid \mathcal{I}, \sigma) \propto P(\mathcal{I} \mid \mathcal{S}, \sigma) P(\mathcal{S} \mid \sigma)$$

where the denominator, which is the marginal probability $P(\mathcal{I} \mid \sigma)$ is a normalizing constant that does not change the relative rankings of different structures.

This score is composed of two main parts: the prior probability of the structure, and the probability of the data given that structure. It turns out that the marginal likelihood is a crucial component, which has the effect of penalizing models with a large number of parameters. Thus, this score automatically balances the complexity of the structure with its fit to the data. In the case where $\mathcal{I}$ is a complete assignment, and we make certain reasonable assumptions about the structure prior, there is a closed form solution for the score.

Structure search. Now that we have a hypothesis space and a scoring function that allows us to evaluate different hypotheses, we only need to provide a procedure for finding a high-scoring hypothesis in our space. For Bayesian networks, we know that the task of finding the highest scoring network is NP-hard [13.3]. As PRM learning is at least as hard as Bayesian network learning (a Bayesian network is simply a PRM with one class and no relations), we cannot hope to find an efficient procedure that always finds the highest scoring structure. Thus, we must resort to heuristic search.

The simplest heuristic search algorithm is greedy hill-climbing search, using our score as a metric. We maintain our current candidate structure and iteratively improve it. At each iteration, we consider a set of simple local transformations to that structure, score all of them, and pick the one with highest score. As in the case of Bayesian networks, we restrict attention to simple transformations such as adding or deleting an edge. We can show that, as in Bayesian network learning, each of these local changes requires that we recompute only the contribution to the score for the portion of the structure that has changed in this step; this has a significant impact on the computational efficiency of the search algorithm. We deal with local maxima using random restarts, i.e., when a local maximum is reached in the search, we take a number of random steps, and then continue the greedy hill-climbing process.

There are two problems with this simple approach. First, as discussed in the previous section, we have infinitely many possible structures. Second, even the atomic steps of the search are expensive; the process of computing the statistics necessary for parameter estimation requires expensive database operations. Even if we restrict the set of candidate structures at each step of the search, we cannot afford to do all the database operations necessary to evaluate all of them.

We propose a heuristic search algorithm that addresses both these issues. At a high level, the algorithm proceeds in phases. At each phase k, we have a set of potential parents $Pot_k(X.A)$ for each attribute $X.A$. We then do a standard structure search restricted to the space of structures in which the parents of each $X.A$ are in $Pot_k(X.A)$. We structure the phased search so that it first explores dependencies within objects, then between objects that are directly related, then between objects that are two links apart, etc. This approach allows us to gradually explore larger and larger fragments of the infinitely large space, giving priority to dependencies between objects that are more closely related. The second advantage of this approach is that we can precompute the database view corresponding to $X.A, Pot_k(X.A)$; most of the expensive computations — the joins and the aggregation required in the definition of the parents — are precomputed in these views. The sufficient statistics for any subset of potential parents can easily be derived from this view. The above construction, together with the decomposability of the score, allows the steps of the search (say, greedy hill-climbing) to be done very efficiently.

13.6 Experimental results

We have tested our learning algorithm in several domains, both real and synthetic. We now describe experimental results on one synthetic dataset and two real ones.

13.6.1 Genetics domain

We begin by presenting our results on a synthetic dataset generated by the genetics example used in this chapter. The goal of these experiments is to test the learning algorithm, showing that it can reconstruct the dependency structure if it is clearly present in the distribution.

The datasets here were generated by a PRM that has the structure shown in Figure 13.5(a). We generated various training sets, of size from 200 to 800, with 10 training sets of each size. We also generated an independent test database of size 10,000. A data set of size n consists of a family tree containing n people, with an average of 0.6 blood tests per person. For each training set, we learned a PRM using the algorithm described in the previous section, and

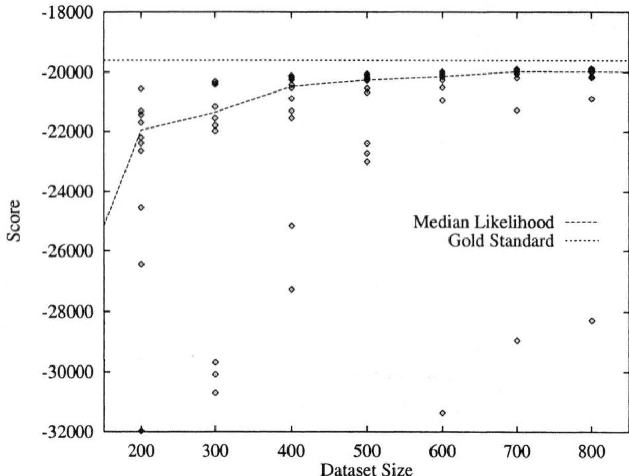

Fig. 13.6. Learning curve showing the generalization performance of PRMs learned in the genetic domain. The x-axis shows the training set size; the y-axis shows log-likelihood of a test set of size 10,000. For each sample size, we show learning experiments on ten different independent training sets of that size. The curve shows median log-likelihood of the models as a function of the sample size.

then tested how well the learned PRM predicts the test data. For measuring the predictive ability, we used the log-likelihood of the test data, the standard measure for evaluating density estimation procedures. Figure 13.6 shows the results for the different training sets. The straight line at the top is the log-likelihood of the test data given the "true" model used to generate the data. The data is presented in the form of a scatter plot, showing the accuracy for each of the training sets, as well as the median log-likelihood of the learned models for each size. We can see that the median is quite reasonable, but there are a few outliers. In most cases, our algorithm learned a model with the correct structure, and scored well; the difference in score is due to the parameter estimation, which is inherently noisy given limited data. However, in a small minority of cases, the algorithm got stuck in local maxima, learning a model with incorrect structure that scored quite poorly.

13.6.2 Tuberculosis patient domain

We also applied the algorithm to various real-world domains. The first of these is drawn from a database of epidemiological data for 1300 patients from the San Francisco tuberculosis (TB) clinic, and their 2300 contacts [13.1, 13.26]. For the Patient class, the schema contains demographic attributes such as age, gender, ethnicity, and place of birth, as well as medical attributes such as HIV status, disease site (for TB), X-ray result, etc. In addition, a sputum sample is taken from each patient, and subsequently undergoes genetic marker analysis. This allows us to determine which strain of TB

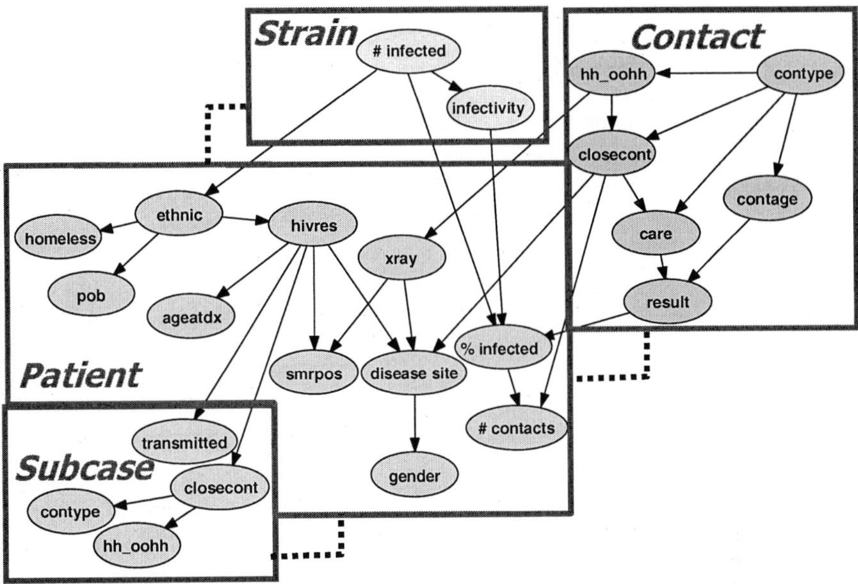

Fig. 13.7. The PRM structure for the TB domain.

a patient has, and thereby create a Strain class, with a relation between patients and strains. Each patient is also asked for a list of people with whom he has been in contact; the Contact class has attributes that specify the type of contact (sibling, coworker, etc.) contact age, whether the contact is a household member, etc.; in addition, the type of diagnostic procedure that the contact undergoes (*Care*) and the result of the diagnosis (*Result*) are also reported. In cases where the contact later becomes a patient in the clinic, we have additional information. We introduce a new class Subcase to represent contacts that subsequently became patients; in this case, we also have an attribute *Transmitted* which indicates whether the disease was transmitted from one patient to the other, i.e., whether the patient and subcase have the same TB strain.

The structure of the learned PRM is shown in Figure 13.7. We see that we learn a rich dependency structure both within classes and between attributes in different classes. We showed this model to our domain experts who developed the database, and they found the model quite interesting. They found many of the dependencies to be quite reasonable, for example: the dependence of age at diagnosis (*ageatdx*) on HIV status (*hivres*) — typically, HIV-positive patients are younger, and are infected with TB as a result of AIDS; the dependence of the contact's age on the type of contact — contacts who are coworkers are likely to be younger than contacts who are parents and older than those who are school friends; or the dependence of HIV status on ethnicity — Asian patients are rarely HIV positive whereas white patients are much more likely to be HIV positive, as they often get TB as a result of having AIDS. In addition, there were a number of dependencies that they found interesting, and worthy of further investigation. For example, the dependence between close contact (*closecont*) and *disease site* was novel and potentially interesting. There are also dependencies that seem to indicate a bias in the contact investigation procedure or in the treatment of TB; for example, contacts who were screened at the TB clinic were much more likely to be diagnosed with TB and receive treatment than contacts who were screened by their private medical doctor. Our domain experts were quite interested to identify these and use them as a guide to develop better investigation guidelines.

We also discovered dependencies that are clearly relational, and that would have been difficult to detect using a non-relational learning algorithm. For example, there is a dependence between the patient's HIV result and whether he transmits the disease to a contact: HIV positive patients are much more likely to transmit the disease. There are several possible explanations for this dependency: for example, perhaps HIV-positive patients are more likely to be involved with other HIV-positive patients, who are more likely to be infected; alternatively, it is also possible that the subcase is actually the infector, and original HIV-positive patient was infected by the subcase and simply manifested the disease earlier because of his immune-suppressed status. Another interesting relational dependency is the correlation between the ethnicity of the patient and the number of patients infected by the strain. Patients who are Asian are more likely to be infected with a strain which is unique in the population, whereas other ethnicities are more likely to have strains that recur in several patients. The reason is that Asian patients are more often immigrants, who immigrate to the U.S. with a new strain of TB, whereas other ethnicities are often infected locally.

13.6.3 Company domain

The second domain we present is a dataset of company and company officers obtained from Security and Exchange Commission (SEC) data.[2] The data set includes information, gathered over a five year period, about companies (which were restricted to banks in the dataset we used), corporate officers in the companies, and the role that the person plays in the company. For our tests, we had the following classes and table sizes: Company (20,000), Person (40,000), and Role (120,000). Company has yearly statistics, such as the number of employees, the total assets, the change in total assets between years, the return on earnings ratio, and the change in return on assets. Role describes information about a person's role in the company including their

[2] This dataset was developed by Alphatech Corporation based on Primark banking data, under the support of DARPA's Evidence Extraction and Link Discovery (EELD) project.

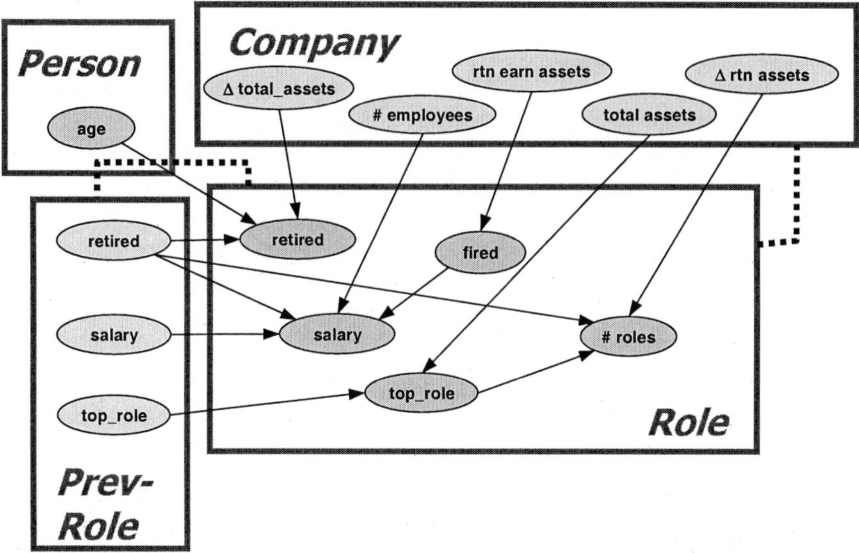

Fig. 13.8. The PRM structure for the Company domain.

salary, their top position (president, CEO, chairman of the board, etc.), the number of roles they play in the company and whether they retired or were fired. Prev-Role indicates a slot whose range type is the same class, relating a person's role in the company in the current year to his role in the company in the previous year.

The structure of the learned PRM is shown in Figure 13.8. We see that we learn some reasonable persistence arcs such as the facts that this year's salary depends on last year's salary and this year's top role depends on last year's top role. There is also the expected dependence between Person.*Age* and Role.*Retired*. A more interesting dependence is between the number of employees in the company, which is a rough measure of company size, and the salary. For example, an employee that receives a salary of $200K in one year is much more likely to receive a raise to $300K the following year in a large bank (over 1000 employees) than in a small one. Again, we see interesting correlations between objects in different relations.

13.7 Discussion and related work

There are clearly many other approaches to learning from data. We can categorize them along three main axes: probabilistic versus deterministic; model learning versus classification; and attribute-based versus relational.

Most approaches to machine learning fall into the category of attribute-based classification, with Naive Bayes falling into the probabilistic category, concept learning into the deterministic category, and approaches such as decision trees and neural networks somewhere in between.

Bayesian networks fall into the category of attribute-based model-learning: the result of the learning is a model of the dependencies within the set of attributes, rather than an attempt to predict one particular attribute using the others. However, as we discussed in Section 13.2, Bayesian networks are attribute-based in nature, which substantially limits their ability to represent complex domains involving multiple entities. One might ask, however, whether they can be nevertheless be applied as a discovery tool for relational data. A standard and fairly obvious solution is to flatten the data, creating a single table that contains all the attributes, and then apply Bayesian network learning. While this approach can be useful, it suffers from three major problems.

One key issue relates to the statistical correctness of this approach. Consider our TB dataset, and imagine flattening the data to a table that contains a patient and a contact. In this case, a patient who has two contacts would appear in the table twice (once with each contact), whereas one who has ten contacts would appear in the table ten times. This has the effect of skewing the data substantially, leading to parameter estimates (and correlations) that are highly non-representative of the true distribution. In more technical terms, Bayesian network learning makes the assumption that the data cases are independent (IID), whereas this assumption is clearly violated when our data set is obtained by flattening a relational database.

A second limitation is that we have to determine in advance which are the relevant attributes to put in the table. Consider, for example, the genetics domain, but where we have only phenotype observations (observable features of the person). We might think to join the table to itself, creating a table that contains the person and his parents. However, this would restrict us to discovering correlations that cross generations. For example, male baldness is inherited from a man's maternal grandfather rather than his father; this type of correlation would be lost if we flattened the data in the obvious way.

Finally, even if a Bayesian network model is learned for such a database, it cannot be used to reach conclusions based on relational dependencies. For example, in the TB domain, we might imagine inferring that a patient has a particularly infectious strain by noticing that he transmitted the strain to many of his contacts; we can then infer that a new contact is likely to have the same strain, and therefore that she is also likely to infect many of her contacts. This type of reasoning is only possible using the relational structure that allows us to use information obtained about one patient to reach conclusions about the strain and from that reach conclusions about another patient entirely.

Other than PRMs (and the related approaches of [13.20, 13.23]), the only relational learning approaches are variants of inductive logic programming (ILP). Most ILP approaches are deterministic classification approaches, which do not attempt to model a probability distribution, but rather only to predict (classify) a particular predicate. However, there are two recent developments within the ILP community that are related to PRMs: stochastic logic programs (SLPs) [13.18, 13.6] and Bayesian logic programs (BLPs) [13.13]. The semantics for these two approaches are quite different, with the BLP semantics being the closest to PRMs. An SLP defines a sampling distribution over logic programming proofs; as a consequence, it induces a probability distribution over the possible ground facts for a given predicate. On the other hand, a BLP consists of a set of rules, along with conditional probabilities and a combination rule; following the approach of knowledge-based model construction [13.25], the BLP essentially specifies a propositional Bayesian network. This approach is very similar to the probabilistic logic programs of [13.20, 13.23].

Learning algorithms for these approaches are being developed. Methods for learning SLPs are described in [13.19]. A maximum likelihood approach is taken for parameter estimation for an SLP which is based on maximizing the posterior probability of the program. The task of learning the structure of an SLP is quite different from learning a PRM structure and is based on more traditional ILP approaches. On the other hand, while BLPs are more closely related to PRMs, and methods for learning BLPs have been suggested in [13.13], learning algorithms have not yet been developed. Methods for learning PRMs may be found to be applicable to learning BLPs.

13.8 Extensions

13.8.1 Structural uncertainty

So far, we have assumed that the skeleton is external to the probabilistic model; in other words, the skeleton is assumed to be part of the input to the PRM. This limitation has two main implications. Most obviously, it implies that the model can only be used in settings where the relational structure is known. Thus, for example, we cannot use it to conclude that a patient is more likely to have a particular strain of TB, based on his demographics and his contacts. A more subtle point is that this restriction can diminish the quality of our model even in cases where the relational structure *is* given, because it ignores interesting correlations between attributes of entities and the relations between them. For example, in a movie domain a "serious" actor is unlikely to appear in many horror movies; hence, we can infer information about an actor's (unknown) attributes based on the Role relation between actors and movies.

The PRM framework can be extended to accommodate uncertainty about the structural relationships between objects as well as about their properties [13.22]. We have extended our learning algorithms to deal with such *structural uncertainty*. We now provide a brief sketch of this extension, which is described in more detail in [13.10].

Suppose we have a simple domain in which we have Movie-Theaters and Movies. Each theater shows some number of movies; this is represented by the class Shows, with reference slots that point to both the theater and the movie. We might know the number of screens that a theater has, but would like to represent a probabilistic model over which movie each theater chooses to show. One way of representing this model is by defining a distribution over the reference slot Shows.Movie: which movie, among the movies currently available, is a theater likely to show. We call this approach to structural uncertainty *Reference Uncertainty.*

Naively, we might think of representing this distribution as a very large multinomial distribution, with a probability for every movie in our domain. This approach is infeasible for two reasons. First, the representation is much too large. Second, our training set will contain one set of movies, but we want to learn a model that can also be applied to other domains, with a different set of movies. In one approach, we can partition the movies into categories, using some set of attributes, either of the movies or of related objects. For example, we might partition on the type of movie (action, horror, comedy), and/or on the origin of the studio which is the source of the movie (Hollywood, independent U.S., or foreign). We then represent the probability that a movie theater shows a movie from a particular category in the partition. As usual, this distribution might depend on some set of parents, e.g., the type of theater (megaplex or art theater). Thus, we might state that an art theater is more likely to play a foreign movie, while a megaplex may be more likely to play a Hollywood movie.

Another approach to structural uncertainty we call *Existence Uncertainty.* Existence uncertainty is a general approach to modeling the probability that a relationship exists between any two entities. To do so, we require that the relationship between the entities is represented by a class; for example, the Registration class in our school domain represents the relationship between a student and a course. We can now represent a probabilistic model that a student will register for a class by introducing a probabilistic model that a given pair (student, class) will appear in the Registration table. More precisely, we specify the probability that such a pair exists in the table given attributes of the student and of the course. For example, it is more likely that a university senior would register for an advanced class, whereas a freshman is more likely to register for an introductory class.

13.8.2 Class hierarchies

We have also investigated the idea of learning PRMs with class hierarchies. Class hierarchies change the language in two important ways. In the non-hierarchical case, all objects in a class must share the same CPDs. Hierarchical models allow us to specialize the CPDs of attributes in different parts of the hierarchy. In our school example, we might choose to partition the Course class according to their intended year, creating the two subclasses Undergraduate-Course and Graduate-Course. We might then *specialize* the CPD for *Difficulty* in each of the subclasses. More interestingly, a hierarchical model allows us to create dependencies that might seem cyclic in the non-hierarchical case. For example, we can now have a student's grades in a graduate course depend on her grades in the undergraduate courses she has taken. Note that this dependency is apparently cyclic at the level of the Course class, since an attribute of one course depends on the same attribute in another course. However, the division of courses into two subclasses resolves this apparent cycle. A preliminary discussion of these issues can be found in [13.9].

13.9 Conclusions

PRMs provide a new approach to relational data mining that is grounded in a sound statistical framework. Algorithms for learning PRMs build on the recent developments in learning Bayesian networks, and extend learning to this rich class of relational models. Because these models do not focus on a single classification task, they are particularly well suited to exploratory data analysis.

There are several directions for future work. Perhaps the most obvious one is the treatment of missing data and hidden variables. We can extend standard techniques (such as Expectation Maximization for missing data) to this task. (See [13.14] for some preliminary work on related models.) However, the complexity of inference on large databases with many missing values make the cost of a naive application of such algorithms prohibitive. Clearly, this domain calls both for new inference algorithms and for new learning algorithms that avoid repeated calls to inference over these very large problems. Even more interesting is the issue of automated discovery of hidden variables. There are some preliminary answers to this question in the context of Bayesian networks [13.7], in the context of ILP [13.17], and very recently in the context of simple binary relations [13.12]. Combining these ideas and extending them to this more complex framework is a significant and interesting challenge. Ultimately, we would want these techniques to help us automatically discover interesting entities and relationships that hold in the world.

Acknowledgments

We thank Benjamin Taskar for his collaboration on several papers that build on this work, and for his help building the software on which our experiments were performed. We also thank Dr. Peter Small and Jeanne Rhee from the Stanford University Medical Center for providing us with the TB data and with their expertise on this topic, and Kendra Moore and Chris White of Alphatech Corporation for providing us with the Company dataset. The work of Lise Getoor, Daphne Koller and Avi Pfeffer was funded by ONR contract N66001-97-C-8554 under DARPA's HPKB program and EELD program, by ONR grant N00014-96-1-0718, and by the generosity of the Sloan Foundation and of the Powell Foundation. Nir Friedman was supported through the generosity of the Michael Sacher Trust and the Sherman Senior Lectureship.

References

13.1 M.A. Behr, M.A. Wilson, W.P. Gill, H. Salamon, G.K. Schoolnik, S. Rane, and P.M. Small. Comparative genomics of BCG vaccines by whole genome DNA microarray. *Science*, 284:1520–1523, 1999.

13.2 J. Breese, D. Heckerman, and C. Kadie. Empirical analysis of predictive algorithms for collaborative filtering. In *Proceedings of the Fourteenth Conference on Uncertainty in Artificial Intelligence*, pages 43–52. Morgan Kaufman, San Francisco, CA, 1998.

13.3 D. M. Chickering. Learning Bayesian networks is NP-complete. In D. Fisher and H.-J. Lenz, editors, *Learning from Data: Artificial Intelligence and Statistics V*, pages 121–130. Springer, Berlin, 1996.

13.4 G. F. Cooper. The computational complexity of probabilistic inference using Bayesian belief networks. *Artificial Intelligence*, 42: 393–405, 1990.

13.5 G. F. Cooper and E. Herskovits. A Bayesian method for the induction of probabilistic networks from data. *Machine Learning*, 9: 309–347, 1992.

13.6 J. Cussens. Loglinear models for first-order probabilistic reasoning. In *Proceedings of the Fifteenth Conference on Uncertainty in Artificial Intelligence*, pages 126–133. Morgan Kaufman, San Francisco, CA, 1999.

13.7 N. Friedman. Learning belief networks in the presence of missing values and hidden variables. In *Proceedings of the Fourteenth International Conference on Machine Learning*, pages 125–133. Morgan Kaufman, San Francisco, CA, 1997.

13.8 N. Friedman, L. Getoor, D. Koller, and A. Pfeffer. Learning probabilistic relational models. In *Proceedings of the Sixteenth International Joint Conference on Artificial Intelligence*, pages 1300–1307. Morgan Kaufman, San Francisco, CA, 1999.

13.9 L. Getoor, D. Koller, and N. Friedman. From instances to classes in probabilistic relational models. In *Proceedings of the ICML-2000 Workshop on Attribute-Value and Relational Learning: Crossing the Boundaries*, pages 25–34. Stanford University, Stanford, CA, 2000.

13.10 L. Getoor, D. Koller, B. Taskar, and N. Friedman. Learning probabilistic relational models with structural uncertainty. In *Proceedings of the AAAI-2000 Workshop on Learning Statistical Models from Relational Data*, pages 13–20. Technical Report WS-00-06, AAAI Press, Menlo Park, CA, 2000.

13.11 D. Heckerman. A tutorial on learning with Bayesian networks. In M. I. Jordan, editor, *Learning in Graphical Models*, pages 301 – 354. MIT Press, Cambridge, MA, 1998.

13.12 T. Hofmann, J. Puzicha, and M. Jordan. Learning from dyadic data. In *Advances in Neural Information Processing Systems 11*, pages 466–472. MIT Press, Cambridge, MA, 1998.

13.13 K. Kersting, L. de Raedt, and S. Kramer. Interpreting Bayesian logic programs. In *Proceedings of the AAAI-2000 Workshop on Learning Statistical Models from Relational Data*, pages 29–35. Technical Report WS-00-06, AAAI Press, Menlo Park, CA, 2000.

13.14 D. Koller and A. Pfeffer. Learning probabilities for noisy first-order rules. In *Proceedings of the Sixteenth International Joint Conference on Artificial Intelligence*, pages 1316–1321. Morgan Kaufman, San Francisco, CA, 1997.

13.15 D. Koller and A. Pfeffer. Probabilistic frame-based systems. In *Proceedings of the Fifteenth National Conference on Artificial Intelligence*, pages 580–587. AAAI Press, Menlo Park, CA, 1998.

13.16 S. L. Lauritzen. The EM algorithm for graphical association models with missing data. *Computational Statistics and Data Analysis*, 19: 191–201, 1995.

13.17 N. Lavrač and S. Džeroski. *Inductive Logic Programming: Techniques and Applications.* Ellis Horwood, Chichester, 1994. Freely available at `http://www-ai.ijs.si/SasoDzeroski/ILPBook/`.

13.18 S.H. Muggleton. Stochastic logic programs. In L. de Raedt, editor, *Advances in Inductive Logic Programming*, pages 254–264. IOS Press, Amsterdam, 1996.

13.19 S.H. Muggleton. Learning stochastic logic programs. In *Proceedings of the AAAI-2000 Workshop on Learning Statistical Models from Relational Data*, pages 36–41. Technical Report WS-00-06, AAAI Press, Menlo Park, CA, 2000.

13.20 L. Ngo and P. Haddawy. Answering queries from context-sensitive probabilistic knowledge bases. *Theoretical Computer Science*, 171:147– 177, 1996.

13.21 J. Pearl. *Probabilistic Reasoning in Intelligent Systems.* Morgan Kaufmann, San Mateo, CA, 1988.

13.22 A. Pfeffer. *Probabilistic Reasoning for Complex Systems.* PhD thesis. Stanford University, Stanford, CA, 2000.

13.23 D. Poole. Probabilistic Horn abduction and Bayesian networks. *Artificial Intelligence*, 64:81–129, 1993.

13.24 P. Spirtes, C. Glymour, and R. Scheines. *Causation, Prediction and Search.* Springer, New York, 1993.

13.25 M.P. Wellman, J.S. Breese, and R.P. Goldman. From knowledge bases to decision models. *The Knowledge Engineering Review*, 7(1): 35–53, 1992.

13.26 M. Wilson, J.D. DeRisi, H.H. Kristensen, P. Imboden, S. Rane, P.O. Brown, and G.K. Schoolnik. Exploring drug-induced alterations in gene expression in Mycobacterium tuberculosis by microarray hybridization. In *Proceedings of the National Academy of Sciences*, 96(22):12833-12838, 1999.

Part IV

Applications and Web Resources

14. Relational Data Mining Applications: An Overview

Sašo Džeroski

Jožef Stefan Institute
Jamova 39, SI-1000 Ljubljana, Slovenia

Abstract

This chapter gives an overview of applications of relational learning and inductive logic programming to data mining problems in a variety of areas. These include bioinformatics, where successful applications come from drug design, predicting mutagenicity and carcinogenicity, and predicting protein structure and function, including genome scale prediction of protein functional class. Other application areas include medicine, environmental sciences and monitoring, mechanical and traffic engineering. Applications of relational learning are also emerging in business data analysis, text and Web mining, and miscellaneous other fields, such as the analysis of musical performances.

14.1 Introduction

Many applications of relational learning and inductive logic programming to data mining problems have emerged in the last five years. When I last wrote a comprehensive overview article on the subject in 1996 [14.20], many of the applications mentioned there were proof-of-the-principle applications and only a handful qualified as real-world applications. Proof-of-the-principle applications formulate a generic task as a relational learning problem and address an illustrative instance of the generic task that which is simpler than real-world instances of the same generic task. Typically, synthetic data would be used to reconstruct a known solution to the problem.

A crucial ingredient of real-world applications is the availability of real-world data collected through the observation of natural or business processes. Another ingredient is the presence of an information or performance need that could be potentially satisfied by analyzing the available data. We take that the latter is manifested by the availability of a domain expert who has some understanding of the processes involved and shows a strong interest in solving the performance/understanding problem in cooperation with a data mining expert.

One can argue that only deployed applications, where data mining results in everyday use have brought tangible financial profit, count as successful real-world applications. While there is a point to this attitude, we should note that applications of this kind are rarely even reported (so as not to jeopardize the competitive advantage of the company involved), let alone described in detail in the scientific literature. Also, the successful discovery of knowledge from scientific data, such as genome data, would not necessarily qualify as a successful application according to this criterion, despite its undisputed potential to affect our everyday lives in the near future.

In this chapter, we will consider applications where relational learning algorithms have been applied to real-world data in cooperation between a data mining and a domain expert. The results have been typically analyzed from a performance viewpoint, as well as for comprehensibility. In some cases, the use of relational learning has enabled the use of data mining on problems that would be difficult to address with single table approaches. In others, measurable improvements have been achieved in comparison to single table approaches, either in terms of performance or comprehensibility.

The applications are grouped by application area. Many successful applications come from life science domains, and in particular bioinformatics. The first three sections describe applications of relational learning and inductive logic programming in drug design, predicting carcinogenicity and mutagenicity, and predicting the structure and function of proteins. The next two sections describe applications from the areas of medicine and environmental science/engineering, including environmental monitoring. Engineering applications come next, including mechanical and traffic engineering. The last three sections summarize applications in business data analysis, text and Web mining, and miscellaneous other domains.

14.2 Drug design

Most pharmaceutical research and development is based on finding slightly improved variants of patented active drugs. In doing this, it is essential to understand the relationships between chemical structure and activity. The first two subsections of this section describe the application of relational learning methods to learning such relationships from data. The remaining three subsections deal with classifying NMR spectra of diterpenes, a class of compounds often used as lead compounds in drug design; finding substructures responsible for the activity of a set of drugs; and recognizing when two drugs with mirror-image structures can be successfully separated.

14.2.1 Pyrimidines, triazines, and tacrine analogues

A central concern of chemistry is understanding the relationships between chemical structure and activity. In most cases, these relationships cannot be derived solely from physical theory — experimental evidence is essential. Such empirically derived relationships are called Structure Activity Relationships (SARs). In a typical SAR problem, a set of chemicals of known structure and activity are given, and the task is to find a predictive theory relating the structure of a compound to its activity. This relationship can then be used to select structures with high or low activity. Typically, knowledge of such relationships is used for devising clinically effective, non-toxic drugs.

The ILP system GOLEM [14.45] was applied to several problems of this kind, including the problem of inhibition of E. Coli Dihydrofolate Reductase by two different groups of drugs (pyrimidines and triazines) [14.34, 14.35]. The structural template for pyrimidines is given in Figure 14.1. The problem of predicting the properties (toxicity, acetocholinesteraze inhibition, etc.) of Tacrine (a drug for treating Alzheimer's disease) derivatives [14.35] has also been addressed.

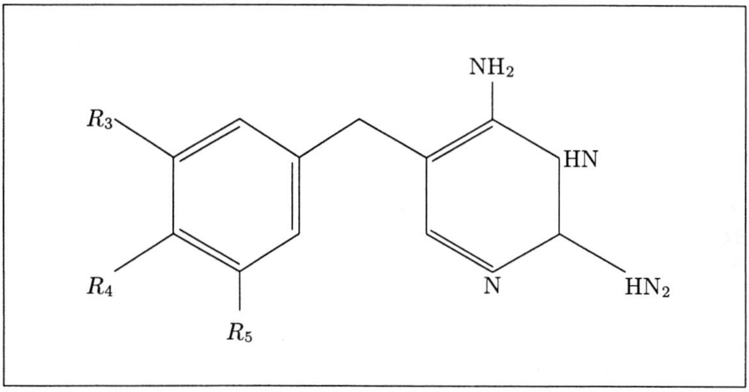

Fig. 14.1. The structure of pyrimidines.

The ILP formulation of the problems compares the properties of pairs of compounds with known activity. The background knowledge contains predicates specifying the chemical structure of the drugs with known activity, as well as properties of some groups of atoms (substituents or radicals). For the problem of inhibition of E. Coli Dihydrofolate Reductase, GOLEM induced nine rules comparing drug activities. The Spearman rank correlation of the drug activity order predicted by GOLEM with the actual order was 0.46 for a testing set of drugs [14.34] as compared to a correlation of 0.42 by a classical SAR approach, taken by Hansch et al. [14.28]. Besides achieving better accuracy than traditional methods, the induced rules also provide a description of the chemical laws governing the problem.

14.2.2 Modulating transmembrane calcium movement

The compounds considered in this study are a class of calcium-channel activators. Their activity is measured as the logarithm of the potency of the compound relative to an accepted standard calcium-channel activator. Initial experiments in the chemical literature used the hydrophobicity and molar reflectivity of the compounds to derive linear models of the SARs in this domain.

PROGOL was applied to this dataset yielding structural concepts that were translated into boolean-valued attributes [14.55]. These were added to the above mentioned attributes and significantly improved the linear model predicting activity. The final model is comparable in accuracy to a much more complex model derived using computational chemistry methods.

14.2.3 Diterpene structure elucidation

Diterpenes are organic compounds of low molecular weight with a skeleton of 20 carbon atoms. They are of significant chemical and commercial interest because of their use as lead compounds in the search for new pharmaceutical effectors. The interpretation of diterpene ^{13}C NMR-spectra normally requires specialists with detailed spectroscopic knowledge and substantial experience in natural products chemistry, more specifically knowledge on peak patterns and chemical structures.

Given a database of peak patterns for diterpenes with known structure, several ILP approaches were applied to discover correlations between peak patterns and chemical structure [14.25]. The approaches used include relational instance based learning, induction of logical decision trees and inductive constraint logic. Performance close to the one of domain experts was achieved, which suffices for practical use.

More specifically, the task addressed was to identify the skeleton (type) of diterpenoid compounds, given their ^{13}C-NMR-Spectra that include the multiplicities and the frequencies of the skeleton atoms. The multiplicity of a carbon atom is the number of hydrogen atoms connected to it. Pre-processed (so-called reduced) multiplicities were used which result from eliminating measurement side-effects. Each molecule was thus represented by a set of facts of the form $red(MoleculeID, Multiplicity, Frequency)$, yielding a nondeterminate representation. Twenty-three different skeleton types are represented in the whole set of 1503 compounds: there are thus 23 possible class values (target predicates).

Several propositional versions of the problem were considered. The simplest represented a molecule by four numbers: the numbers of atoms of each of the four possible multiplicities. The most complex represented each molecule by 860 attributes. All of these representations yield an accuracy of approx. 80% on unseen cases as estimated by cross-validation.

Using the relational representation in combination with the four features mentioned above, the ILP systems RIBL [14.26] and TILDE [14.1] yield accuracies of over 90%. This is in the range of the accuracy with which experts classify diterpenes into skeleton types given ^{13}C NMR spectra only. That number can actually only be estimated since it is expensive to have an expert carry out a statistically significant number of structure predictions without using other additional information that often becomes available from heterogeneous sources (such as literature, and ^{1}H NMR spectra).

14.2.4 Pharmacophore discovery for ACE inhibition

The task in this domain [14.47, 14.27] was to identify the structure (pharmacophore) responsible for the activity of ACE (Angiotensin-converting enzyme) inhibitors. Given were the structures of 28 molecules that display the activity of ACE inhibition, described by atom and bond information, including the 3D positions of the atoms. Background knowledge about atom groups and the distances between pairs of groups was available.

Pharmacophores are typically described in terms of types of atoms (e.g., hydrogen donors) or functional groups and the pairwise distances among them. The pairwise distances define the geometric arrangement of the atoms or groups that is necessary for them to lock into a particular site of a protein such as ACE. P-PROGOL [14.54] discovered a four-piece pharmacophore with one zinc-binder and three hydrogen acceptors present in all molecules. According to expert opinion, the discovered pharmacophore is equivalent to the generally accepted pharmacophore for ACE inhibition.

14.2.5 Characterizing successful enantioseparations

Isomers are chemicals with the same chemical formula but different three-dimensional molecular structure. Enantiomers are mirror image isomers. Enantioseparation is the process of separating two enantiomers: a third chemical is used (so-called chiral selector) which has a preference for reacting with one of the two enantiomers in the pair as a consequence of its stereochemistry. This is achieved by selecting a chiral stationary phase (CSP), hence the third chemical is referred to as a CSP chiral selector.

Bryant [14.6] applied PROGOL to analyze data from a study [14.9] that investigated the ability of seven CSP chiral selectors to separate enantiomeric drugs. The dataset contained data on 197 separations, involving 50 drugs whose structure varied widely. Rules were induced by PROGOL on when a given chiral selector will successfully separate enantiomeric drugs based on the structural properties of the drug and in particular the distance of particular structural groups from the chiral center of the drug.

14.3 Predicting mutagenicity and carcinogenicity

Prevention of environmentally induced cancers is a health issue of unquestionalble importance. An increasing number of chemicals are in use in large amounts every day, while very few of them are evaluated for toxic effects like carcinogenicity (the capacity to cause cancer) or mutagenicity (the capacity to cause changes in genetic material). It is thus important to be able to predict such effects directly from the structure of the chemicals.

14.3.1 Predicting mutagenicity

Srinivasan et al. [14.59, 14.60] applied the ILP system PROGOL [14.43] to induce theories for predicting the mutagenicity of a set of 230 aromatic and heteroaromatic nitro-compounds. The prediction of mutagenicity is important as it is relevant to the understanding and prediction of carcinogenicity. The compounds used in this study are more heterogeneous structurally than those used in the drug design domains and can only be fully represented in a first-order setting.

Of the 230 compounds, 138 have positive levels of log mutagenicity, these are labeled "active" and constitute the positive examples: the remaining 92 compounds are labeled "inactive" and constitute the negative examples. The target relation is in this case $active(C)$, stating that compound C has positive log mutagenicity. The background knowledge contains the structure of the compounds represented as a list of atoms and bonds that can be found in each compound. The predicate $atm(C, A, E, T, Charge)$ states that atom A in compound C is an atom of element E (e.g., carbon), of type T (e.g., aromatic carbon) with charge $Charge$. The predicate $bond(C, A1, A2, BT)$ states that there is a bond of type BT (e.g., aromatic bond) between atom $A1$ and atom $A2$ of compound C. The facts for these predicates were generated by the molecular modeling program QUANTA, where the compounds were entered through a chemical editing facility.

In addition, four attributes are provided for analysis of the compounds. These can be used directly by both propositional and ILP learners. They are: (1) the hydrophobicity of the compound (termed logP); (2) the energy level of the lowest unoccupied molecular orbital (termed LUMO); (3) a boolean attribute identifying compounds with 3 or more benzyl rings (termed indicator variable I1); and (4) a boolean attribute identifying a sub-class of compounds termed acenthryles (termed indicator variable Ia). The last two are pre-selected structural features that incorporate chemical expert knowledge.

Generic structural knowledge was used as background knowledge in some experiments with PROGOL. It includes definitions of the concepts of methyl groups, nitro groups, aromatic rings, heteroaromatic rings, connected rings, ring length, and the three distinct topological ways to connect three benzene rings. These definitions are generic to the field of organic chemistry.

The 230 compounds are divided into two sets: 188 compounds that could be fitted using linear regression (regression-friendly set), and 42 compounds that could not (regression-unfriendly set). In the PROGOL experiments, accuracies of theories constructed for the 188 compounds were estimated by a 10-fold cross-validation and the accuracy of theories for the 42 compounds were estimated by a leave-one-out procedure.

In summary, PROGOL produced theories that perform as well as linear regression or neural networks on the regression-friendly set and much better (88%, 20 percentage points) on the regression-unfriendly set. In comparison with CART [14.4], no significant differences in performance exist when the pre-selected structural features are available. However, when these features are not available, PROGOL performs much better on the regression-friendly set (88 % vs 83 %). PROGOL also performed better than FOIL [14.50].

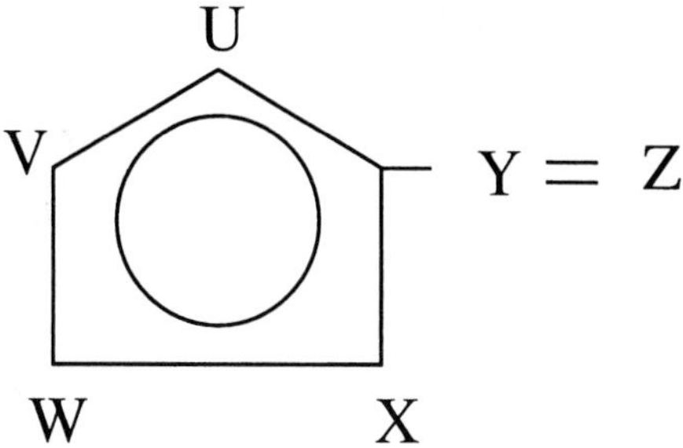

Fig. 14.2. A new structural alert for mutagenicity discovered by PROGOL.

It is worth noting that PROGOL generated a single rule for the regression-unfriendly set. The structure indicated by this rule is a new structural alert for high mutagenicity in chemical compounds. This alert is depicted un Figure 14.2.

14.3.2 Prediction of rodent carcinogenicity bioassays

The problem addressed by King and Srinivasan [14.58] was to predict the carcinogenicity of a diverse set of chemical compounds, learning from a dataset that originates from the US National Toxicology Program. The data on carcinogenicity was obtained by testing the chemicals on rodents, each trial taking several years and hundreds of animals (long term rodent bioassays). The training set consisted of 291 compounds, of which 161 carcinogens, while the testing set consisted of 39 compounds, of which 22 carcinogens.

An accuracy of 64% on unseen cases was achieved by PROGOL. No SAR models (of either human or machine origin) were significantly more accurate than this. PROGOL was also the most accurate method that did not use data from biological tests on rodents (these data were not available). A set of structural alerts for carcinogenicity was generated automatically and the chemical rationale for them investigated. Unlike other SAR methods, the PROGOL alerts are statistically independent of those available from an existing carcinogenesis test based on Salmonella mutagenicity.

In a competitive comparative study [14.57, 14.56], named the Predictive Toxicology Evaluation (PTE) challenge, a number of different prediction methods were applied to the above problem. In this evaluation, ILP approaches turned out to perform surprisingly well in terms of quantitative/predictive performance. A new comparative study of this kind is underway at the time of writing, which would consider four separate prediction problems: predicting the outcomes of carcinogenicity tests for rats and mice, for male and female individuals [14.29].

14.4 Predicting protein structure and function

Predicting the three-dimensional shape of proteins from their amino acid sequence is widely believed to be one of the hardest unsolved problems in molecular biology. It is also of considerable interest to pharmaceutical companies since a protein's shape generally determines its function as an enzyme. This section describes the application of relational learning to predicting protein structure (secondary structure and protein folds) and function (recognizing neuropeptide precursor proteins and predicting the protein functional class for all proteins in the genomes of two microorganisms).

14.4.1 Predicting protein secondary structure

A protein is basically a string of amino acids (or *residues*). Predicting the three-dimensional shape of proteins from their amino acid sequence is widely believed to be one of the hardest unsolved problems in molecular biology. It is also of interest to the pharmaceutical industry since the shape of a protein determines its function.

The sequence of amino acids is called the *primary structure* of the protein. Spatially, the amino acids are arranged in different patterns (spirals, turns, flat sections, etc.). The three-dimensional spatial shape of a protein is called the *secondary structure.* A limited version of the problem of predicting the shape (*secondary structure*), involving only the α-helix shape, was considered by Muggleton et al. [14.46]. The ILP system GOLEM used information on the sequence of residues and properties of the individual residues to predict whether a given residue will belong to an α-helix. The induced rules achieved an accuracies of 81% on unseen cases, which, at the time, was better than the best previously reported result of 76% [14.36].

14.4.2 Classifying domains into three-dimensional folds

Turcotte et al. [14.61] consider the problem of predicting the three-dimensional structure of proteins. The task considered here is the classification of proteins into folds of the classification scheme SCOP, which a classification done manually by a world-expert on protein structure [14.5]. Proteins in the same fold share the same secondary structure and interconnections. Folds are grouped into classes, based on the overall distribution of their secondary structure elements. Example folds are 'Globin-like' (which belongs to the all-α class) and 'β/α (TIM) barrel' (which belongs to the α/β class).

The basic unit of this classification is a domain, a structure or substructure that is considered to be folded independently. Small proteins have a single domain, while larger ones can have several. Secondary structure is taken as input in this task.

PROGOL was used to induce rules that characterize each of 20 folds, selected from the 334 folds of the four main SCOP protein classes (all-α, all-β, α/β, and $\alpha+\beta$). In addition to global properties of domains (such as the length/number of residues of the domain, number of α-helices and number of β-strands), PROGOL also takes into account structural properties, such as the adjacency of secondary structures within a given domain, the length of the loop between them, the length, average hydrophobicity and the hydrophobic moment of individual secondary structures, and the presence of proline therein. In addition to yielding higher accuracy than propositional decision trees, PROGOL produced some rules that represent expert-like knowledge and rely on the relational information about the secondary structures. One rule, for example, states that a domain that begins by a long helix followed by another helix belongs to the fold '4-helical cytokines'.

14.4.3 Recognising neuropeptide precursor proteins

Muggleton et al. [14.44] addressed the problem of identifying proteins that belong to a class of proteins known as human neuropeptide precursors (NPPs). These proteins have considerable therapeutic potential and are of widespread interest in the pharmaceutical industry. Unlike enzymes and other structural proteins, NPPs tend to show lower sequence similarity. They are highly variable in length and undergo specific enzymatic degradation before the biologically active short peptides (neuropeptides) are released.

The 3910 human sequences available at the time of the experiment from the SWISS-PROT annotated protein sequence database were used by Muggleton et al. [14.44]. Of these, 44 are known to be NPPs, but there is no guarantee that all of the NPPs in the database have been properly identified. Thus, PROGOL was applied to learn from positive examples only (the sequences of the known NPPs) and derive a grammar describing the NPP sequences.

A group of features was derived from the grammar. These were then used together with other features: the length of the sequence; proportions of the number of residues in the sequence which are a specific amino-acid or have a specific property of an amino-acid; and the results of SIGNALP [14.48], the pre-eminent automated method for predicting the presence and location of signal peptides. Learning to classify sequences as NPPs/non-NPPs from these features makes the search for NPPs in SWISS-PROT much more efficient than random search (1 in 2048 randomly selected sequences is expected to be an NPP, as compared to 1 in 22 of those that are classified as NPP by the learned classifier). Using the features from the grammar drastically improved the recognition of NPPs.

14.4.4 Genome scale prediction of protein functional class

This application addresses the task of automating the analysis of genomic data. It is in the area of functional genomics, concerned with determining the function of genes with unassigned function. In particular, the task is to predict protein functional class from sequence information only.

A combination of inductive logic programming and rule learning was applied to the *M. tuberculosis and* E. coli genomes [14.33, 14.32]. The basic unit of classification here is an open reading frame (ORF), which is derived from the genome sequence and is very similar to an actual protein sequence, except that it may in fact turn out not to code for a protein. Protein sequences (and ORFs) are classified into a functional hierarchy: the hierarchy is organism specific and has four levels for *M. tuberculosis* and three for *E. coli.* For illustration, the first level of the *M. tuberculosis* hierarchy contains the classes *Small-molecule metabolism*, *Cell processes*, *Macromolecule metabolism* and *Others.* The second level entries of the hierarchy under *Small-molecule metabolism* include *Lipid biosynthesis*, *Energy metabolism* and *Degradation*, while the third level entries under *Degradation* include (among others) *Carbon compounds* and *Fatty acids.*

Classification at each of the hierarchical levels was addressed, separately for each of the organisms. ORFs with known classifications are used for training and the learned rules are applied to ORFs with unassigned function. ORFs are described with the properties of their sequences, such as the sequence length, proportions of specific residues and pairs thereof, the predicted secondary structure for the sequence, as well as information on similarity (homology) to proteins in the SWISS-PROT annotated protein sequence database.

WARMR was applied to find frequently occurring patterns on the ORFs with assigned function, which were then used as features for propositional rule induction with C4.5/C5. Different sets of rules were learned, one for each of the respective level of the functional hierarchy. The learned rules were then applied to the ORFs with unassigned function. The rules predict 65% of the ORFs with no assigned function in *M. tuberculosis* and 24% of those in *E. coli*, with an estimated accuracy of 60-80%, depending on the level of functional assignment. The rules also provide insight into the evolutionary history of *M. tuberculosis and* E. coli.

14.4.5 mRNA signal structure detection

Although not strictly a task of predicting protein structure or function, this application ([14.30], see also Chapter 9) is related to the above as it is concerned with identifying subsequences in mRNA that are responsible for biological functions. mRNA is a sequence of nucleotic acids (guanine, adenine, uracil, cytosine). The secondary structure of an mRNA contains special subsequences called *signal structures* that are responsible for special biological functions, such as RNA-protein interactions and cellular transport.

Signal structures are grouped into classes, such as IRE (Iron Responsive Element) that have the same (or similar) biological function. Horvath et al. [14.30] used RIBL2 to classify signal structures with unknown classes. They achieve highest accuracy using a term-based representation and similarity measure.

14.5 Medical applications

Several applications of relational learning methods to medical domains exist. These include

- Learning rules for early diagnosis of rheumatic diseases [14.39].
- Identifying glaucomatous eyes from ocular fundus images [14.41]
- Classifying X-ray angiograms of a patient's cerebral vasculature [14.52]
- Modeling the therapeutic effects of drugs on patients in intensive care [14.42]
- Analysis of epidemiological data from the San Francisco tuberculosis clinic (see Chapter 13)

Lavrač et al. [14.39] applied the ILP system LINUS to the problem of diagnosis he early stage of rheumatic diseases. In addition to the patient records available, a specialist for rheumatic diseases provided his knowledge about the typical co-occurrences of symptoms. This was used as background knowledge by LINUS and was found to improve the predictive performance and noise robustness of the learning process.

Mizoguchi et al. [14.41] addressed the problem of identifying glaucomatous eyes from ocular fundus images. The images are divided into circular segments, each of which is is marked as normal or abnormal by a domain expert. The problem of diagnosing the individual segments is of a relational nature, since neighbouring segments are likely to have the same diagnosis. By applying their ILP system GKS, the authors achieved human-level performance.

Sammut and Zrimec [14.52] applied ILP to the problem of medical image understanding, in particular classifying X-ray angiograms of a patient's cerebral vasculature. Relational machine learning methods were applied to automatically build rules for classifying types of blood vessels (e.g. ICA = Internal Carotid Artery). Their learning system can use complex Prolog programs as background knowledge, in particular knowledge needed for classifying features in an image.

Morik et al. [14.42] used relational learning to revise knowledge on the therapeutic effects of drugs on patients in intensive care. The knowledge provided by a medical expert predicts the effects of administering a drug to a patient. This is matched with actual intervention records and a revision of the knowledge is attempted where inconsistencies arise.

Finally, Getoor et al. (see Chapter 13) analyzed epidemiological data from the San Francisco tuberculosis (TB) clinic. Probabilistic relational models were learnt from a multi-relation database containing information on TB patients, their contacts, and TB strains. Several relational dependencies were discovered: for example, there is a dependency between the ethnicity of the patient and the number of patients infected by the same TB strain as that patient.

14.6 Environmental applications

This section describes three environmental applications of relational learning. The first is a SAR problem and concerns the prediction of biodegradation rates of chemicals from their chemical structure. The second is concerned with the biological classification of British rivers and the third with finding relationships between the physical/chemical and biological properties of water quality in Slovenian rivers.

14.6.1 Predicting biodegradation rates

Džeroski et al. [14.19] studied the problem of predicting biodegradation rates for compounds from their chemical structure. They used a database of 328 structurally diverse and widely used (commercial) chemicals described in a handbook of degradation rates. Complete data on the structure of the chemicals (SMILES notation) was available, as well as data on the overall, biotic and abiotic degradation rates in four environmental compartments (soil, air, surface water and ground water). Models were built for biotic degradation in surface water, predicting the logarithm of the half-life time of aqueous biodegradation. Half-life times were measured for some compounds and estimated by experts for others: in the latter case, an upper and a lower bound were given and the arithmetic mean of these was taken.

Several propositional and ILP methods, including FFoil, ICL, S-Cart, and Tilde, were used for decision tree, regression tree and rule induction. In addition to a few global features, such as molecular weight, the main information used for learning was the data on the structure of compounds, i.e., the atoms within a molecule and the connections/bonds between them. Domain knowledge about a variety of functional groups and substructures was used. ILP systems use this data directly, while propositional systems use features derived from it, which represent the compounds' structure approximately, but not completely. Several of the derived models perform better (correlation 0.7) than a state-of-the-art biodegradability prediction system (correlation 0.6) based on linear regression.

Table 14.1. A relational rule for predicting the biodegradability of a compound.

A compound M degrades fast IF
M *contains* an atom A1 and
atom A1 is a nitrogen atom and
atom A1 *is connected to* atom A2 with bond B and
bond B is an aromatic bond and
the molecular weight of M is less than 110 units and
the logP value (hydrophobicity) of M is greater than zero .

An example rule for predicting biodegradability is given in Table 14.1. Note that this rule is relational, since it makes use of the relations *contains* between a compound and its components (in this case an atom) and *is connected to* between atoms.

Two issues speak in favor of using ILP approaches in this domain. The first is the need to handle structural information, i.e., information on the structure of chemicals. The natural representations of chemical structures are not straightforward to squeeze into a fixed-width-table. The second is the need for prior/domain knowledge: chunks of knowledge defining functional groups and substructures are essential for good performance. ILP methods provide facilities for using both types of information directly.

14.6.2 Biological classification of British rivers

The task addressed by Džeroski et al. [14.22] is to interpret benthic samples of macro-invertebrates in water quality terms. In other words, given are samples of the river beds at different sites and their classification into one of five quality classes. The task is to learn general rules for the classification of samples into water quality classes. The study used 292 field samples of benthic communities collected from British Midlands rivers, classified by an acknowledged expert river ecologist (H. A. Hawkes). The samples come from the database maintained by the National River Authority of the United Kingdom, where the results of monitoring the environmental quality of British rivers are stored.

Two ILP systems were applied to the problem of classification of biological samples. GOLEM [14.45] (see also Chapter 3) works in the normal ILP setting, where the task is to find classification rules that explain the training examples, while CLAUDIEN [14.15] (see also Chapter 5) works in the nonmonotonic ILP setting, where the task is to find valid rules that are confirmed by the training examples. A propositional learning system, CN2 [14.8, 14.21] was also applied to learn classification rules.

For the normal ILP setting, the problem was formulated as follows: for each class a separate ILP problem was created, where the positive examples are the samples classified in that class, and all the other samples are negative examples. The background knowledge consisted of eighty predicates of the form $family(X, A)$, each denoting that $family$ is present in sample X at abundance level A. (In several cases, identification was carried to levels other than family, e.g., species or genera level. For simplicity, we will use the term family throughout, regardless of the taxonomic identification level.) Predicates of this kind include $tipulidae(X, A)$, $asellidae(X, A)$, etc. In addition, the background predicate $greater\_than(A, B)$ was available, stating that abundance level A is greater than abundance level B.

Example rules induced by GOLEM are the following. The rule $b1a(X) \leftarrow leuctridae(X, A)$ states that a sample belongs to the best water quality class if Leuctridae are present. This rule covers forty-three positive and four negative examples, and agrees with expert knowledge; the family Leuctridae is an indicator of good water quality. Another good rule is the following: $b1b(X) \leftarrow ancylidae(X, A), gammaridae(X, B), hydropsychidae(X, C), rhyacophilidae(X, D), greater\_than(B, A), greater\_than(B, D)$. Gammaridae in abundance is a good indicator of class B1b, along with the other families present.

14.6.3 Relating physical and chemical parameters of water quality in Slovenian rivers

Physical and chemical properties give a specific picture of river water quality at a particular point in time, while the biota (living organisms) act as continuous monitors and give a more general picture of water quality over a period of time. This has increased the relative importance of biological methods for monitoring water quality. The problem of inferring the chemical properties from the biota is practically relevant, especially in countries where extensive biological monitoring is conducted. Regular monitoring for a very wide range of chemical pollutants would be very expensive, if not impossible. On the other hand, the state of the biota can reflect an increase in pollution and indicate likely causes/sources.

Džeroski et al. [14.23] used data on biological and chemical samples from Slovenian rivers collected through the monitoring program of the Hydrometeorological Institute of Slovenia. Pairs of biological and chemical samples taken at the same site at approximately the same time were used: 1061 such pairs were collected over six years. Data on biological samples list all the species/taxa present at the site and their abundances. Chemical samples contain the measured values of 16 physical an chemical parameters: biological oxygen demand (BOD), chlorine concentration (Cl), CO_2 concentration, electrical conductivity, chemical oxygen demand COD ($K_2Cr_2O_7$ and $KMnO_4$), concentrations of ammonia (NH_4), NO_2, NO_3 and dissolved oxygen (O_2), alkalinity (pH), PO_4, oxygen saturation, SiO_2, water temperature, and total hardness.

First, regression tree induction was used to learn predictive models for each of the 16 parameters separately [14.23]. The models for the most important indicators of pollution (ammonia, biological oxygen demand, chemical oxygen demand) had the best predictive power. TILDE was next applied to learn first-order logical decision trees, which can be also used as clustering trees [14.2] (see also Chapter 5). These were then to predict the values for all 16 parameters at the same time [14.3]: this actually improved the accuracy as compared to individual predictions for each of the 16 parameters.

Three relational data mining issues were raised in this application:

1. Varying length data records: biological samples list all the species present. Depending on the site and water quality, the number of taxa present can vary. Methods for handling structural information or careful feature selection are thus needed.
2. Aggregating data: we used detailed data, where organisms were identified to species level, and aggregated data, where species from the same family were grouped together. Domain knowledge on the taxonomy of river water organisms was used.
3. Making multiple predictions: most KDD methods for prediction only deal with one target variable. In many cases, however, it might be beneficial to try to predict several interrelated variables simultaneously.

14.7 Mechanical engineering applications

This section describes three applications of relational learning in mechanical engineering and two in traffic engineering. These include finite-element mesh design, steel grinding and electrical discharge machining.

14.7.1 Finite-element mesh design

The datasets on finite element mesh design [14.17] come from the area of mechanical engineering where finite element meshes are used to analyze the behavior of structures under different kinds of stress. The problem addressed here is to determine an appropriate resolution of a finite element mesh for a given structure so that the corresponding computations are both accurate and fast.

The resolution of a finite element (FE) mesh is determined by the number of elements on each of its edges. It depends on the geometry of the body studied and on the boundary conditions. Given are descriptions of ten structures for which experts have determined an appropriate mesh resolution. An object to be partitioned is represented as (1) a set of edges, (2) the properties of the edges, and (3) relations among the edges.

These properties and relations are represented as part of background knowledge by predicates, such as: *short*(*Edge*), *loaded*(*Edge*), *not_loaded*(*Edge*), *two_side_fixed*(*Edge*), *neighbour_xy*(*Edge*1, *Edge*2), etc. The target relation to be learned is *mesh*(*Edge*, *N*), where *Edge* is the name of an edge in the structure, and *N* is the recommended number of finite elements along this edge. The task is thus to learn rules that determine an appropriate resolution of a FE mesh (i.e., an appropriate resolution for each given edge) from the geometry of the body, the types of edges, boundary conditions and loadings.

A number of ILP systems, including FOIL, GOLEM, and CLAUDIEN, have been applied to this problem. Two example rules generated from the data are given in Table 14.2. The first rule says that partitioning an *Edge* into 7 elements is appropriate if *Edge* has a neighbour *EdgeY* in the XY-plane so that *EdgeY* is fixed at both ends, and *Edge* has another neighbour *EdgeZ* in the XZ-plane so that *EdgeZ* is not loaded. The second rules is recursive and observes that an edge's partition can be determined by looking for an edge of the same length and shape positioned similarly in the same object.

Table 14.2. Two relational rules for finite-element mesh design.

```
mesh(Edge, 7) ←
    usual_length(Edge),
    neighbour_xy(Edge, EdgeY), two_side_fixed(EdgeY),
    neighbour_zx(EdgeZ, Edge), not_loaded(EdgeZ)

mesh(Edge, N) ←
    equal(Edge, Edge2), mesh(Edge2, N)
```

14.7.2 Electrical discharge machining

In electrical discharge machining (EDM), the workpiece surface is machined by electrical discharges occurring in the gap between two electrodes – the tool and the workpiece. The process consists of numerous monodischarges generating a crater-textured surface. There are three kinds of pulses associated with the monodischarges: A (empty), B (effective), and C (arc) pulses. Some parameters can be controlled during the process (gap, flow), some require process interruption (duration of discharges), and some of them are inherent to the particular machining task. For a standard set of workpiece types there exist predefined sets of values for the process parameters which guarantee a certain degree of generated surface quality. However, the settings are very conservative and do not yield really good performance in terms of the time needed to accomplish the task. Therefore, a human operator is normally employed to control the process parameters and minimize machining time. Our task in this domain was to model the operator's behavior.

The ILP system FORS [14.31] was applied to traces of operator behavior (including control actions). Gap and flow were chosen to be the controlled parameters (class), while the mean values and the deviations of the observed quantities (A, B, C and I - electric current) for the last 5 and last 20 seconds were chosen as the attributes. To enable detection of the rate of change of attributes, the predicate smaller-than ($<$) was defined as the background knowledge.

Since we tried to model control by two variables (gap and flow) the learning task was decomposed into two separate learning tasks – learning gap control and learning flow control. For each control variable three actions were possible: increase the variable (the action was assigned a numerical value +1.0), no action (0.0), and decrease the variable (-1.0).

The domain expert qualitatively defined the relation between the effectiveness of the process and the control parameters – gap and flow. The state of the process can be represented in a two dimensional diagram. The diagram defines two states of the process: stable state and arcing state. The goal of the control is to guide the process as close as possible to the boundary between the arcing and the stable region. The model was evaluated by plotting suggested actions into the state diagram of the process. Most of the vectors point towards the boundary between the regions – the optimal working region. It thus seems that the model is correctly modeling the operator's performance.

14.7.3 Steel grinding

The task in the steel grinding domain [14.31] is to determine the roughness of the workpiece from the properties of the sound produced during the process of steel grinding. Data were obtained during an experiment in which vibration signals generated by the grinding wheel and the workpiece were detected by an accelerometer sensor and processed by a spectrum analyzer. From the obtained spectra predefined spectral features were extracted: total spectrum area (SpArea), frequency of the maximum area peak (MaxAreaX), and frequency of the spectrum area central point (AreaCX). Simultaneously, workpiece surface roughness was measured. When the ILP system FORS [14.31] was applied to these data, comparisons between the frequencies were also allowed through the predicates smaller-than-or-equal-to ($\leq$) and greater-than-or-equal-to ($\geq$).

While the use of background knowledge did not bring any significant improvement in predictive performance, the newly induced models frequently contained background knowledge literals. The domain experts considered this a significant improvement because the newly induced models are more general than the models without background knowledge. For example, without using background knowledge, a literal such as MaxAreaX $\leq$ 6125 typically appeared, saying that the frequency of the maximum area peak is less than 6125 Hz. In the particular setting (grinding regime, choice of the tools) this means that the maximum area peak is in the lower part of the spectrum.

If we use a different grinding wheel speed, however, the whole spectrum would shift to higher or lower frequencies, thus making the above literal useless. The use of background knowledge typically yielded the literal MaxAreaX $\leq$ AreaCX, which directly states that the maximum peak must lie in the lower part of the spectrum, regardless of the frequency area in which the spectrum is situated. This enables the rule to be used in a much broader class of working regimes.

14.8 Traffic engineering applications

The traffic engineering applications addressed by ILP include the detection of traffic problems from sensor data and classifying road accidents as being caused by young male drivers or otherwise.

Džeroski et al. [14.24] applied ILP to the task of detecting traffic problems, such as accidents and congestions, in the urban-ring of the city of Barcelona. Overall, two kinds of input were available to the learning process. The first type is background knowledge on the road network, which is present in an existing traffic management system. An object oriented representation is used to capture the different types of road sections, the relations among them, and the placement of sensors on individual road sections. The second type is sensor readings on three basic quantities describing traffic behavior: speed, flow and occupancy. The goal of the learning process is to identify critical sections (where problems have occurred) by using sensor readings and road geometry. Technically speaking, a critical section is a section of the road which constrains the road capacity the most, e.g., because an accident has occurred just after this section in the immediate past.

Three ILP systems (ICL, PROGOL, and TILDE) were applied to a dataset containing examples of accidents and congestions. Sensor data (obtained from a high-quality traffic simulator) as well as background knowledge on the road network geometry were used. Sensors provide us with a continuous stream of information, sending five readings each minute that refer to the last minute and each of the four minutes preceding it. Typically, flow (number of cars that passed the sensor in the last minute) and occupancy (the proportion of time the sensor is occupied, in thousandths) are measured. Some sensors (which are actually double sensors) also measure the average speed of the cars that passed the sensor during the last minute. The measurements of sensors related to a single section are aggregated: flow is summed across lanes, while occupancy and velocity are averaged across lanes. Saturation is a derived quantity defined as the ratio between the flow and the capacity of a section: the latter depends on the number of lanes and is part of the background knowledge.

In a comparison to a propositional learning systems (C4.5), all three ILP systems perform much better, showing that it is essential to take the (variable) road geometry into account. Two rules generated by TILDE clearly demonstrate this. The first states there is an accident at section `A` if the saturation at `A` is low and the occupation at the section preceding `A` is high. The second states there is an accident at section `A` if the saturation is low and the occupation is high at `A`, whereas the velocity at the next section is high.

Roberts et al. [14.51] address the problem of classifying road accidents as being caused by young male drivers (a class of interest to traffic experts) or otherwise. Namely, young male drivers have many more accidents per mile than other drivers. The purpose of the investigation was to find out if they were overrepresented in a particular type of accident, or if they just have more accidents. The latter conclusion was supported: young males appear to have accidents due to inexperience, without any mitigating circumstances.

The data used consisted of factual details of road traffic accidents, such as road surface conditions, weather details, casualties, and vehicle maneuvering details, as well as more subjective information concerning the accident such as who caused the accident, what were the underlying causes and what precipitated the accident. Two ILP systems, PROGOL, and TILDE, and a propositional learning system (C4.5) were applied to a dataset of 1413 accident records, yielding comparable performance.

14.9 Text mining, Web mining, and natural language processing

Relational learning algorithms have also been used in a number of applications to analyze and classify text, and more recently hypertext.

Information extraction aims at finding specific pieces of information from in a natural language document. Califf and Mooney [14.7] use a relational learning system RAPIER to extract database entries for a jobs database from newsgroup postings. The relational representation used allows for constraints on words, POS (part-of-speech) tags and semantic classes of the extracted phrase and the surrounding context to be taken into account.

Information retrieval is concerned with identifying small sets of relevant documents in very large document collections. ILP has been used to learn the concept of relevant documents from example documents marked as relevant (and irrelevant) by the user. TILDEwas used to learn to identify relevant medical documents in the English or Slovene language from examples [14.16]. Loggie [14.40] used PROGOL to learn to identify relevant documents concerning telecommunications in the INSPEC database, using the hierarchical orderings of INSPEC codes and terms as background knowledge.

Text categorization is the task of classifying text into one of several predefined categories. Cohen [14.11] applies the ILP systems FOIL and FLIPPER, as well as two propositional learning systems, to ten text categorization problems. Positional relations between words (near, after) are used by the relational systems and prove to be useful in generating high precision classifiers.

Craven and Slattery [14.13] apply a combination of statistical and relational learning in the context of extracting knowledge bases from the web. They address two tasks: the problem of classifying Web pages (which corresponds to recognizing instances of knowledge base classes, such as students, faculty, courses, etc.) and the problem of learning prototypical patterns of hyperlink connectivity among pages. The relational knowledge used specifies which pages are linked through hyperlinks, what words can be found on a page, what words are in the anchor or neighbourhood of a hyperlink, and such like.

Natural language processing will be of increasing importance for text mining, since pre-processed and enriched text holds more potential for knowledge discovery. Natural language tasks that have been addressed by relational learning included the learning of morphology (rules of word formation), learning to perform POS (part-of-speech) tagging, learning grammars, learning semantic relations from corpora, learning semantic parsers and learning in the context of machine translation. For an overview we refer the reader to Cussens and Džeroski [14.14].

14.10 Business data analysis

The discovery challenge [14.53] at the PKDD-2000 conference included the task of analyzing data from a multi-relation relational database describing the operation of a Czech bank. It describes the operations of 5369 clients holding 4500 accounts. The eight tables in the database are: *account*, *client*, *disposition*, *permanent order*, *transaction*, *loan*, *credit card*, and *demographic data*. The first three tables describe client and account information, the next four describe the usage of products, and the remaining table contains demographic information about 77 Czech districts.

Many data mining tasks can be defined on these data. One of these is determining the loan quality of an account, where the table *account* is the target table. There are thus 4500 examples, but the dataset contains a total of 1.079,680 records. Note that there is a one to many association between the account and transaction relations. Knobbe et al. [14.37] applied a propositionalization technique based on aggregation to this problem in conjunction with the C5.0 decision tree induction system. The aggregation turned out to be especially useful for the large *transactions* table, which contains 1.056,320 records. A very high accuracy on this problem was reported.

Another application of relational data mining to the analysis of business data is reported in Chapter 13. Data on companies, corporate officers and their roles is stored in three relations and analyzed to learn probabilistic relational models, a relational extension of Bayesian networks. Among other, an interesting dependency has been found of a person's salary and the number of employees in their company.

14.11 Miscellaneous applications

Miscellaneous other applications of relational learning exist. The one in adaptive system management deserves a special mention, as it is included in a software product for the management of large computer networks. Other application areas mentioned in this section are music, software engineering, and the modeling, control and design of dynamic systems.

14.11.1 Adaptive system management

Knobbe et al. [14.38] addressed the task of identifying the causes for low performance in complex computer networks. This is critical for upholding service level agreements (SLAs), where a minimum performance level is prescribed for certain services (e.g., maximum database access time). Data from performance monitors on individual components in the network (gathered at regular time intervals) are used, together with background knowledge on the structure of the network (connections between the components). A typical induced rule predicts low performance if any of the NFS (Network-File-Servers) servers in the network has a monitor (such as network load, usage, etc.) with a high value. This rule is a generalization of a number of rules generated by propositional learning, which involve a particular NFS server (e.g., server 11) and a particular monitor (e.g., high usage of disk 14 on server 11).

14.11.2 Music

Dovey [14.18] applied ILP to analyze Rahmaninoff's piano performances. Van Baelen and De Raedt [14.62] addressed the task of analysis and prediction of piano performances. CLAUDIEN (see Chapter 5) was used to induce theories for predicting MIDI files from the musical analysis of Mendelssohn's Lied Ohne Worte.

Pompe et al. [14.49] tried to automate composing the two-voice counterpoint. ILP was applied to learn rules that help generate the counterpoint melody from the cantus firmus melody. The learned rules together with some rules from musical theory produced counterpoints that professional musicians and nonmusicians judged to be at the level of an acceptable junior composer.

14.11.3 Software engineering

ILP approaches have been applied with varying degrees of success to a number of software engineering problems. These include program development from high level specification through data reification, program construction using a strong declarative bias, construction of invariants for proving program correctness, and generation of test cases. For an overview, we refer the reader to Džeroski and Bratko [14.20].

Two applications of relational learning to software engineering deserve special mention. Cohen [14.10] used relational learning to recover an abstract specification of a large software system. He also applied FOIL and FLIPPER [14.12] to the problem of software fault prediction. The task addressed was to predict whether a fault occurrs in a C++ class, given background knowledge on coupling relationships to other C++ classes.

14.11.4 Modeling, control, and design

Preliminary applications of ILP have been investigated in the area of dynamical systems and qualitative physics. The tasks addressed include learning qualitative models of dynamic systems, learning control rules from behavioral traces of operators/controllers, and design of devices from first principles. For an overview, we refer the reader to Džeroski and Bratko [14.20].

14.12 Summary and discussion

This chapter presented an overview of data mining applications of relational learning. The most successful applications come from the area of bioinformatics and include drug design (note that there exists a data mining company that specializes in applying relational learning to such problems) and genome scale prediction of the functional class of proteins. Other applications come from the areas of medicine, environmental sciences/monitoring, engineering, and business data analysis.

The relational learning systems that have been most successful in application settings so far have been GOLEM, PROGOL and P-PROGOL (ALEPH). This has been due to alarge extent to the flexibility of the latter to accommodate new forms of knowledge constraints into the learning process. The importance of the good cooperation between the system developers and experts in the application domains cannot be emphasized enough.

In more recent applications, relational instance-based approaches (such as RIBL) have shown their potential. Also the discovery of relational association rules with WARMR and the use of WARMR to generate good features for propositional learning has yielded good results in several application domains. Other ILP systems (e.g., PROGOL) have been used in the same general setup. Finally, relational decision tree induction systems, such as S-CART and TILDE have strong potential for practical applications due to their efficiency.

We hope that this book has made the concept of relational data mining accessible to a large audience and will thus stimulate interest in the field. We hope that this chapter has illustrated the kind of applications where relational data mining can make a difference (due to the representation of the available data and domain knowledge) and will thus stimulate further applications of relational data mining.

References

14.1 H. Blockeel and L. De Raedt. Top-down induction of first order logical decision trees. *Artificial Intelligence*, 101(1-2): 285–297, 1998.
14.2 H. Blockeel, L. De Raedt, and J. Ramon. Top-down induction of clustering trees. In *Proceedings of the Fifteenth International Conference on Machine Learning*, pages 55–63. Morgan Kaufmann, 1998.
14.3 H. Blockeel, S. Džeroski, and J. Grbović. Simultaneous prediction of multiple chemical parameters of river water quality with TILDE. In *Proceedings of the Third European Conference on Principles of Data Mining and Knowledge Discovery*, pages 15–18. Springer, Berlin, 1999.
14.4 L. Breiman, J.H. Friedman, R.A. Olshen, and C.J. Stone. *Classification and Regression Trees*. Wadsworth, Belmont, CA, 1984.
14.5 S.E. Brenner, C. Chothia, T.J. Hubbard, and A.G. Murzin. Understanding protein structure: Using SCOP for fold interpretation. *Methods in Enzymology*, 266: 635–643, 1996.
14.6 C. Bryant. Data mining via ILP: The application of PROGOL to a database of enantioseparations. In *Proceedings of the Seventh International Workshop on Inductive Logic Programming*, pages 85–92. Springer, Berlin, 1997.
14.7 M.E. Califf and R. Mooney. Relational learning of pattern match rules for information extraction. In *Proceedings of the Sixteenth National Conference on Artificial Intelligence*, pages 328–334. AAAI Press, Menlo Park, CA, 1999.
14.8 P. Clark and R. Boswell. Rule induction with CN2: Some recent improvements. In *Proceedings Fifth European Working Session on Learning*, pages 151–163. Springer, Berlin, 1991.
14.9 T. Cleveland. Pirkle-concept chiral stationary phases for the HPLC separation of pharmaceutical racemates. *Journal of Liquid Chromatography*, 18(4): 649–671, 1995.
14.10 W. Cohen. Recovering software specifications with inductive logic programming. In *Proceedings of the Twelfth National Conference on Artificial Intelligence*, MIT Press, Cambridge, MA, 1994.
14.11 W. Cohen. Learning to classify English text with ILP methods. In L. De Raedt, editor, *Advances in Inductive Logic Programming*, pages 124–143. IOS Press, Amsterdam, 1996.
14.12 W. Cohen and P. Devanbu. A Comparative Study of Inductive Logic Programming Methods for Software Fault Prediction. In *The Fourteenth International Conference on Machine Learning*, pages 66–74. Morgan Kaufmann, San Francisco, CA, 1997.
14.13 M. Craven and S. Slattery. Relational learning with statistical predicate invention: Better models for hypertext. *Machine Learning*, 43: 97–119, 2001.
14.14 J. Cussens and S. Džeroski, editors. *Learning Language in Logic*. Springer, Berlin, 2000.
14.15 L. De Raedt and M. Bruynooghe. A Theory of Clausal Discovery. In *Proceedings of the Thirteenth International Joint Conference on Artificial Intelligence*, pages 1058–1063. Morgan Kaufmann, San Mateo, CA, 1993.
14.16 J. Dimec, S. Džeroski, L. Todorovski, and D. Hristovski. WWW search engine for Slovenian and English medical documents. In *Proc. Fifteenth International Congress for Medical Informatics*, pages 547-552. IOS Press, Amsterdam, 1999.
14.17 B. Dolšak, I. Bratko, and A. Jezernik. Applications of machine learning in finite element computation. In R.S. Michalski, I. Bratko, and M. Kubat, editors, *Machine Learning, Data Mining and Knowledge Discovery: Methods and Applications*, pages 147–171. John Wiley and Sons, Chichester, 1997.

14.18 M. J. Dovey. Analysis of Rachmaninoff's piano performances using inductive logic programming. In *Proceedings of the Eighth European Conference on Machine Learning*, pages 279–282. Springer, Berlin, 1995.
14.19 S. Džeroski, H. Blockeel, B. Kompare, S. Kramer, B. Pfahringer, and W. Van Laer. Experiments in Predicting Biodegradability. In *Proceedings of the Ninth International Workshop on Inductive Logic Programming*, pages 80–91. Springer, Berlin, 1999.
14.20 S. Džeroski and I. Bratko. Applications of inductive logic programming. In L. De Raedt, editor, *Advances in Inductive Logic Programming*, pages 65–81. IOS Press, Amsterdam, 1996.
14.21 S. Džeroski, B. Cestnik and I. Petrovski. Using the m-estimate in rule induction. *Journal of Computing and Information Technology*, 1(1): 37–46, 1993.
14.22 S. Džeroski, L. Dehaspe, B. Ruck and W. Walley. Classification of river water quality data using machine learning. In *Proceedings of the Fifth International Conference on the Development and Application of Computer Techniques to Environmental Studies, Vol. I: Pollution modelling*, pages 129–137. Computational Mechanics Publications, Southampton, 1994.
14.23 S. Džeroski, J. Grbović, and D. Demšar. Predicting chemical parameters of river water quality from bioindicator data. *Applied Intelligence*, 13(1): 7–17, 2000.
14.24 S. Džeroski, N. Jacobs, M. Molina, C. Moure, S. Muggleton, and W. Van Laer. Detecting traffic problems with ILP. In *Proceedings of the Eighth International Conference on Inductive Logic Programming*, pages 281–290. Springer, Berlin, 1998.
14.25 S. Džeroski, S. Schulze-Kremer, K. Heidtke, K. Siems, D. Wettschereck, and H. Blockeel. Diterpene structure elucidation from ^{13}C NMR spectra with Inductive Logic Programming. *Applied Artificial Intelligence*, 12: 363–383, 1998.
14.26 W. Emde and D. Wettschereck. Relational Instance-Based Learning. In *Proceedings of the Thirteen International Conference on Machine Learning*, pages 122–130. Morgan Kaufmann, San Francisco, CA, 1996.
14.27 P. Finn, S. Muggleton, C.D. Page, and A. Srinivasan. Pharmacophore discovery using the inductive logic programming system PROGOL. *Machine Learning*, 30: 241–271, 1998.
14.28 C. Hansch, R. Li, J. Blaney, and R. Langridge. Comparison of the inhibition of escherichia coli and lactobacillus casei dihydrofolate reductase by 2,4-diamino-5-(substituted-benzyl) pyrimidines: Quantitative structure-activity relationships, X-ray crystallography, and computer graphics in structure-activity analysis. *J. Med. Chem.*, 25: 777–784, 1992.
14.29 C. Helma, R.D. King, S. Kramer, and A. Srinivasan. The predictive toxicology challenge 2000-2001. *Bioinformatics*, 17: 107–108, 2001. Web pages at `http://www.informatik.uni-freiburg.de/~ml/ptc/`.
14.30 T. Horváth, S. Wrobel, and U. Bohnebeck. Relational instance-based learning with lists and terms. *Machine Learning*, 43(1/2): 53–80, 2001.
14.31 A. Karalič and I. Bratko. First order regression. *Machine Learning*, 26(2/3): 147–176, 1997.
14.32 R.D. King, A. Karwath, A. Clare, and L. Dehaspe. Accurate prediction of protein functional class in the M. tuberculosis and E. coli genomes using data mining. *Yeast (Comparative and Functional Genomics)*, 17: 283–293, 2000.
14.33 R.D. King, A. Karwath, A. Clare, and L. Dehaspe. Genome scale prediction of protein functional class from sequence using data mining. In *Proceedings of the Sixth International Conference on Knowledge Discovery and Data Mining*, pages 384–389. ACM Press, New York, 2000.

14.34 R.D. King, S. Muggleton, R. Lewis, and M.J.E. Sternberg. Drug design by machine learning: The use of inductive logic programming to model the structure-activity relationships of trimethoprim analogues binding to dihydrofolate reductase. *Proc. of the National Academy of Sciences of the USA* 89(23): 11322–11326, 1992.

14.35 R.D. King, A. Srinivasan, and M.J.E. Sternberg. Relating chemical activity to structure: An examination of ILP successes. *New Generation Computing*, 13: 411–433, 1995.

14.36 D. Kneller, F. Cohen, and R. Langridge. Improvements in protein secondary structure prediction by an enhanced neural network. *J. Mol. Biol.*, 214: 171–182, 1990.

14.37 A.J. Knobbe, M. de Haas, and A. Siebes. Propositionalization and aggregates. In *Proceedings of the Fifth European Conference on Principles of Data Mining and Knowledge Discovery*. Springer, Berlin, 2001.

14.38 A.J. Knobbe, B. Marseille, O. Moerbeek, and D. van der Wallen. Results in data mining for adaptive system management. In *Proceedings of the Eighth Belgian-Dutch Conference on Machine Learning*, ATO-DLO, Wageningen, The Netherlands.

14.39 N. Lavrač, S. Džeroski, V. Pirnat, and V. Križman. The utility of background knowledge in learning medical diagnostic rules. *Applied Artificial Intelligence*, 7:273–293, 1993.

14.40 W.T.H. Loggie. Using inductive logic programming to assist in the retrieval of relevant information from an electronic library system. In *Notes of the Workshop on Data Mining, Decision Support, Meta Learning and ILP* held at *The Fourth European Conference on Principles of Data Mining and Knowledge Discovery*, Lyon, France, September 2000. Available at `http://eric.univ-lyon2.fr/~pkdd2000/Download/#Workshops`.

14.41 F. Mizoguchi, H. Ohwada, M. Daidoji, S. Shirato. Using inductive logic programming to learn classification rules that identify glaucomatous eyes. In N. Lavrač, E. Keravnou, B. Zupan, editors, *Intelligent Data Analysis in Medicine and Pharmacology*, pages 227–242. Kluwer, Boston, 1997.

14.42 K. Morik, P. Brockhausen, and T. Joachims. Combining statistical learning with a knowledge-based approach — A case study in intensive care monitoring. In *Proceedings of the Sixteenth International Conference on Machine Learning*, pages 268–277. Morgan Kaufmann, San Francisco, CA, 1999.

14.43 S. Muggleton. Inverse entailment and Progol. *New Generation Computing*, 13: 245–286, 1995.

14.44 S.H. Muggleton, C.H. Bryant, and A. Srinivasan. Learning Chomsky-like grammars for biological sequence families. In *Proceedings of the Seventeenth International Conference on Machine Learning*, pages 631–638. Morgan Kaufmann, San Francisco, CA, 2000.

14.45 S. Muggleton and C. Feng. Efficient induction of logic programs. In *Proceedings of the First Conference on Algorithmic Learning Theory*, pages 368–381. Ohmsma, Tokyo, Japan, 1990.

14.46 S. Muggleton, R. D. King, and M. J. E. Sternberg. Protein secondary structure prediction using logic-based machine learning. *Protein Engineering*, 5(7): 647–657, 1992.

14.47 S. Muggleton, C.D. Page, and A. Srinivasan. An initial experiment into stereochemistry-based drug design using inductive logic programming. In *Proceedings of the Sixth International Workshop on Inductive Logic Programming*, pages 25–40. Springer, Berlin, 1997.

14.48 H. Nielsen, J. Engelbrecht, S. Brunak, and G. von Hejne. Identification of prokaryotic and eukaryotic signal peptides and prediction of their cleavage sites. *Protein Engineering*, 10: 1–6.
14.49 U. Pompe, I. Kononenko, and T. Makše. An application of ILP in a musical database: Learning to compose the two-voice counterpoint. In *Proceedings of the MLnet Workshop on Data Mining with ILP*, pages 1–11. University of Bari, Italy, 1996.
14.50 J.R. Quinlan. Learning logical definitions from relations. *Machine Learning*, 5: 239–266, 1990.
14.51 S. Roberts, W. Van Laer, N. Jacobs, S. Muggleton, and J. Broughton. A comparison of ILP and propositional systems on propositional traffic data. In *Proceedings of the Eighth International Conference on Inductive Logic Programming*, pages 291–299. Springer, Berlin, 1998.
14.52 C. Sammut and T. Zrimec. Learning to classify X-ray images using relational learning. In *Proceedings of the Tenth European Conference on Machine Learning*, pages 55–60. Springer, Berlin, 1998.
14.53 A. Siebes and P. Berka. *Discovery Challenge.* Notes of the workshop held at *The Fourth European Conference on Principles of Data Mining and Knowledge Discovery*, Lyon, France, September 2000. Available at `http://eric.univ-lyon2.fr/~pkdd2000/Download/#Challenge`.
14.54 A. Srinivasan. The Aleph Manual. Technical Report, Computing Laboratory, Oxford University, 2000. Available at `http://web.comlab.ox.ac.uk/oucl/research/areas/machlearn/Aleph/`
14.55 A. Srinivasan and R.D. King. Feature construction with inductive logic programming: A study of quantitative predictions of biological activity aided by structural attributes. In *Proceedings of the Sixth International Workshop on Inductive Logic Programming*, pages 89–104. Springer, Berlin, 1997.
14.56 A. Srinivasan, R.D. King, and D.W. Bristol. An assessment of ILP-assisted models for toxicology and the PTE-3 experiment. In *Proceedings of the Ninth International Workshop on Inductive Logic Programming*, pages 291–302. Springer, Berlin, 1999.
14.57 A. Srinivasan, R.D. King, and D.W. Bristol. An assessment of submissions made to the predictive toxicology challenge. In *Proceedings of the Sixteenth International Joint Conference on Artificial Intelligence*, pages 270–275. Morgan Kaufmann, San Francisco, CA, 1999.
14.58 A. Srinivasan, R.D. King, S. Muggleton, and M.J.E. Sternberg. Carcinogenesis prediction using inductive logic programming. In N. Lavrač, E. Keravnou, B. Zupan, editors, *Intelligent Data Analysis in Medicine and Pharmacology*, pp. 243–260. Kluwer, Boston, 1997.
14.59 A. Srinivasan, S.H. Muggleton, R.D. King, and M.J.E. Sternberg. Mutagenesis: ILP experiments in a non-determinate biological domain. In *Proceedings of the Fourth International Workshop on Inductive Logic Programming*, pages 217–232. GMD, Sankt Augustin, Germany, 1994.
14.60 A. Srinivasan, S. Muggleton, R. D. King, and M. J. E. Sternberg. Theories for mutagenicity: A study of first-order and feature based induction. *Artificial Intelligence*, 85(1,2): 277–299, 1996.
14.61 M. Turcotte, S.H. Muggleton, and M.J.E. Sternberg. The effect of relational background knowledge on learning of protein three-dimensional fold signatures. *Machine Learning*, 43(1/2): 81–96, 2001.
14.62 E. Van Baelen and L. De Raedt. Analysis and prediction of piano performances using Inductive Logic Programming. In *Proceedings of the Sixth International Workshop on Inductive Logic Programming*, pages 55–71. Springer, Berlin, 1996.

15. Four Suggestions and a Rule Concerning the Application of ILP

Ashwin Srinivasan

Computing Laboratory, Oxford University
Wolfson Building, Parks Road, Oxford OX1 3QD, UK

Abstract

Since the late 1980s there has been a sustained research effort directed at investigating the application of Inductive Logic Programming (ILP) to problems in biology and chemistry. This essay is a personal view of some interesting issues that have arisen during my involvement in this enterprise. Many of the concerns of the broader field of Knowledge Discovery in Databases manifest themselves during the application of ILP to analyse bio-chemical data. Addressing them in this microcosm has given me some directions on the wider application of ILP, and I present these here in the form of four suggestions and one rule. Readers are invited to consider them in the context of a hypothetical Recommended Codes and Practices for the application of ILP.

15.1 Introduction

Since 1986, an on-going collaboration between researchers now variously located at the Universities of Edinburgh, Louisville, Oxford, Wales, and York; the Imperial Cancer Research Fund (ICRF); Pfizer UK; and Smith-Kline Beecham has resulted in applications of symbolic machine-learning to problems in molecular biology and biochemistry [15.6, 15.8, 15.7, 15.14, 15.13, 15.9, 15.10, 15.11, 15.12, 15.24, 15.26, 15.31, 15.33]. Much of this has been accomplished within the setting of Inductive Logic Programming (ILP: see [15.21]). This is an anecdotal account of some practical guidelines that I have found useful during the course of the applied work. That they have had a role to play in my thinking about the biological applications of ILP may not be apparent from formal accounts in the literature. This is not due to any special deviousness – it merely follows from the nature and form of technical papers, as memorably portrayed by Medawar ("Is the scientific paper a fraud?" in [15.16]):

> As to what I mean by asking 'is the scientific paper a fraud?'– I do not of course mean 'does the scientific paper misrepresent facts', and I do not mean that the interpretations you find in a scientific paper are wrong or deliberately mistaken. I mean the scientific paper may be a fraud because it misrepresents the processes of thought that accompanied or gave rise to the work that is described in the paper. That is the question, and I will say right away that my answer to it is 'yes'.

This essay contains not so much the 'processes of thought' but some of the signposts that were important in guiding my thought processes during the course of the applications of ILP, namely:

- Clearly evaluate when and why a first-order representation is useful.
- First encode background knowledge as declaratively as possible. Once correct, re-code background knowledge to be efficient.
- Check whether the user's cost function matches that of the ILP system.
- Comprehensibility of theories is vital.
- Maintain records of all experiments.

That these were especially noteworthy is a purely personal point of view, and it would be interesting to see the extent to which this is reflected in experience elsewhere.

15.2 Background

The application gained early impetus from discussions initiated by Donald Michie (Turing Institute) with Walter Bodmer (ICRF) on the possibility of machine discovery of biological function from molecular structure. In 1986, at a meeting between Stephen Muggleton and Ross King (Turing Institute) and Michael Sternberg (Birkbeck College, and later ICRF), it was decided that protein secondary structure prediction was a worthwhile task for machine analysis. This topic – the subject of Ross King's subsequent PhD dissertation – led in 1990 to a publication in the *Journal of Molecular Biology* [15.14], and in 1992 to the application of the ILP program GOLEM [15.24]. It was around this time that rational drug design was identified as an area where ILP theories could provide chemical insight. It was also at this time that I became involved in the development and application of ILP.

A central concern of drug-chemistry is understanding the relationships between chemical structure and activity. These relationships cannot be derived solely from physical theory and experimental evidence is essential. Such empirically derived relationships are called Structure Activity Relationships (SARs). In a typical SAR problem, a set of chemicals of known structure and activity are given, and the problem is to construct a predictive theory relating the structure of a compound to its activity. This relationship can then be used to select for structures with high or low activity. Typically, knowledge of such relationships form the basis for devising clinically effective, non-toxic drugs. With continued scientific interest from Michie, Bodmer, Sternberg, and later from Paul Finn (Pfizer) and Chris Rawlings (Smith-Kline Beecham), a number of ILP applications to obtaining SARs have resulted. Each uses increasingly sophisticated representation schemes and algorithms, and the reader is referred to [15.6, 15.13] for a summary of the results, details of which would introduce an unnecessary actuarial flavor here and mask the fact that such application work can only flourish against a backdrop of enlightened academic and industrial support over a long period.

15.3 When and why ILP?

Much of the application work to biology has used a first-order representation. It is evident from the results reported in [15.13] that predictivity of theories could not be the only motive. There, it was found that the predictivity of ILP theories was often no greater than some propositional learners. However, as in earlier results from chess databases [15.23], the first-order representation had much to commend itself in terms of expressive power and compactness. Thus, it is evident that while a hypothesis such as

> Every molecule that contains a bonded pair of aromatic rings is mutagenic.

can be expressed straightforwardly as a first-order statement within an ILP system, a method employing a propositional representation would need precoding of the feature "contains a bonded pair of aromatic rings". In some cases, precoding all possible features may not be tractable, as in the application described in [15.6], which is an example of the "multiple-instance problem" [15.4]. The representation concerns raised by these problems would appear to be naturally addressed within the first-order setting of an ILP learner. However, this expressive power comes at a computational cost, and the worst-case time-complexity for an ILP learner is usually of a much higher order than a propositional learner. Thus, while in principle, the effect of all other propositional methods (including statistical ones) could be obtained within the logical setting of ILP, practical constraints placed on ILP programs ensure that it is rarely the case. It has therefore been worthwhile investing some effort in identifying precisely where a first-order representation is really required, and in cultivating a temperament that sees no dishonour in a symbiotic relationship between ILP and other well-established analysis methods. Indeed, the results in [15.13, 15.31] could be summarised as follows:

- When good attributes are available, just using a first-order representation rarely resulted in higher accuracies;
- When good attributes are unavailable, using a first-order representation has done reasonably well; and
- Using propositional and first-order learning together gains advantage from both.

Other application work in ILP has also shown promise in this direction, and I present two examples drawn from experience of this form of 'focussed' ILP. In the experiments to identify mutagenic structures [15.10, 15.33], the bulk of the data were known to be 'linearisable' – that is, explainable with linear regression on chemist-specified attributes. For this data, the ILP theory obtained was comparable in classification accuracy to the linear model. But this is uninteresting, and unsurprising. Of far more interest to the chemical community was a new structural explanation for a subset of the non-linearisable data. This same data were used in [15.31] where a regression model was constructed using 'new' attributes found by an ILP program, and in [15.30] where an ILP program used regression as background knowledge. In both cases, it became possible to predict the actual numerical activity to a much better degree of precision than had been possible earlier. Thus, while a first-order representation is natural when reasoning with complex structured objects, and can enable the construction of compact theories, it is probably worth investigating thoroughly the extent to which statistical and other machine-learning methods can aid the task.

15.4 Encoding background knowledge

The use of background knowledge to construct explanations is a distinctive feature of ILP. Results from experiments concerned with learning simple logic programs for list processing [15.27] suggest that the performance of an ILP system can be sensitive to the type and amount of background knowledge provided. Background knowledge that contains large amounts of information that is irrelevant to the problem being considered can, and typically does, hinder an ILP system in its search for an adequate explanation of the data. On the other hand, in [15.32], the addition of relevant predicates was found to improve performance (both accuracy and time for theory construction). The two questions that arise naturally are: (1) What should be encoded? and (2) How should it be encoded? I have chosen only to address the second question here. The choice of predicates that appear in the background knowledge is in some sense the most important, and appears to be the least understood step in an ILP application. While ILP theory and implementation do provide clear directions on the enumeration and testing of hypotheses, the process of constructing the background knowledge appears to be a combination of imagination and informed judgement that has counterparts in most scientific disciplines. All that can be said of the applications cited is that they have relied largely on the competence of the domain-expert and on extensive literature searches. Thus, for cutting-edge biological problems, it has been possible to find a consensus opinion on what is likely to be relevant – what is unknown is how these pieces fit together.

In addressing the second question, it is evident that the nature of an ILP program lends itself naturally to the use of background knowledge that is 'largely' declarative. The qualification is needed here because Prolog programs can be written in a procedural fashion. For example, in [15.6], one part of the background knowledge required the computation of potential sites for a zinc atom given the descriptions of chemical groups that can bind to zinc.

This portion of the background knowledge was highly procedural, although written in Prolog, and included code for intersecting planes and spheres, for Gaussian elimination etc. Fortunately, this did not require interaction with the expert chemist.

By contrast, the part of the background knowledge that defined chemical groups that can bind to zinc was highly declarative, being almost a direct translation of pictorial descriptions provided by the chemist. As a result, where ambiguities or questions arose, or where debugging was necessary, these issues were addressed by actually going through that portion of the background knowledge with the chemist. The experiments in [15.6] involved at least two major debugging phases. The first resulted when the ILP algorithm PROGOL [15.22] was not able to re-discover a description (called a "pharmacophore") common to a given set of molecules. The problem was tracked down to a possible assumption in the chemical background knowledge. This was corrected, and the experiment repeated. A second phase of debugging ensued when PROGOL's result was found to be overly-specific. The original "fix" was now found to be incorrect, and the true error in the background knowledge was uncovered. This form of iterative experimentation would have been extremely difficult had the background knowledge been non-declarative.

The debugging that inevitably accompanies the early stages of an application appears to benefit from a declarative encoding of domain-specific knowledge. This was most apparent in some of the toxicity applications cited earlier. For example, several errors in the translation of chemical ring structures were found relatively easily by a non-expert. Conversely, the use of opaque procedural encoding of a regression program caused considerable discomfort and required an elaborate sequence of checkpoints and assertions during the initial stages of the work in [15.30]. The process of settling on correct and adequate background knowledge should not be underestimated. I have been asked about a breakdown of the different activities that comprise a typical application – the underlying thinking being, no doubt, that such a question could be answered quite easily by someone who maintained extensive records of all experimental work. If I had always practised what I preach in Section 15.7, I would indeed be able to supply such breakdowns from my files. As things are, I present in Table 15.1 what I believe to be a reasonable reconstruction in the form of a hypothetical 100-day application. The precision is of course illusory, but as with all myths, there is an element of truth in the tabulation – namely, the phases of data collection and early experimentation with representation are far longer than the final phase of obtaining results.

Thus, the appropriate response to the question of how the background knowledge should be encoded would appear to be: 'as declaratively as possible.' However the utility of a declarative encoding is in ensuring correctness. In final experimental runs, mundane concerns of efficiency were usually

Table 15.1. A hypothetical 100-day ILP application.

Phase	Activity	Days	Sub-total
I. (Data collection)	1. Identify problem	1–5	
	2. Locate data	5–15	
	3. Convert data	15–40	40 days
II. (Early experiments)	4. Write background	40–55	
	5. Pilot results	55–60	
	6. Change background and repeat 5 if needed	60–80	40 days
III. (Final results)	7. Results and evaluation	80–90	10 days
IV. (Dissemination)	8. Write-up	90–100	10 days

found to dominate. For an ILP algorithm that employs best-first search using the operations of selecting clauses, refining clauses, scoring clauses, and sorting them according to their score, the step of scoring is usually the rate-limiting one. As an illustration, Table 15.2 (from [15.29]) shows the order – in time-complexity terms – of each of these steps for some well-known ILP applications.

For the debugged background knowledge, concerns of speeding-up the "score" step have often translated into pre-computation and representation as ground facts or bit-sets, definitions that make a clever use of the argument-indexing mechanisms of the Prolog engine, dynamic storage of proofs, or representations that are amenable to more efficient proof procedures. For example, in many large-scale natural language applications it is usually more efficient to use a bottom-up "chart-parsing" technique, rather than the standard Prolog top-down theorem prover. All background definitions are then re-written to reflect this other style of theorem proving (J. Cussens, personal communication).

Table 15.2. A comparison of time-complexities for search steps for the ILP implementation described in [15.29]. Here $|E|$ refers to the size of the example-set.

Problem	$\|E\|$	Select	Refine	Score	Sort
Trains [15.17]	10	10^0	10^1	10^2	10^3
Pharmacophore [15.25]	28	10^0	10^2	10^3	10^3
Mutagenesis [15.33]	188	10^0	10^2	10^5	10^4
Mesh [15.5]	2500	10^0	10^2	10^7	10^4
Tagging [15.1]	6000	10^0	10^1	10^{12}	10^2
KRK [15.23]	10000	10^0	10^3	10^6	10^2

15.5 Utility mismatch

Heuristics built-in to a general purpose learning program sometimes do not match particular problem domains. This makes it possible for the program to look as though it has returned "incorrect" or "poor" answers, simply because it has been minimising the wrong cost function. Examples of this problem abound. For example, for the problem in [15.6], chemical intuition seeks to find the most complex pharmacophore common to all active molecules. This was in direct conflict with the ILP program that used a simple "compression" measure that depended on the number of literals in a clause. Similar conflicts arose when the task required minimising a numerical cost function (like mean-square-error: [15.30]), or learning rules in a labeled logic [15.2]. Initial attempts to address this problem relied on manual intervention – either by forcing the program to learn more complex solutions or by carefully selecting the examples. In the event, it was found to be much more satisfactory to devise an algorithm that accepted, as part of a general search-specification, the cost function of actual interest to the user [15.30]. The idea of using a custom-built preference criterion to select amongst consistent solutions is, in itself, not very dramatic, but it can make a considerable difference in thinking about a problem and the acceptability of the solutions obtained by an ILP program.

15.6 Comprehensibility

Despite early and repeated recognition of the importance of comprehensibility in Machine Learning [15.17, 15.18, 15.20], the emphasis in the testing methodology has been on predictive accuracy. While accurate theories are expected, the overriding factor in the successful application of ILP to biological problems has been the attempts made to make the results comprehensible to domain experts. This has made it possible to abstract general biochemical principles worthy of publication in journals in the field. It is worth citing Medawar again in this context ("Two conceptions of science" in [15.15]):

> Here then are some of the criteria used by scientists when judging their colleagues' discoveries and the interpretations put upon them. Foremost is their *explanatory value* – their rank in the grand hierarchy of explanations and their power to establish new pedigrees of research and reasoning.

Given this, it is not surprising that it was found very quickly with the applications to biology that "explanatory value" rarely resulted from just the Prolog rules returned by an ILP program. These were usually viewed with some suspicion and distaste. Whatever explanatory value was present only appeared to come into focus with English translations of the rules or molecular visualisations of instances entailed by them. This made it very clear that it was of paramount importance to communicate the results obtained in a language acceptable to the user. Ten years ago Michie [15.19] formulated a distinction between "weak" and "strong" criteria for Machine Learning. In the weak criterion, a system uses sample data to generate an updated basis for improved performance on subsequent data. In the strong criterion, the system satisfies the weak Criterion and can communicate its internal updates in explicit symbolic form. If we restrict the term "knowledge" to what is explicit, it can be argued that the strong criterion is essential to achieve "knowledge discovery". Few, I think, would argue with the position that "Accurate + Comprehensible" is better than "Accurate + Incomprehensible" when dealing with automated assistants.

15.7 From nursery slopes to Darwin's rule

After hours of hard labour I have begun to harbour the suspicion that in experimental work, almost anything that was once thought useless usually becomes exactly the opposite. The maxim of keeping routine (and extensive) logs is commonplace in experimental science, being drilled into budding scientists on nursery slopes. Even then, it is a long haul to the Olympian mark set by Charles Darwin [15.3]:

> I had, also, during many years followed a golden rule, namely, that whenever a published fact, a new observation or thought came across me, which was opposed to my general results, to make a memorandum of it without fail and at once; for I had found by experience that such facts and thoughts were far more apt to escape from the memory than favorable ones. Owing to this habit, very few objections were raised against my views which I had not at least noticed and attempted to answer.

This is not just "more log-book stuff" – it identifies an important category of observations that are often lost in the quest for scientific knowledge. Speaking for myself, a New Year's resolution is needed simply to Record Everything and Destroy Nothing. Paradise awaits those who have surmounted this particular hillock and can adhere to Darwin's golden rule.

Acknowledgments

This essay is based on a talk delivered at a CompulogNet Area Meeting following the Seventh International ILP Workshop [15.28]. As a journeyman in the craft of ILP development and application, I have learned much from the guidance of others, particularly the teachings and example of Donald Michie in experimental method. It is also a pleasure to acknowledge the counsel of Claude Sammut, Stephen Muggleton, Michael Sternberg, Paul Finn, Chris Rawlings, Doug Bristol and members of the ILP groups at the Turing Institute, Glasgow, and the Computing Laboratory, Oxford.

References

15.1 J. Cussens. Part-of-speech tagging Using Progol. In *Proceedings of the Seventh International Workshop on Inductive Logic Programming*, pages 93–108, Springer, Berlin, 1997.

15.2 J. Cussens, A. Hunter and A. Srinivasan. Generating explicit orderings for non-monotonic logics. In *Proceedings of the Eleventh National Conference on Artificial Intelligence*, pages 420–425. MIT Press, Cambridge, MA, 1993.

15.3 C. Darwin. *Autobiography.* Collins, London, 1958.

15.4 T. Dietterich, R. Lathrop, and T. Lorano-Perez. Solving the multiple-instance problem with axis-parallel rectangles. *Artificial Intelligence*, 89: 31–71, 1997.

15.5 B. Dolšak and S. Muggleton. The application of inductive logic programming to finite element mesh design. In S. Muggleton, editor, *Inductive Logic Programming*, pages 453–472. Academic Press, London, 1992.

15.6 P. Finn, S. Muggleton, D. Page, and A. Srinivasan. Pharmacophore Discovery using the Inductive Logic Programming system Progol. *Machine Learning*, 30: 241–270, 1998.

15.7 J. D. Hirst, R. D. King, and M. J. E. Sternberg. Quantitative structure-activity relationships by neural networks and inductive logic programming. I. The inhibition of dihydrofolate reductase by triazines. *Journal of Computer-Aided Molecular Design*, 8: 421–432, 1994.

15.8 J. D. Hirst, R. D. King, and M. J. E. Sternberg. Quantitative structure-activity relationships by neural networks and inductive logic programming. II. The inhibition of dihydrofolate reductase by pyrimidines. *Journal of Computer-Aided Molecular Design*, 8: 421–432, 1994.

15.9 R. D. King, S. Muggleton, A. Srinivasan, C. Feng, R. A. Lewis and M. J. E. Sternberg. Drug design using inductive logic programming. In *Proceedings of the Twenty-sixth Hawaii International Conference on System Sciences*, IEEE Computer Society Press, Los Alamitos, 1993.

15.10 R. D. King, S. Muggleton, A. Srinivasan, and M. J. E. Sternberg. Structure-activity relationships derived by machine learning: The use of atoms and their bond connectivities to predict mutagenicity by inductive logic programming. *Proceedings of the National Academy of Sciences*, 93: 438–442, 1996.

15.11 R. D. King, S. Muggleton, and M. J. E. Sternberg. Drug design by machine learning: The use of inductive logic programming to model the structure-activity relationships of trimethoprim analogues binding to dihydrofolate reductase. *Proceedings of the National Academy of Sciences*, 89(23): 11322–11326, 1992.

15.12 R. D. King and A. Srinivasan. Prediction of rodent carcinogenicity bioassays from molecular structure using inductive logic programming. *Environmental Health Perspectives*, 104(5): 1031–1040, 1996.

15.13 R. D. King, A. Srinivasan, and M. J. E. Sternberg. Relating chemical activity to structure: an examination of ILP successes. *New Generation Computing*, 13(3,4), 1995.

15.14 R. D. King and M. J. E. Sternberg. A machine learning approach for the prediction of protein secondary structure. *Journal of Molecular Biology*, 216: 441–457, 1990.

15.15 P. B. Medawar. *Pluto's Republic.* Oxford University Press, Oxford, 1984.

15.16 P. B. Medawar. *The Strange Case of the Spotted Mice and other classic essays on science.* Oxford University Press, Oxford, 1996.

15.17 R. S. Michalski. A theory and methodology of inductive learning. In R. S. Michalski, J. Carbonnel, and T. Mitchell, editors, *Machine Learning: An Artificial Intelligence Approach*, pages 83–134. San Mateo, CA, 1983.
15.18 D. Michie. The superarticulacy phenomenon in the context of software manufacture. In *Proceedings of the Royal Society of London*, A 405: 185–212, 1986.
15.19 D. Michie. Machine learning in the next five years. In *Proceedings of the Third European Working Session on Learning*, pages 107–122. Pitman, London, 1988.
15.20 D. Michie, D. J. Spiegelhalter, and C. C. Taylor, editors. *Machine Learning, Neural and Statistical classification*. Ellis Horwood, New York, 1994.
15.21 S. Muggleton. Inductive logic programming. *New Generation Computing*, 8(4): 295–318, 1991.
15.22 S. Muggleton. Inverse entailment and Progol. *New Generation Computing*, 13: 245–286, 1995.
15.23 S. Muggleton, M. E. Bain, J. Hayes-Michie, and D. Michie. An experimental comparison of human and machine learning formalisms. In *Proceedings of the Sixth International Workshop on Machine Learning*. Morgan-Kaufmann, San Mateo, CA, 1989.
15.24 S. Muggleton, R. D. King, and M. J. E. Sternberg. Predicting protein secondary structure using inductive logic programming. *Protein Engineering*, 5: 647–657, 1992.
15.25 S. Muggleton, C. D. Page, and A. Srinivasan. An initial experiment into stereochemistry-based drug design using ILP. In *Proceedings of the Sixth International Workshop on Inductive Logic Programming*. Springer, Berlin, 1996.
15.26 S. Muggleton, A. Srinivasan, R. D. King, and M. J. E. Sternberg. Biochemical knowledge discovery using Inductive Logic Programming. In *Proceedings of the First Conference on Discovery Science*. Springer, Berlin, 1998.
15.27 J. R. Quinlan. FOIL: A midterm report. In *European Conference on Machine Learning*, pages 3–20. Springer, Berlin, 1993.
15.28 A. Srinivasan. Five lessons in representation based on the application of ILP. In *Proceedings of the CompulogNet Area Meeting on Representation issues in Reasoning and Learning*. Czech Technical University, Prague, 1997.
15.29 A. Srinivasan. A study of two sampling methods for analysing large datasets with ILP. *Data Mining and Knowledge Discovery*, 3(1): 95–123, 1999.
15.30 A. Srinivasan and R. C. Camacho. Numerical reasoning with an ILP program capable of lazy evaluation and customised search. *Journal of Logic Programming*, 40(2-3): 185-213, 1999.
15.31 A. Srinivasan and R. D. King. Feature construction with inductive logic programming: A study of quantitative predictions of biological activity aided by structural attributes. *Data Mining and Knowledge Discovery*, 3(1): 37–57, 1999.
15.32 A. Srinivasan, R. D. King, and S. Muggleton. The role of background knowledge: using a problem from chemistry to examine the performance of an ILP program. *Technical report*. Oxford university, Oxford, 1999.
15.33 A. Srinivasan, S. Muggleton, R. D. King, and M. J. E. Sternberg. Theories for mutagenicity: A study of first-order and feature based induction. *Artificial Intelligence*, 85: 277–299, 1996.

16. Internet Resources on ILP for KDD

Ljupčo Todorovski[1], Irene Weber[2], Nada Lavrač[1], Olga Štěpánkova[3], Sašo Džeroski[1], Dimitar Kazakov[4], Darko Zupanič[1], and Peter Flach[5]

[1] Jožef Stefan Institute
Jamova 39, SI-1000 Ljubljana, Slovenia

[2] Institut für Informatik, Universität Stuttgart
Breitwiesenstr. 20–22, D-70565 Stuttgart, Germany

[3] Faculty of Electrical Engeneering, Department of Cybernetics
Czech Technical University
Technicka 2, 166 27 Prague 6, Czech Republic

[4] Department of Computer Science, University of York
Heslington, York YO10 5DD, UK

[5] Department of Computer Science, University of Bristol
Merchant Venturers Building, Woodland Rd, Bristol BS8 1UB, UK

Abstract

The aim of this chapter is to review ILP resources available on the Internet. These are especially important due to the growing interest for applying ILP methods to knowledge discovery in databases (KDD) and data mining problems. They also play an important role in connecting the ILP research community with potential end users of ILP methods. Most of the chapter is dedicated to the ILP Internet resources gathered and maintained in the ILPnet and ILPnet2 projects. Some other ILP and KDD related Internet resources are briefly reviewed as well.

16.1 Introduction

This chapter outlines some of the inductive logic programming resources available on the Internet. The focus is on resources that are of interest from the viewpoint of knowledge discovery in databases (KDD) and data mining (DM). Since these are two very active areas, there are numerous academic and commercial Internet resources offering relevant information (e.g., `http://www.kdnuggets.com/`). However, these resources usually do not provide sufficient information on ILP systems and their applications. This lack of information is now being compensated by specialized repositories of ILP-related Internet resources, such as the ones reviewed in this chapter.

Recently, there has been a growing interest for applying ILP methods to various KDD and DM problems. Both researchers and end users are interested in the application of ILP methods to real-world problems. Relevant Internet resources can play a very important role in connecting these two communities – users with specific problems on one hand, and ILP researchers as solution providers on the other.

Initially, ILP-related information was gathered and organized in Internet repositories covering the wider area of machine learning. A first systematic effort to build ILP-related Internet resources was undertaken in the ILPnet (European Inductive Logic Programming Scientific Network) project from 1993 to 1996. The work in the ILPnet project resulted in a WWW repository of ILP systems, data sets and bibliography. This work has been continued in the successor of ILPnet, named ILPnet2 (Inductive Logic Programming Network of Excellence). The major part of this chapter is dedicated to the Internet resources gathered and organized in the ILPnet2 project.

The chapter is organized as follows. First, a brief history of ILP Internet resources is presented in Section 16.2. A detailed presentation of ILP-related Internet resources maintained in the ILPnet2 project is given in Section 16.3. Sections 16.4 and 16.5 briefly review other ILP and KDD Internet resources, respectively. Finally, links to the most important Internet resources reviewed in this chapter are collected in Section 16.6.

16.2 Brief history on ILP Internet resources

One of the earliest attempts to regularly collect and organize information about machine learning systems and applications on the Internet was the ML archive. It was established and maintained by the GMD Artificial Intelligence Research Division and was available at `http://www.gmd.de/ml-archive/`. This WWW repository gathered information about machine learning systems and their applications, including many ILP systems and applications. The repository also collected useful information about the machine learning research community. The GMD ML Archive is no longer available on the WWW, but the information gathered in the archive is now part of the MLNet Open Information Service at `http://www.mlnet.org/`.

The UCI (University of California, Irvine) Machine Learning repository is another very important and well known Internet resource related to the area of machine learning, which includes data sets, domain theories and data generators. It is widely used in the machine learning community as a source of benchmark problems for performance evaluation of machine learning systems, and is regularly maintained. Recently, its scope has been broadened to include the area of KDD by setting up the UCI Knowledge Discovery in Databases Archive, a collection of large data sets which encompasses a wide variety of data types, data analysis tasks, and application domains.

The first coordinated attempt to build ILP-related Internet resources occurred in the ILPnet (European Inductive Logic Programming Scientific Network) project. This network was coordinated by Nada Lavrač at the Jožef Stefan Institute in Ljubljana, and was funded by the European Commission from 1993 to 1996. One of the aims of ILPnet was to provide an infrastructure for dissemination of results of ongoing European research in ILP, most of which was performed in the Esprit ILP project: http://www.cs.kuleuven.ac.be/~ml/esprit.6020.ilp.html.

One of the results of the ILPnet project was a WWW repository of ILP related information, available at: http://www-ai.ijs.si/ilpnet/. The ILPnet WWW repository included a bibliography of ILP-related publications. Besides books and regular articles or papers published in journals and conference proceedings, the bibliography included bibliographical information about ILP-related technical reports and PhD theses including their abstracts. Another part of the ILPnet WWW repository included descriptions of ILP systems and also links to their entries in the GMD ML archive. Links to ILP data sets in the GMD ML archive were also included. The publication of an electronic ILP Newsletter also started in the ILPnet project.

In 1996 the research project ILP was succeeded by the ILP2 project (http://www.cs.kuleuven.ac.be/~ml/esprit/esprit.ilp2.20237.html), and in 1998 the ILPnet2 Network of Excellence in Inductive Logic Programming continued where ILPnet had left off in 1996. In particular, maintenance and upgrade of the ILP-related Internet resources, established in the ILPnet project, is continued in the ILPnet2 project (http://www.cs.bris.ac.uk/~ILPnet2/). These resources are reviewed in the next section.

16.3 ILPnet2 Internet resources

The six most important categories of resources available at the ILPnet2 website are: ILP systems, applications and datasets, educational materials, online library, calendar of events, and a regular newsletter. These categories are reviewed in the sections below.

16.3.1 ILP systems: Descriptions and portability

This section of the ILPnet2 Internet resources gives useful information about ILP systems and their availability on and portability to different hardware and software platforms. It consists of two parts: system descriptions and portability.

A first version of the list of links to ILP system descriptions was collected in the ILPnet WWW site. Most of the links point to entries in the ML archive and were dedicated to the systems developed by the partners of the ILP project as well as systems developed in the wider ILP community (which includes e.g., the Machine Learning Research Group at the University of Texas at Austin). Currently, ILPnet2 list of systems contains entries for more than 40 ILP systems.

Each entry in the list of ILP system descriptions contains at least one of two possible links. One link points to a short description of the system and another one to the WWW page of the system, typically provided by the system developer. Short descriptions of ILP systems provide concise and uniformly structured information, such as specification of capabilities and features of the system, details about its implementation (e.g., programming language and current version) and references to relevant papers and/or technical reports.

The list of ILP systems covers interactive ILP systems, that can learn from small data sets and require user interaction during the learning process, as well as empirical ILP systems capable of learning from large data sets requiring little or no user guidance and thus suitable for KDD and DM. The current list of ILP systems is given in Table 16.1.

An important aspect of ILP systems availability is the question whether they are portable to different hardware and software platforms. The issue of portability of ILP systems is also covered in the ILPnet2 Internet resources. Compatibility tests have been carried out for a number of ILP systems by the Research Group on Artificial Intelligence at the Hungarian Academy of Sciences. The results of these tests are regularly reported in the ILP system portability page, which currently provides information about the portability of the following five ILP systems: CPROGOL, FOIL, MFOIL, PPROGOL and SPECTRE.

16.3.2 ILP applications and data sets

The information about applications of ILP systems to different real-world data sets is even more relevant to potential users. It is also useful for the developers of ILP systems, because they can use these data sets to test the performance of systems on real-world problems.

This section of the ILPnet2 Internet resources contains a collection of short descriptions of ILP applications. This information was collected in the ILP and ILP2 projects. Some of the application descriptions are available only on the ILPnet2 WWW page and have not been extensively published elsewhere.

Short descriptions of applications, similar to the ones for systems, provide concise and uniformly structured information about ILP applications. Each description includes a short digest of the background knowledge from the application domain, a description of the data set with detailed information about its format and size, a report on experiments performed with different ILP systems used in the particular application, comparison with non-ILP approaches (where available) and references to related papers, articles and techical reports. Some of the short descriptions also provide links pointing to more detailed information prepared by the researchers involved in the application.

Currently, this section of ILPnet2 Internet resources provides information about more than 40 applications of ILP systems. They address different real-world problems from various application domains, such as engineering, life sciences and environmental sciences, natural language processing and market

Table 16.1. The list of ILP systems (as of 01.01.2001) provided in the ILPnet2 Internet resources. **S/H** denote the availability of a short description/home page.

Name	S	H	Reference	A very short description
1BC		•	[16.9]	A first-order naive Bayesian classifier, implemented in the context of TERTIUS.
ACL	•		[16.15]	Learns abductive theories by first learning rules, then learning integrity constrains.
ALEPH		•	[16.26]	Based on inverse entailment. Supersedes PPROGOL. Closely related to CPROGOL.
CHILLIN	•	•	[16.29]	Combines bottom-up and top-down induction to compress the examples.
CLAUDIEN		•	Chapter 5	Induces integrity constraints in the framework of learning from interpretations.
CLINT		•	[16.6]	A system for interactive concept learning and theory revision.
CPROGOL	•	•	Chapter 7	Induces logic programs by inverting entailment. Implemented in C.
DOLPHIN		•	[16.28]	Optimizes logic programs by combining analytical and inductive techniques.
FDEP		•	[16.11]	Induces functional dependencies from relations in a database.
FFOIL	•	•	Chapter 12	A derivative of FOIL specialized for learning functional relations.
FILP		•	[16.2]	Induces functional logic programs by querying the user.
FLIP		•	[16.14]	Induces functional logic programs from facts. Based on the reversal of narrowing.
FOCL	•	•	[16.22]	Integrates an explanation-based (analytical) learning component within the inductive learning approach of FOIL.
FOIDL	•	•	[16.18]	Learns first-order decision lists (ordered sets of rules).
FOIL	•	•	Chapter 12	Learns logical definitions (logic programs) from relations (in a database).
FORC			Chapter 9	k-means clustering of multi-relational data.
FORS	•	•	[16.16]	Learns to predict a real-valued class variable from examples and background knowledge (relational regression).
FORTE	•	•	[16.24]	Revises function-free first-order Horn clause theories (theory revision).
GOLEM	•	•	[16.21]	Learns pure logic programs by least-general generalization (see Chapter 3).
HYDRA	•	•	[16.1]	Extends FOCL to handle noisy data by learning rules for each class and attaching likelihood ratios to each rule.

Table 16.1. Continued.

Name	S	H	Reference	A very short description
ICL	•	•	Chapter 5	Learns classification rules in the framework of learning from interpretations.
ILP-R	•		[16.23]	Probabilistic first-order classifier with a non-myopic search heuristic and a weak language bias.
INDEX		•	[16.8]	Decomposes relations by predicate invention based on discovered functional dependencies.
LILP		•	[16.17]	Lambda ILP. Learns predicate definitions from positive-only examples.
LINUS	•	•	Chs. 3, 11	Transforms ILP problems to propositional form.
MARKUS	•	•	[16.13]	A derivative of MIS [16.25] that learns non-incrementally, uses optimal refinement operators and iterative deepening search.
MERLIN	•	•	[16.3]	A multiple predicate learning system suitable for learning recursive hypotheses.
MFOIL	•	•	[16.7]	Extends FOIL with noise-handling capabilities.
MIDOS			Chapter 4	Subgroup discovery in multi-relational data.
MILES		•	[16.27]	A flexible environment for testing ILP methods and operators.
MOBAL		•	[16.19]	An integrated knowledge acquisition toolbox, including some ILP systems/facilities.
PROGOL				See entries for CPROGOL and PPROGOL.
PPROGOL		•		Based on inverting entailment. Implemented in Prolog. Superseded by ALEPH.
RDBC			Chapter 9	Performs hierarchical agglomerative clustering for multi-relational data.
REGAL	•	•	[16.12]	Uses a genetic algorithm to learn first-order concept descriptions.
RIBL, RIBL2	•		Chapter 9	Extends nearest-neighbor classification to the relational/first-order framework.
SPECTRE	•	•	[16.4]	Specializes overly general initial theories that cover all positive and some negative examples.
S-CART			Chapter 6	Learns (structural) classification and regression trees in first-order logic.
STILL	•		[16.20]	Uses stochastic matching of examples to rules. Keeps all consistent rules that cover a positive example. Classifies nearest-neighbor style.
TERTIUS		•	[16.10]	Performs confirmation-based unsupervised discovery of first-order rules.
TILDE	•	•	Chapter 5	Learns logical decision trees in the framework of learning from interpretations.
WARMR			Chapter 8	Discovers relational association rules.

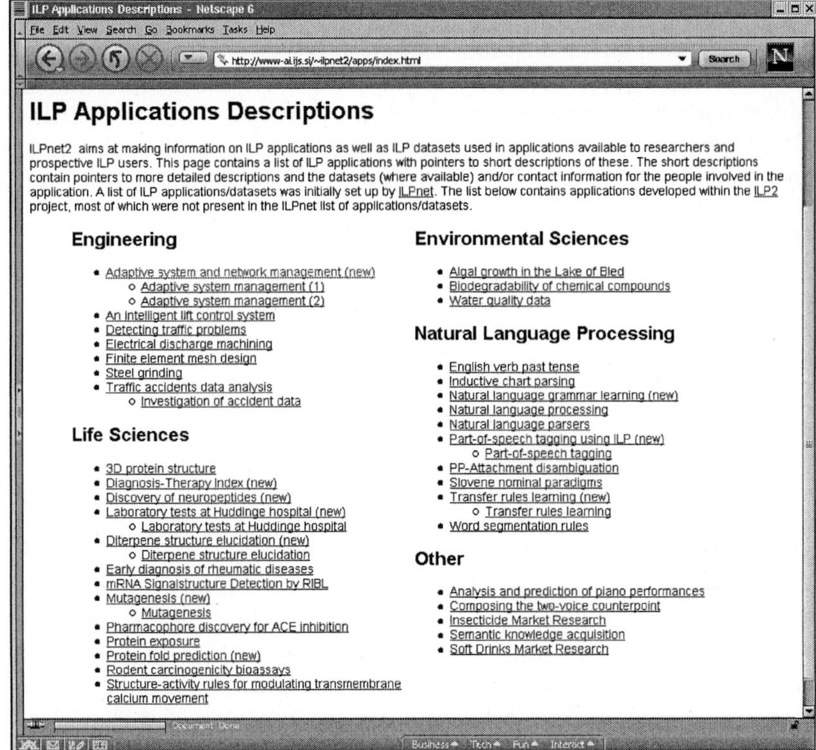

Fig. 16.1. A snapshot of the list of ILP applications and data sets, grouped by application domain.

analysis. Let us mention here that most successful ILP applications come from the area of life sciences and that the natural language domain holds a lot of promise and challenges for successful ILP applications [16.5]. For a more detailed overview of ILP applications we refer the reader to Chapter 14.

Applications of ILP have been also investigated in the EU-funded ALADIN industrial research project. Embedded applications of ILP in the areas of information retrieval and adaptive system management have been considered. Information on the project partners, the project and its results, as well as some information on ILP relevant to practical applications can be found at the project WWW site (`http://www.aladin-eu.com`).

16.3.3 ILP educational materials

This section of the ILPnet2 website is dedicated to the dissemination of ILP-related educational materials. Such materials can be used for self-training by potential end users of ILP methods. Using these materials, they can get familiar with the basic ILP methodology, necessary for planning potential future application of ILP methods to their problems.

The materials are collected from different ILP-related university courses, seminars and summer schools, some of them organized in the ILPnet and ILPnet2 projects. Usually, they are provided in the form of on-line versions of lecture slides or syllabi of courses and seminars. These include the on-line school on ILP & KDD, a collection of materials from the *International Summer School on Inductive Logic Programming and Knowledge Discovery in Databases*, held in Prague, Czech Republic, 15-17 September 1997 (`http://www-ai.ijs.si/SasoDzeroski/ILP2/ilpkdd/`).

16.3.4 On-line library of ILP-related references

The on-line library contains ILP-related references from 1970 onwards. It was developed from the ILP bibliography for the period 1970-1996 that was compiled by ILPnet. A number of references over 1997 and 1998 were added courtesy of the ILP2 project. The database is maintained by ILPnet2 and contains at the time of writing 900 entries by 477 different authors.

Each entry in the on-line library is listed on its own Web page listing full bibliographic details of the publication, and an abstract if available. Many publications also include a link to an on-line version of the paper, or a Web page where more information can be found. BibTEX records can be obtained for individual publications, for all publications in a year, or for all publications in the library. The user can interactively browse through the references by author name, keywords (see Figure 16.2), type of publication (journal article, conference paper, book or any other publication type in BibTEX) and publication year. A global search on any term can be performed from the on-line library's homepage.

New entries can be added either through on-line forms, or by submitting BibTEX entries to the ILPnet2 librarian. This is open to anybody with ILP-related publications. ILPnet2 members can also include links to their personal WWW pages, which then appears in all their references in the database. Thus, the database also provides a list of contacts to people in the ILP research community.

16.3.5 ILP-related events

This WWW page is included in the ILPnet2 Internet resources in order to track the information about forthcoming and past events related to the field of ILP. Four categories of events are collected: conferences, workshops, seminars and summer schools. Each entry of the list of forthcoming events contains short information about the event: location, important dates, link to WWW page of the event and other important information (like call for papers), where available. The second part of the list includes reports on past events with highlights of the aspects related to ILP.

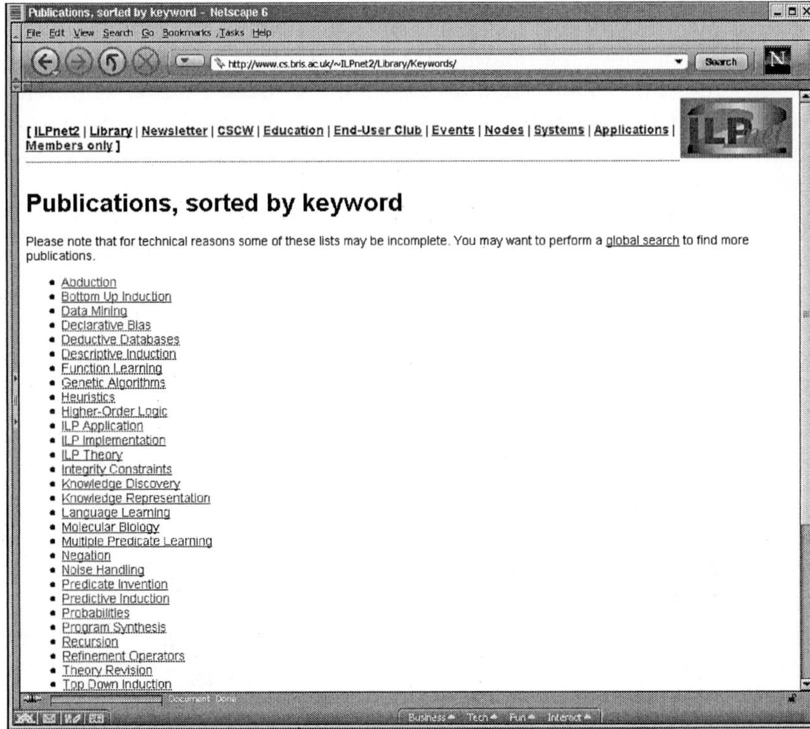

Fig. 16.2. A snapshot of an interactive browse through the ILP library references sorted by keywords.

16.3.6 ILP newsletter

The ILP newsletter was initiated by ILPnet in 1993. It was regularly published until ILPnet expired in 1996. The Newsletter has been revived in 1999. Currently, it is being sent to 235 subscribers. The ILP Newsletter includes material relevant to ILPnet2 and ILP in general, including a calendar of ILP events, conference reports (from the ILP perspective), book reviews, etc. The archive of all published issues of ILP newsletter can be accessed from the ILPnet2 website.

16.4 Other ILP-related Internet resources

This section reviews Internet resources provided by three Networks of Excellence, and also some of the ILP-related resources provided by research groups dealing with development and applications of ILP systems.

The MLNet Online Information Service (`http://www.mlnet.org`), the successor of the ML Archive at GMD, is probably the most complete Internet resource that provides information about machine learning and related research fields. The site includes information about the fields of knowledge discovery, case-based reasoning, knowledge acquisition, and data mining. It provides information about different learning systems, their availability and examples of applications to various real-world problems. In addition, MLNet OIS includes contact information for research groups and individual researchers in the machine learning community. Links to WWW sites dedicated to events and projects related to machine learning are also included. In the MLNet OIS, there is a lot of information related to the field of ILP.

Compulog Net is a Network of Excellence in computational logic providing networking and WWW infrastructure for the purpose of research, development and application of logic-based techniques and methods. Therefore, their WWW pages provide a lot of materials and links relevant for ILP as a logic-based machine learning technique. Their WWW site is available at `http://www.compulog.org/`. CoIL is another Network of Excellence integrating four communities that perform research, development and application in four different areas: Erudit (fuzzy logic), EvoNet (evolutionary computing), MLNet (machine learning) and NEuroNet (neural networks): `http://www.dcs.napier.ac.uk/coil/`.

A number of WWW sites of research groups working in the field deserve to be mentioned in this review. Some of these, listed in alphabetical order of the city/town, are reviewed in the remainder of this section.

The WWW site of the Machine Learning Research Group at the University of Texas at Austin (`http://www.cs.utexas.edu/users/ml/`) provides information about ILP systems and applications (especially in the domains of natural language processing and theory refinement) developed by the group and links to other ILP-related WWW pages.

Several ILP systems and datasets are available from the WWW page of the Machine Learning Group at the Department of Computer Science of the Katholieke Universiteit Leuven (`http://www.cs.kuleuven.ac.be/~ml/`).

The WWW site of the Machine Learning Group at Oxford University (`http://web.comlab.ox.ac.uk/oucl/research/areas/machlearn/`) covers different aspects of ILP. It includes an introduction to the theory of ILP with a brief outline of important state-of-the-art research issues and also presents applications of ILP systems in the areas of bioinformatics and medicine, especially drug design.

The Machine Learning Group at the University of York maintains a WWW site (`http://www.cs.york.ac.uk/mlg/`) with an introduction to the theoretical background of ILP and an interesting collection of ILP data sets. It also provides information on ILP systems developed by the group members and their projects and publications, many of them related to ILP.

16.5 KDD related Internet resources

In this section we briefly review two KDD related Internet resources. We believe that they are good representatives of the vast number of KDD related resources available. They are, at least, good starting points with many further links.

The first is KDnuggets (`http://www.kdnuggets.com/`), an exhaustive source of links and information related to data mining (DM) and knowledge discovery in databases (KDD). The first version of the KDnuggets site appeared in 1994, and the site has been regularly maintained and updated since then. The information is organized in sections dealing with the following aspects of KDD and DM: software, data sets, solutions, companies, job offers, events (courses, meetings and conferences), publications and links to related WWW sites. Subscription to KDnuggets News, an electronically published newsletter, can be arranged on-line and an archive of past issues is available.

There are not many references to ILP in KDnugets WWW site. Links to a few ILP systems are provided in the software section under different subsections such as classification, suites or association rules. The ILPnet2 WWW site is listed in the related sites section. Also, there are several references to ILP in past issues of the KDnuggets News newsletter.

KD Central (`http://www.kdcentral.com/`) is a WWW portal and resource center providing links to KDD related WWW sites. Currently, it provides about a thousand links clustered in twenty-two categories. Most of the links are accompanied with a short review of their contents.

16.6 Conclusion

A substantial body of Internet resources on ILP is available, a large portion of which has been produced by the ILP and ILP2 projects and further organized by the ILPnet and ILPnet2 networks. They include information on ILP systems, applications and datasets, as well as ILP-related educational materials, events and publications. This chapter has given an overview of these resources and some essential pointers to KDD resources. The links mentioned in this chapter are collected in Table 16.2.

Acknowledgments

We would like to thank the partners of the ILP and ILP2 projects, as well as the member nodes of the ILPnet and ILPnet2 networks for contributing to the resources described in this chapter. These projects and networks have been or are still funded by the European Commission. The participation of Jožef Stefan Institute in the above projects has been funded also by the Slovenian Ministry of Science and Technology.

Table 16.2. Links to ILP and KDD related Internet resources.

ILPnet and ILPnet2	
ILPnet	http://www-ai.ijs.si/ilpnet.html
ILPnet2	http://www.cs.bris.ac.uk/~ILPnet2/
ILPnet2 @ IJS	http://www-ai.ijs.si/~ilpnet2/
systems	http://www-ai.ijs.si/~ilpnet2/systems.html
applications	http://www-ai.ijs.si/~ilpnet2/apps/
edu. materials	http://www-ai.ijs.si/~ilpnet2/education/
on-line library	http://www.cs.bris.ac.uk/~ILPnet2/Library/
related events	http://www-ai.ijs.si/~ilpnet2/events/
newsletter	http://www-ai.ijs.si/~ilpnet2/newsletter/

EU projects: ILP, ILP2, ALADIN
http://www.cs.kuleuven.ac.be/~ml/esprit.6020.ilp.html
http://www.cs.kuleuven.ac.be/~ml/esprit/esprit.ilp2.20237.html
http://www.aladin-eu.com/

Repositories of data sets	
UCI ML	http://www.ics.uci.edu/~mlearn/MLRepository.html
UCI KDD	http://kdd.ics.uci.edu/

Networks of Excellence	
MLnet	http://www.mlnet.org/
CompuLog	http://www.compulog.org/
CoIL	http://www.dcs.napier.ac.uk/coil/

Machine Learning Research Groups	
U of Texas	http://www.cs.utexas.edu/users/ml/
U of York	http://www.cs.york.ac.uk/mlg/
KU Luven	http://www.cs.kuleuven.ac.be/~ml/
Oxford U	http://oldwww.comlab.ox.ac.uk/oucl/groups/machlearn/

KDD related	
KDnuggets	http://www.kdnuggets.com/
KD Central	http://www.kdcentral.com/

References

16.1 K. M. Ali and M. J. Pazzani. Hydra: A noise-tolerant relational concept learning algorithm. In *Proceedings of the Thirteenth International Joint Conference on Artificial Intelligence*, pages 1064–1071. Morgan Kaufmann, San Mateo, CA, 1993.

16.2 F. Bergadano and D. Gunetti. Functional inductive logic programming with queries to the user. In *Proceedings of the Sixth European Conference on Machine Learning*, pages 323–328. Springer, Berlin, 1993.

16.3 H. Bostrom. Theory-guided induction of logic programs by inference of regular languages. In *Proceedings of the Thirteenth International Conference on Machine Learning*, pages 46–53. Morgan Kaufmann, San Francisco, CA, 1996.

16.4 H. Bostrom and P. Idestam-Almquist. Specialization of logic programs by pruning SLD-trees. In *Proceedings of the Fourth International Workshop on Inductive Logic Programming*, pages 31–48. GMD, Sankt Augustin, Germany, 1994.

16.5 J. Cussens and S. Džeroski, editors. *Learning Language in Logic.* Springer, Berlin, 2000.

16.6 L. De Raedt and M. Bruynooghe. An overview of the interactive concept-learner and theory revisor CLINT. In S. Muggleton, editor, *Inductive Logic Programming*, pages 163–192. Academic Press, London, 1992.

16.7 S. Džeroski. Handling imperfect data in inductive logic programming. In *Proceedings of the Fourth Scandinavian Conference on Artificial Intelligence*, pages 111–125. IOS Press, Amsterdam, 1993.

16.8 P. A. Flach. Predicate invention in inductive data engineering. In *Proceedings of the Sixth European Conference on Machine Learning*, pages 83–94. Springer, Berlin, 1993.

16.9 P. A. Flach and N. Lachiche. 1BC: A first-order Bayesian classifier. In *Proceedings of the Ninth International Workshop on Inductive Logic Programming*, pages 92–103. Springer, Berlin, 1999.

16.10 P. A. Flach and N. Lachiche. Confirmation-guided discovery of first-order rules with Tertius. *Machine Learning*, 42(1-2): 61–95, 2001.

16.11 P. A. Flach and I. Savnik. Database dependency discovery: a machine learning approach. *AI Communications*, 12(3):139–160, 1999.

16.12 A. Giordana and F. Neri. Search-intensive concept induction. *Evolutionary Computation Journal*, 3(4):375–416, 1996.

16.13 M. Grobelnik. Induction of Prolog programs with MARKUS. In *Proceedings of the Third International Workshop on Logic Program Synthesis and Transformation*, pages 57–63. Springer, Berlin, 1993.

16.14 M. Hernández-Orallo and M. J. Ramírez-Quintana. A strong complete schema for inductive functional logic programming. In *Proceedings of the Ninth International Workshop on Inductive Logic Programming*, pages 116–127. Springer, Berlin, 1999.

16.15 A. C. Kakas and F. Riguzzi. Learning with abduction. In *Proceedings of the Seventh International Workshop on Inductive Logic Programming*, pages 181–188. Springer, Berlin, 1997.

16.16 A. Karalič and I. Bratko. First order regression. *Machine Learning*, 26(2/3): 147–176, 1997.

16.17 Z. Markov. A functional approach to ILP. In *Proceedings of the Fifth International Workshop on Inductive Logic Programming*, pages 267–280. Department of Computer Science, Katholieke Universiteit Leuven, Belgium, 1995.

16.18 R. J. Mooney and M. E. Califf. Induction of first-order decision lists: Results on learning the past tense of English verbs. *Journal of Artificial Intelligence Research*, 3:1–24, 1995.
16.19 K. Morik, S. Wrobel, J.-U. Kietz, and W. Emde. *Knowledge Acquisition and Machine Learning: Theory, Methods and Applications.* Academic Press, London, 1993.
16.20 M. Sebag and C. Rouveirol. Polynomial time learning in logic programming and constraint logic programming. In *Proceedings of the Sixth International Workshop on Inductive Logic Programming*, pages 105–126. Springer, Berlin, 1997.
16.21 S. Muggleton and C. Feng. Efficient induction of logic programs. In *Proceedings of the First Conference on Algorithmic Learning Theory*, pages 368–381. Ohmsha, Tokyo, 1990.
16.22 M. J. Pazzani and D. Kibler. The utility of knowledge in inductive learning. *Machine Learning*, 9(1):57–94, 1992.
16.23 U. Pompe and I. Kononenko. Naive Bayesian classifier within ILP-R. In *Proceedings of the Fifth International Workshop on Inductive Logic Programming*, pages 417–436. Department of Computer Science, Katholieke Universiteit Leuven, Belgium, 1995.
16.24 B. L. Richards and R. J. Mooney. Refinement of first-order Horn-clause domain theories. *Machine Learning*, 19(2):95–131, 1995.
16.25 E. Shapiro. *Algorithmic Program Debugging.* MIT Press, Cambridge, MA, 1983.
16.26 A. Srinivasan. The Aleph Manual. Technical Report, Computing Laboratory, Oxford University, 2000. Available at
`http://web.comlab.ox.ac.uk/oucl/research/areas/machlearn/Aleph/`
16.27 I. Stahl and B. Tausend. MILES - A Modular Inductive Logic Programming Experimentation System. Technical Report, Fakultät Informatik, Universität Stuttgart, 1994.
16.28 J. M. Zelle and R. J. Mooney. Combining FOIL and EBG to speed-up logic programs. In *Proceedings of the Thirteenth International Joint Conference on Artificial Intelligence*, pages 1106–1111. Morgan Kaufmann, San Mateo, CA, 1993.
16.29 J. M. Zelle, R. J. Mooney, and J. B. Konvisser. Combining top-down and bottom-up techniques in inductive logic Programming. In *Proceedings of the Eleventh International Conference on Machine Learning*, pages 343–351. Morgan Kaufmann, San Francisco, CA, 1994.

Author Index

Subject Index

Druck: Strauss Offsetdruck, Mörlenbach
Verarbeitung: Schäffer, Grünstadt